THE SCIENCE OF ENGINEERING DESIGN

Percy H. Hill

TUFTS UNIVERSITY

*"Some men see things as they are and ask 'Why?'
I dream of things that never were and ask 'Why not?'"*
GEORGE BERNARD SHAW

Holt, Rinehart and Winston, Inc.
New York   Chicago   San Francisco   Atlanta   Dallas
Montreal   Toronto   London   Sydney

Illustrations by Visa Directions

Copyright © 1970 by Holt, Rinehart and Winston, Inc.
All Rights Reserved
Library of Congress Catalog Card Number 71-121639
SBN: 03-081390-5
Printed in the United States of America
0 1 2 3   17   9 8 7 6 5 4 3 2 1

*To
the
patience
of*

**EH²
MAH
CH²**

PREFACE

This book was written for the engineer and the manager, student as well as professional. It will give the engineer some insight into the problems facing management when it comes to risks of launching a new product and a respect for decisions which must be made to insure company growth and survival. At the same time the book will give management an understanding of the engineering design process and creative engineering activity necessary to conceive of and produce a new product to measure up to management's analysis of the market situation.

Eight years with a senior level course termed Inventive Design, now a graduate level requirement, as well as long experience with freshman students has convinced the author that material of the type presented in this book must be conveyed to the student as early as possible in the engineering curriculum and continually throughout his learning period. Students of engineering need a real-world experience with such variables as economics, environment, people, company policy, reliability, optimum function, and possibly limited conditions of time and space imposed upon their course work at some point before entering professional practice. This book attempts to do just that. It does not concern itself with the science of solid mechanics, fluid mechanics, integrated circuitry, thermodynamics, and the like, since this material is more than adequately covered in the literature. This book focuses the reader's attention on the process by which ideas originate and are developed into workable products. To convey this methodology of design, the book is divided into four interrelated parts:

PART I (Chapters 1, 2, 3) is concerned with the problems of management in seeking out new-product ideas, the creative process for generating ideas, and the design process involving the reader with the methodology of engineering design.

PART II (Chapters 4, 5, 6) gives the reader the necessary experience and provides the principles to refine the design in light of material optimization and use of devices by the human operator.

PART III (Chapters 7, 8, 9, 10) involves both the engineer and the manager with techniques of scheduling, decision making, and presenting (selling) the design idea.

PART IV includes a total of forty-five (45) open-end type design problems, projects, and cases. These are based on the assumption that wisdom cannot be told; the learning process, to be effective, does not depend on the simple process of telling, but rather calls for something dynamic taking place within the learner. This method does not require "pat" answers of the student but instead calls for an involvement with a situation that stimulates creativity and requires the individual to think for himself and to make judgements and decisions within the framework of real-life situations.

Finally, this book was written for students of engineering design who must avail themselves of all techniques, methods, and approaches to the subject so they may practice the discipline in the most effective way possible.

Percy H. Hill

March 1970
Medford, Massachusetts

CONTENTS

PROLOGUE   xi

1 NEW PRODUCTS AND THE CORPORATE ENVIRONMENT   1

New products; new product evaluation; engineering the prototype; problems and pitfalls; exercises.

2 THE CREATIVE PROCESS   15

Creativity; imagineering; ideation; functional visualization; idea diagram; idea matrix; brainstorming; synectics; creative process; barriers to creativity; exercises.

3 THE DESIGN PROCESS   33

Engineering design; scientific method; design process; stages and techniques; morphological approach; optimization; reliability; exercises.

4 CASE HISTORIES   61

Case 001—Rod Light; Case 002—Cable Stabilization Unit; Case 003—Automotive Ignition Shut-Off; exercises.

5 GUIDE TO MATERIAL SELECTION   115

Properties of metals; properties of plastics; composition and density of selected materials; properties of sheet metal; properties of selected materials; choice of metals for forming; finishes and coatings; joining and fastening; making the selection; exercises.

6 HUMAN FACTORS IN DESIGN   135

Da Vinci and Gilbreth; the man-machine system; human factors; human factors data; some examples of human engineering; analog of the human body as a dynamic system; design considerations for powered exoskeletons; the industrial robot; exercises.

7 C.P.M./P.E.R.T.   173

Program control techniques; historical background; glossary of terms; calculating the critical path; managing the critical path; preparing a P.E.R.T. network; CPM/PERT for the IBM 1130; project management with PERT/CPM; exercises.

8 VALUE ENGINEERING   199

The value engineering process; value engineering case history; pricing and the break-even point; pricing policies for new products; linear programming; exercises.

9 PATENTS, PRINCIPLES, AND PROTECTION   219

Copyrights; trademarks; patents; condition of patentability; protection and disclosure; self-interrogation before filing; basic steps to obtain a patent; claims; proprietary information and trade secrets; patent infringement; promoting the invention; exercises.

10 THE DESIGN CRITIQUE   237

Criticism; design review; the engineer as an entrepreneur; how to sell ideas; the design report; planning and preparing a presentation; presentation check lists; exercises.

BIBLIOGRAPHY WITH ABSTRACTS   255

Engineering design; creativity and innovation; design methods; human factors in design; background materials; management technique; patents and inventions.

APPENDIX A-1 DESIGN PROBLEMS   277

APPENDIX A-2 DESIGN PROJECTS   287

APPENDIX A-3 CASE STUDIES   311

INDEX   369

PROLOGUE

History of technology is filled with the accounts of men of practical genius and serendipity who are in part responsible for today's technological advances. This book is not intended for this type but instead for the engineer or engineer-to-be who has a stick-to-it attitude, a creative bent, and wishes to solve problems in a unique way. If the individual has these attributes and follows the sequential steps involved in the creative and design processes described in the chapters to follow, he will produce a design worthy of consideration. This is not to say that his design will be award-winning, for excellence in design depends upon experience, but his design will be better due to the sequential steps described here.

This, then, is the Science of Engineering Design. Science usually means a generalized and structured body of knowledge. Science is a method that once applied to a problem will produce an answer, and when applied again will produce a comparable answer again, again, and again. The same is true with the Science of Engineering Design. Once the individual is convinced to follow the process of sequential steps, he will succeed in producing a successful design again, again, and again. The value of the design improves with experience as competence in traditional disciplines improves with acquired knowledge. This is explained in detail in Chapter 3 when the Scientific Method is compared to the Design Method.

Design, as explained in the following chapters, is revealed in terms of a number of fundamental principles and relationships known as the creative and design processes. These processes when integrated and practiced systematically will evolve within the individual as the Science of Engineering Design. Particular attention must be paid to the separate phases and their interrelationships and to self integration to realize the full potential of this science. One must realize that this differs from the physical sciences in that design focuses more attention on the individual to structure his thinking along guided and hopefully productive paths (the process of creativity and design).

1 NEW PRODUCTS AND THE CORPORATE ENVIRONMENT

> *"New products are basic to company growth and survival—today more than ever before."*
>
> *"The evolution of new products is not an abstract mystery. It is a practical business function that can best be described as a management process."*
>
> *"The new product program of a company must be organized and controlled if it is to be managed effectively."*
>
> Source: Booz, Allen & Hamilton, Inc., Management Consultants

NEW PRODUCTS

All profit-oriented industries deal, in one way or another, with a product or products in either the software or hardware lines and depend upon the sales of these products for survival and growth. The full attention, talent, and abilities of management, engineering, production, inspection, advertising, marketing, sales, and servicing, are focused on causing the product to return a profit for the company and in turn for the stockholders. Unfortunately, sooner or later every product is pre-empted by another or else degenerates into profitless price competition. Changes in technology and style are strong forces in product displacement. We can quickly recall such items as the ice box, ice cream freezer, 78 rpm phonograph, steam locomotive, and airplane propeller. Even dry ink carbon paper is being replaced by the nonsmudge type produced through the science of microencapsulation. These are but a few of the products that have been displaced or are in danger of being displaced because something better, more convenient, or more attractive has come along. There are also products which were virtually unknown a few years ago yet are profitable industries today, for example Xerox, Polaroid, lasers, Telstar, and the ball point pen.

For an industry to survive in today's world, it must continue to grow; it cannot afford to remain static. This growth, throughout history, has been heavily built on new products. Ford, General Motors, Eastman Kodak, Du Pont, and AT&T are good examples of this. Industry's search for new-product ideas has been of increasing importance in recent years and will reach even greater emphasis in the future as competition continues to intensify and the flood of new products shortens the life span of existing ones.

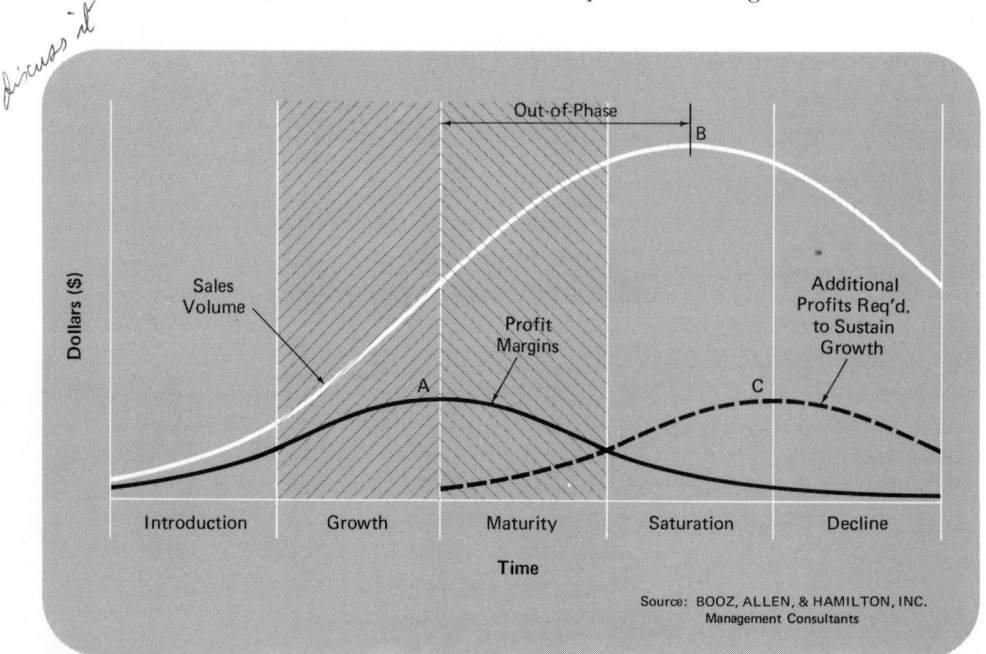

Figure 1-1 **BASIC LIFE CYCLE OF NEW-PRODUCTS**

New products have a characteristic life cycle pattern in sales volume and profit margins, as illustrated in Fig. 1-1. A product (if any good at all) will after introduction rapidly increase (A) in sales during its growth and maturing periods, peak out (B) when it has saturated the market, and then begin to decline. Although the shape of the two curves is similar, studies have shown that they are out of phase with a decrease in profit margins (from A) occurring before a decrease in sales volume. This is understandable since the maximum profit margin (A) occurs at the point of maximum slope of the sales volume curve. As the rate of sales volume begins to decrease, one can expect a decrease in profit margin. Historically, market strategy has been based on the shape of the sales curve, but it is obvious here that product strategy is planned best around the profit curve since by doing so predictions can be made earlier in time. This is easier said than done as most companies do not chart profit margins until the decline period. Use of the computer is beginning to improve this situation. The timing of the introduction of a new product as model change to sustain growth differs among products as well as industries and is shown here only to emphasize the need for continued introduction. As a general rule, the closer a company is to consumer goods and the marketplace, the shorter the cycle of its products; conversely, the closer a company is to basic industry or producer's goods, the longer the cycle.

Now that the need for new products to insure corporate growth has been firmly established, it is obvious that an industry must seek out and promote a flow of new-product ideas. Statistics show that between 55 and 60 good ideas must be found to obtain one successful new product valuable enough to have a significant effect on sales and profit. These ideas can be found from *internal sources* as well as *external*. *Internal sources* consist of:

1. Improvement to present products;
2. Peripheral products, such as a manufacturer of boats beginning to sell outboard motors, boat trailers, anchors, or accessories;
3. Organized creativity among design and development groups;
4. A program of rewarding creative ideas received from salesmen, production workers, laboratory technicians, shipping clerks, secretaries, or middle management personnel.

Idea sources *external* to the company include the following:

1. Knowledge of what the competition is doing;
2. Activities and innovations among companies outside the immediate market area;
3. Ideas from present and potential customers;
4. Inventors and patent application listings;
5. Patent search consultants;
6. Consulting scientists and engineers;
7. Universities, sponsored research, private laboratories, and foundations;
8. Technical, trade, and other publications;
9. Professional society and trade association meetings;
10. Government advisory agencies, such as the Small Business Administration.

NEW-PRODUCT EVALUATION

When a company selects and develops a product it is taking a one-in-five risk that it will return a profit. The company, through the product, is determining which customers, competitors, suppliers, facilities, skill needs, and socioeconomic environment will form the perimeter of its opportunity for success. The only assurance of an organization's success is that it has defined and thoroughly understands the company's goals over a two-to-five or ten-year period. There must also be an intelligent division of responsibility from the new-product staff to engineers to marketing specialists to the sales force that are "in" on what the company is trying to accomplish.

The process by which a product comes into being can be broken down into a series of manageable stages for purposes of planning and control, as illustrated in Fig. 1-2.

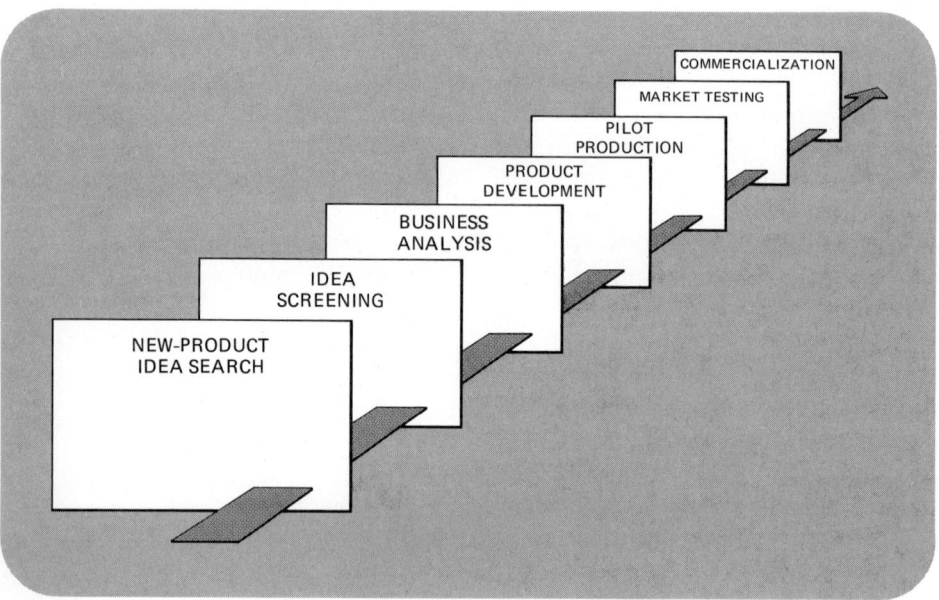

Figure 1-2   **STAGES OF NEW-PRODUCT EVOLUTION**

new-product idea search.

The seeking out of product ideas both from internal and external sources as described earlier.

idea screening.

Identifying product ideas obviously unsuited to the goals and capabilities of the company. This should be done at a high level in the organization by individuals with the broadest overview of company goals and objectives.

business analysis.

An expansion of the product idea through creative analysis into a concrete business recommendation including an initial program for the product. This initial program consists of testing the product idea against the competition and establishing an estimate of sales potential, manufacturing requirements, channels of distribution, type of customers, services, facilities to construct product, and financial outlay to finally launch the product.

product development.

Designing and testing from the idea-on-paper to a product-in-hand. Here the product is closely defined as to size, shape, weight, function, workability, materials, method of fabrication, performance, patentability, tooling, cost of production, and method of servicing. Management groups often prefer to call this stage R & D (research and development). This is unfortunate, for there is relatively little place for research in a new-product development program. Research usually results in knowledge, while the result of development is a product.

pilot production.

A relatively small number of units are constructed on a fabricated production line in order to test jigs and fixtures, personnel skills, production time, quality control and inspection, and to verify estimated costs.

market testing.

This analysis phase requires a sample market area to be located where products constructed during the pilot production stage are placed on sale. The market specialist is interested in customer reaction, distributor opinion, how well the product sells, how it competes with similar lines, if it seems to find a place in the market, and who purchases the product. These commercial experiments are necessary to verify earlier business judgments.

commercialization.

Here all previous stages come into play. After a complete review of results obtained from previous stages a plan is formalized to launch the product into full-scale production and sale. This commits the company's reputation and

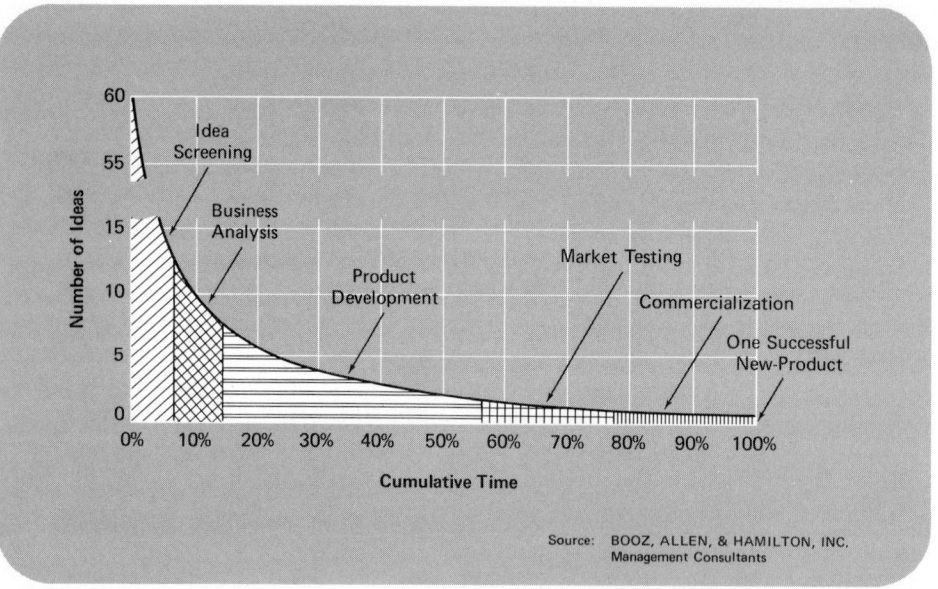

Figure 1-3 **MORTALITY OF NEW-PRODUCT IDEAS**

resources to the new product which earlier was only an idea on paper. During this period, management must continue to check results which may lead to improvements in the product, its manufacturing, or its sales procedures.

We learned earlier that it takes between 55 and 60 ideas to yield one successful new product. Figure 1-3 shows the decay curve of product ideas and projects by stages of product evolution. The process of new-product evolution involves a series of important management decisions. In each stage decisions become more difficult to make since they become progressively more expensive as measured in expenditures of both time and money. The rate at which expense dollars are spent as time accumulates for the average product by stages of product evolution is shown in Fig. 1-4.

This figure represents an industry average; the upper line shows an all-industry average of capital concentrated in the last three stages of product evolution. It is of extreme importance that companies do the best job of screening and business analysis during the early stages of product evolution, thereby eliminating ideas of limited potential before they reach the more expensive stages.

It is now quite clear that management's objective is to choose the best ideas in which to invest available new product time and money. Unfortunately, there are more high-risk than low-risk products and more low-payout than high-payout. On an industry average, for an idea to have a high payout, it must be of high risk. Management's charge is to beat this probability by selecting those rare

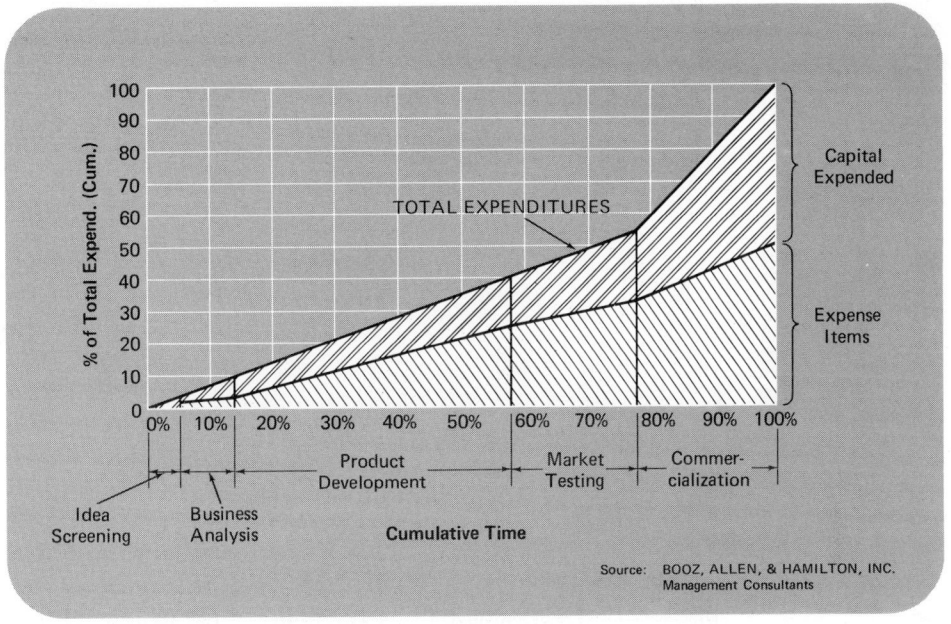

Figure 1-4   **CUMULATIVE EXPENDITURES AND TIME**

ideas that are both low risk and high payout. Certainly this combination leads to successful business, for it is the key to maximum yield with available manpower and resources. Management's effectiveness in product selection can be measured by examining the degree to which projects in the company combine low risk and high payout.

ENGINEERING THE PROTOTYPE

The translation of an idea into a successful new product that conforms to business criteria, is of the highest quality and lowest production cost, and is delivered on time to be shipped to the customer depends upon the talent and creative ability of the engineering staff. The amount of time and dollars required to design and develop a new product is directly proportional to the technical content of the idea and the degree to which familiar materials and fabrication techniques may be applied.

Product development (which is wholly an engineering responsibility) is the necessary first step in arriving at a new product. This is followed by *process development*, shown diagrammatically in Fig. 1-5. Product development consists of *design, analysis,* and *testing. Design* is probably the most demanding of the three

8 THE SCIENCE OF ENGINEERING DESIGN

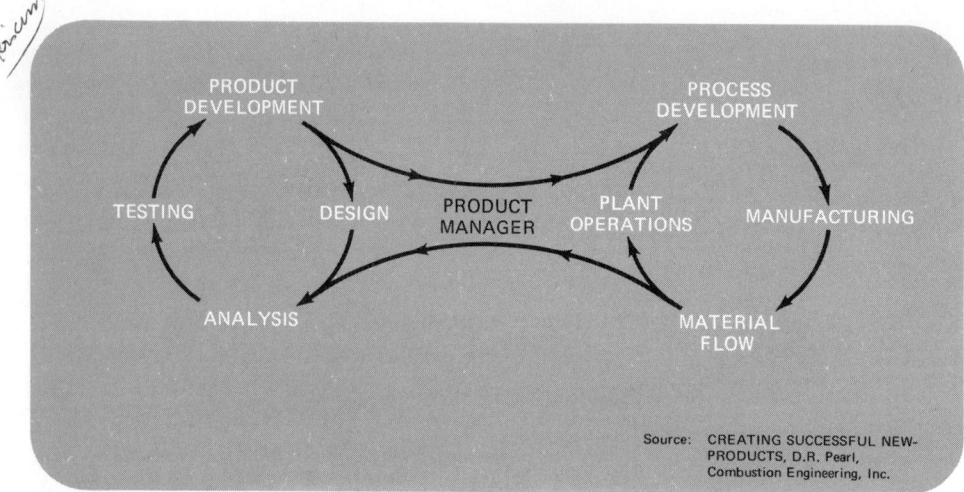

Figure 1-5 **THE ENGINEERING PROCESS**

since it specifies the physical shape and performance characteristics of the product that will demonstrate feasibility of the assigned idea task. Many companies have met with unsuccessful new-product programs by leaving design decisions to draftsmen working from hurried verbal instructions or back-of-the-envelope sketches from engineers in an attempt to economize in this phase. A successful new product program places this area in the hands of a professional design engineer who has a knowledge and understanding of (and in smaller industries must assume the role of) industrial designer and styler, technical analyst, detail draftsman, model maker, test engineer, cost analyst, materials engineer, and manufacturing engineer. He must have a close working relationship with production workers, the purchasing agent, quality control, the salesman, the shipper, the field service worker, the customer, and the investor. Any of these people may cause a good product design to succeed or fail. Analysis is required for purposes of verification and proof of workability, and can make the difference between intuitive judgments (which are often costly because of time-consuming trial-and-error development periods), and formal evaluation based on the optimum use of materials and labor to determine if the product will perform as predicted. Testing is the most costly phase since it requires especially designed space and instrumentation. This phase must be carefully planned to confirm or deny predicted performance, to reveal deficiencies under normal operating conditions, and to prove adequate durability.

The second step in engineering the product, *process development*, also consists of three activities: *manufacturing, material flow, and plant operations*. The *manufacturing* engineer in consultation with the designer will develop the method of fabricating the product and controlling its quality as well as the tools, equipment, and facilities necessary to satisfy design and marketing specifications. Material flow is responsible for procurement, receiving, and handling of all materials related to the product as well as storing and shipping the completed unit. One of the most difficult activities to predict in terms of performance is that of plant operations which must select, train, assign, and supervise the labor force, combining the materials and machines to produce the product according to plan.

Material flow and plant operations in engineering a new product are usually coordinated by a *product manager* who is responsible for interaction between the two groups. Product development and design must have feedback from process development activities to insure savings in cost and time for improvements in present products and model changes as well as for conceptualization of new products to fit present shop facilities. At the same time Process Development is most effective when completely informed of innovations, model changes, and improvements originating in the Product Development Section.

PROBLEMS AND PITFALLS

Once a company recognizes the importance of new products to its total business, understands that there is a process of new-product evolution, and that this process can be managed (organized, directed, and controlled), it will spend sizeable amounts of capital funds to develop an on-going product program. These funds are vital to future growth of the company which can expect returns through successful new products. Any business will prosper through continuing new product successes and can survive occasional product failures so long as they do not occur at frequent intervals. However, the company is "out of business" if it experiences successive failures. Some of the common problem areas in new-product strategy are outlined here in an attempt to aid management to ward off these pitfalls.

organization

Companies new to the new product game often attempt to manage new product programs in the same way they do programs of more familiar items that have proven their worth in sales and acceptance. This often leads to problems of coordinating people to work together constructively and creatively. As stated earlier, it takes a bit more dedication, information transfer, and *esprit de corps* when dealing with the unknown (new products) than with items known to all personnel. In order to decrease or possibly eliminate conflicting objectives among the many functional units involved in new product activity, top management must clearly assign responsibilities for product evolution so as to reduce communication breakdowns and improve working and reporting relationships. The writing of job descriptions with a scheduled (for example, monthly) review in the early stages of reorganization is always a good idea.

control and follow-up

Many companies lack knowledge and/or experience in the initial planning stages of new product evolution and find the product is often late in reaching the commercialization stage and at a higher cost than originally estimated. The process of scheduling events and activities so they will occur on time to insure planned product delivery (thereby keeping costs at a minimum), requires management to initiate adequate control and follow-up methods. PERT (Program Evaluation and Review Technique) has proven to be an effective method of controlling the development of new products from the initial planning and idea stages. This method is briefly discussed as a management tool in the next article and thoroughly explored in theory and technique in Chapter 7.

poor idea screening

New products are often doomed to failure when ideas are selected through glamour, hunches, or emotional belief. There are companies which have learned a hard lesson by marketing a product that was selected on the basis of its emotional appeal rather than screening criteria, or was allowed to pass the screening stage as certain governing rules were bent. Screening must be based on a clear statement in specific terms of corporate objectives and must take into account what price the company is willing to pay in terms of the product. This price is based on personnel, space, availability of materials, marketing, sales force, and whether the company wishes to parallel its present product line or service or diversify it.

inadequate business analysis

It is important that a company determine the business opportunity that exists for a new product before it invests time and money to evolve the product. Products fail when the business analysis program only skims the surface, due to the company's wish to save money on this stage or when the analysis neglects to cover in detail areas that are unfamiliar. This latter failure is quite common because new products are also new to the company producing them and rarely have a business history with which to be compared. The more novel the product, the more difficult the business analysis. The analysis requires a thorough study of the total market, its trend, competition, volume, profit opportunities, consumer needs, and some indication of consumer feeling for the new product idea. If the company is inexperienced in collecting and analyzing data in any of these areas, it is wise to employ a reliable consultant group to handle one or more items.

new ideas and creativity

When a company becomes new product oriented and maintains a program of routinely presenting a successful product on an annual basis, it will experience at some point the leveling-off and ultimate decline of idea generation. This will eventually affect company growth and have a deciding effect on profits. Successful companies guard against this decline in creativity through a continuing evaluation of the rate and relevance of new product ideas, establishing the proper incentive and environment for creative activity, and maintaining an awareness of technological changes, market and population trends, government interests, and other factors affecting opportunities for continued growth.

faulty performance in product development, market testing, and commercialization

Although not a great problem among experienced companies, faulty performance in the last and most expensive stages of new product evolution can decide the difference between product success or failure. Since Product Development is responsible for design and prototype construction of the product, it must work from well-written and carefully thought-out specifications. If these specifications are poorly conceived and do not wholly reflect the stages that precede product development, the product itself will not benefit from the thinking and planning that has already occurred.

Market Testing can cause a product to fail if the testing plan is not well designed to seek out the right questions. Due to the money invested in *developing* a product, management often fails to believe test data and, rather than redesign the product, it attempts to break even by commercializing an item already proven unacceptable.

Commercialization must be well mapped out and represent the complete marketing plan. This stage can cause product failure by presenting the product to the consumer at the wrong time (too early, too late, wrong season, before its time), without proper announcement (advertising), with too little attention to salesman- and consumer-feedback, and by using the consumer to test workability instead of the laboratory.

The fact that the evolution of a new product can be managed does not suggest that one single formula will work for all companies. It does mean, however, that if a company wishes to have some chance of success with new products, it must formulate a plan of operation to insure an optimal use of time and funds in developing them.

Management's responsibility to the new product program is primarily one of decision-making consisting of:

Defining company objectives, both short- and long-range, which will serve as criteria for new product screening

Deciding on which products will enter the product development stage

Deciding when the new product is to be market-tested and deciding the scope of the test

Approving a commercialization plan and a decision on the size of the market effort

Deciding on product changes and market strategy based on feedback information once the product is on the market

Decision-making and budget allocations at any stage of product evolution is most strongly influenced by top management, marketing, and design and development groups. Information and factual data to form the basis for decisions are affected by numerous activities throughout the company. For information to be accurate and to arrive at the right time requires a well-organized effort with assigned responsibility, delegation of authority, and an understanding among personnel of the new-product program. Such an organization, proven effective among many successful companies, consists of the following group.

product planning department

This is a relatively small department of from four to five members whose backgrounds are in the areas of market research, design and development, and administration. This group reports directly to top management and has the responsibility of setting up a program for insuring a continual flow of new product ideas, recommending new product objectives, idea screening, preliminary business analysis, product specifications, coordinating interdepartmental activities to insure cooperation between departments, and precise planning of new product evolution. One of the most successful tools for planning and managing new product development is PERT (Program Evolution and Review Technique), mentioned earlier and discussed in more detail in Chapter 7. This technique assures that the project is thoroughly thought out before execution. It identifies requirements for project review and forces decisions to keep work on schedule.

It clearly indicates individual responsibility and coordinates requirements. PERT also identifies activities that can delay a project and shows those that are most critical to getting the job done on time, as well as indicating risks of altering a program to complete work earlier than scheduled.

product teams

Once a product idea has been accepted, analyzed, and is ready to enter the product development stage, a product team should be formed to coordinate activities among departments and to be actively involved with all aspects of the product through its evolutionary stages. The team should have representatives from each of the major stages, including design and development, manufacturing, marketing and sales, as well as a finance officer. The project team usually reports directly to the project planning department.

new-product committee

If management finds that it has no difficulty in supervising present product lines, has the time to come to grips with crises as they arise and still keep in close contact with its new-product program, there is no need for a new product committee. But, once an industry finds that time normally allotted to new product management is being robbed in favor of more pressing problems, a new product committee can be most effective. This committee is used by top management as a senior advisory board to assist in the review and evaluation of new product plans and, in some cases, set product objectives. The staffing of such a committee is often a difficult task. One effective way is to employ reputable consultants on a retainer basis, to use newly-retired executives from the company as well as the competition, and to use key managers of divisions within the company who have creative ideas. A combination of the personnel suggested (from five to six) will often result in a most effective committee.

EXERCISES

1. Name three industries that change a product model at a frequency of at least once each year.
2. List as many products as you can think of that have undergone a slow evolutionary change but have increased sales appeal through innovations in packaging.
3. Name five new products in the appliance or household line that have been introduced in the past six years.

4. Name a company that has had a new product or model failure and bounced back through success with a new product or model. What was the product or model?
5. Write down on paper a product you are familiar with. List three ways in which you could enhance its sales appeal.
6. List one new product you think would have sales appeal in each of the following consumer areas, even though the product has not yet been invented:
    a. The automotive parts industry.
    b. The vacation-camper industry.
    c. The kitchen appliance and/or utensil industry.
7. What, in your opinion, does planned obselesence mean in the new product game?
8. What criteria would you impose on a new product idea to have it accepted by top management?
9. List as many reasons for new product failures as you can think of in five minutes.
10. Construct an organization chart to effectively manage a new product program.

2 THE CREATIVE PROCESS

> "Imagination is more important than knowledge, for knowledge is limited, whereas imagination embraces the entire world—stimulating progress, giving birth to evolution. . . ."
>
> Source: Albert Einstein

> "I submit that creativity will never be a science—in fact, much of it will always remain a mystery—as much of a mystery as 'what makes our heart tick?' At the same time, I submit that creativity is an art—an applied art—a workable art—a teachable art—a learnable art—an art in which all of us can make ourselves more and more proficient, if we will."
>
> Source: Alex F. Osborn, from an address at M.I.T.

CREATIVITY

Creativity can be defined as a successful step across the borderline of knowledge. It is an addition to knowledge—a defining of things previously unknown to one's self. Creative engineering devices are often a combination of known components or principles in a novel and unique way. Engineering creativity is more akin to inventiveness than research. The creative person has a driving curiosity, a willingness to explore the nonconventional, to emphasize the unique as opposed to the traditional, to continually seek out the need for a device or product, and he believes that a truly unique solution exists to the problem at hand.

The best preparation for creative ideas is *experience*. Firsthand experience provides the richest preparation, for it is apt to stay with one and may be recalled when needed. Such preparation may be classified as *active*. *Passive* preparation results from secondhand experience, such as reading, listening, or speculating. Therefore, the creative person spends much time in gaining personal experience through observation, tinkering with devices, and investigating the multitude of products we have today to find out how they work. He is continually aware of the needs of society for time-saving, labor-saving, and comfort devices.

Creativity is a personal thing requiring self-discipline. The personal trait whereby an individual sticks with a difficult problem, becomes obsessed with it, and does not leave it until a solution is in hand, will usually result in a creative act. For this reason the creative person is usually unconventional, has intense faith in himself, prefers to work alone, and is often disturbed by rules and regulations.

IMAGINEERING

Creative discoveries are more likely to occur when one lets his imagination soar and then engineers it back to earth. The Aluminum Company gave birth to a word that aptly describes this activity—imagineering. One can imagineer only if he is able to cast off the traditional, build up an immunity to the "it won't work" critics, and accept the challenge to achieve the "impossible." Possible solutions to the difficult, messy, unsolved problems around us will never be met with present engineering techniques. Historically, solutions offered through present engineering know-how are only temporary and often cause additional problems at a later date. We are now experiencing some of these problems with super highways, air pollution which is a byproduct of the industrial state, and unemployment through automation. Through the technique of imagineering, which is simply allowing one's imagination to think "way out" and then engineering the thought back to reality, we may find lasting solutions to many of today's unsolved problems, including poverty, slums, segregation, automation, resource conservation, air and waste pollution, transportation, education, and urbanization.

IDEATION (produce ideas)

Almost every creative idea is selected from among a large number of less significant ones. If enough alternatives are listed, the mathematical likelihood of finding a really creative solution is increased. The process through which this is achieved is known as *ideation* and requires alertness, imagination, and a bit of self-discipline to generate truly worthwhile alternatives to a given task. One source of idea-building is through *association*. Association of ideas, the phenomenon by which one imagines something else when seeing, hearing, tasting, feeling, or smelling what exists at the moment, can be used most effectively if one lets his imagination soar to other ideas, allowing one idea to build on another.

It has been proven that the person capable of producing a large number of ideas per unit of time has a greater chance to produce truly significant ones. The skill of coming up with new ideas requires constant practice and can be increased through an awareness of nature, objects, and everyday occurrences.

The personal game of thinking up alternative solutions to tasks can be most productive when the mind is unoccupied, for instance, when mowing the lawn, painting, washing windows, waiting for transportation or an appointment, or doing anything not requiring mental activity. In fact the game of idea-thinking can often make these things bearable. If one gets an idea that is truly unique or seems worthwhile, it is wise to write it down as soon as possible, for it may be forgotten once the mind starts working on something else.

A more direct approach to ideation requires that characteristics of the problem be analyzed according to the following suggested criteria:

what's wrong with it?

Make a list of all the things you feel are wrong with the present product, idea, or task. (Twist-off caps are replacing pry-off types.)

how can it be improved?

Forgetting feasibility, list all the ways you would improve the present product, idea, or task. (The addition of steel studs to snow tires increases their traction.)

other uses.

What other uses does it have in its present form? What other uses are there if the idea is modified? Can it perform a function that was not originally intended? (The collapsible spare tire was first designed for aircraft use.)

modify.

Change trim, shape, description, weight, sound, form, contours. (The yearly automobile model change uses the criteria that the public's task continually changes.)

magnify.

Make larger, higher, longer, wider, heavier, stronger. (The giant economy size).

minify.

Make smaller, shorter, narrower, lighter, subtract something, miniaturize. (The introduction of micro switches has created a whole new industry, to say nothing about the mini-skirt.)

adapt.

Is there something similar? What can be copied? Can it be associated with something else? Is there something in stock or surplus that can be used? (By gluing together chips of wood, usually discarded by lumber mills, companies are now producing chipboard that has some advantages over plywood.)

reverse.

Try a new twist: opposites, upside-down, around, rearrange, opposite pattern, opposite sequence. (A classic example is the fisherman's spinning reel which is just the opposite in design from the bait-casting reel.)

new look.

Change color, form, or style, streamline, use a new package or new cover. (The princess telephone as well as novel packaging techniques attract public attention).

old look.

Copy a period, antique, parallel a previous winner, look for prestige features, trade on "they don't build them like that anymore." (Many lighting fixtures follow this trend.)

rearrange.

Try a different order, interchange components, piece together differently, change places. (Cars with engines in the rear are selling quite well and the front wheel drive seems to be catching on.)

substitute.

What can take its place: plastic for metal, metal for plastic, light instead of dark, round instead of square. What other process, principle, theory, or method can be used? (The use of plastic instrument panels on automotive dashboards has reduced costs as well as pitting problems from chrome-plated metal.)

EMPATHY : put yourself in the place of something else.

combine.

Combine ideas, principles, methods, groups, components, hardware, issues. (Computerized stabilizing fins on ships are a good example of combining engineering principles.)

convenience.

Make it easier, less work, easier to reach, disposable, simpler to use, quicker, automatic. (The electric can opener, knife, disposal, dishwasher, and others have made the kitchen an enjoyable area for the housewife and created many profitable industrial divisions.)

safety.

What devices, properties, controls, or sensors can be added to prevent injury, accident, explosion? Make it nontoxic, nonpoisonous, nonflammable, stainproof, nonbreakable. Is it safer than a competitor's product, less harmful, less easily cut or torn? (The electric razor industry as well as the manufacturers of plastic bottles for laundry solvents, toiletries, and gasoline containers use safety to promote their products.)

Another method of idea generation recently proven successful for troubleshooting an especially sticky problem is that of *role-playing*. This requires the idea generator to place himself within the situation and through self-interrogation ask what he would do if he were the thing, the idea, the device, or the issue. This method can also be used effectively to test the feasibility of an idea when one "becomes" the idea and others ask leading questions, for and against it. To test market acceptance, a number of engineers or executives can become customers and try to think of ways of tearing the product down or reasons for not purchasing it. Thus role-playing can be used individually or among groups to produce effective ideas. This technique requires a bit of experimentation, however, to determine the medium that is most productive.

FUNCTIONAL VISUALIZATION

Creative design solutions to problems can be stimulated by the manner in which the problems are defined. Too often the engineer develops outstanding solutions to the wrong or an insignificant problem. The technique of functional visualization, developed by Don Taylor, Senior Value Consultant, Harbridge House, Inc., serves as an interface between problem definition and problem solution which prepares the mind and broadens the frame of reference.

The primary purpose of functional visualization is definition of the problem so as to maintain proper emphasis on the functional concept and to ensure a smooth, thoughtful transition from function to the concept best answering the requirements. An example of the type of problem definition that is typically placed before the designer is: Design a unique or novel lawn mower. The designer would naturally visualize the present two- or four-wheel rotary or reel-type lawn mower when attempting to conceive of a novel design. He would probably come up with a solution to improve what already exists instead of inventing something new. A problem statement that embraces the same objective but emphasizes function would be: Design a method or device for shortening grass. Here the designer will automatically think of the function involved and place hardware on a secondary level. Hardware should be thought of as the means of accomplishing the function. Another example is to charge a designer with: Design a novel can opener. Using the functional visualization technique of problem definition, the following might result: Design a method of removing the contents from a can.

In summary, functional visualization involves picturing the function to be accomplished, devising methods of achieving the function, and then assembling the hardware necessary to support these methods.

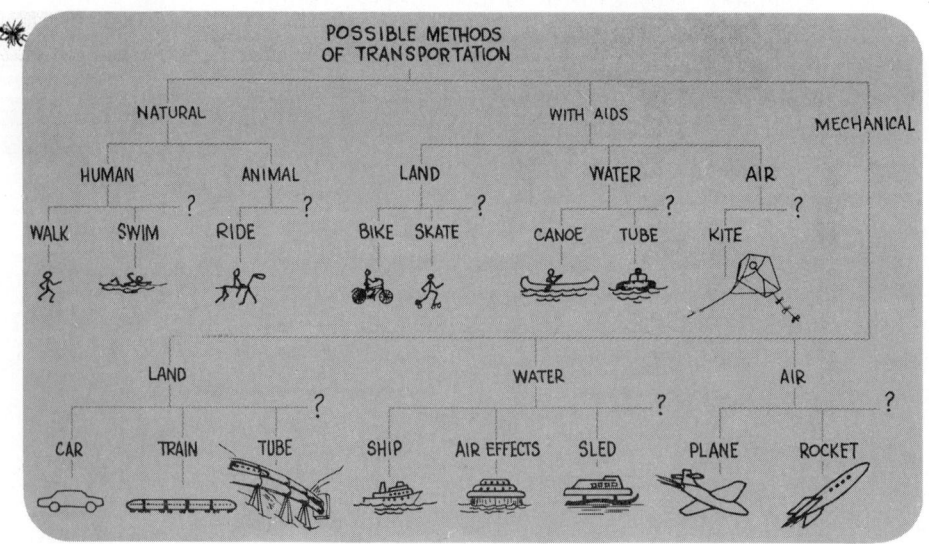

Figure 2-1 **TYPICAL IDEA DIAGRAM**

IDEA DIAGRAM

It is extremely difficult for the mind to handle more than one thought at a time, while the eye can accept an almost infinite variety of things simultaneously and readily distinguish one from another. The method of idea generation through the use of an *idea diagram* takes advantage of the visual media to add a dimension of flexibility to the mind in seeking alternatives to a problem or task.

An example of the use of the idea diagram is illustrated in Fig. 2-1, where the designer is seeking ideas for a novel transportation system. The diagram begins with a listing of general headings, then subheadings, sub-subheadings, and so on. The more fragmented the diagram, the more the likelihood of significant ideas. It is good practice to sketch some of the ideas for a clearer understanding since it forces the designer to offer positive evidence of his thinking. The diagram shown is incomplete to allow the reader to involve himself with this method (See Exercise 8 at the end of this chapter).

IDEA MATRIX

A more exacting procedure of idea generation consists of a morphological (pertaining to form or structure) analysis of the independent variables related to the problem or task. Each variable is subdivided into parameters, types, features, or methods, and listed separately to form a matrix grid. The intersection of listed descriptions represents alternative ideas or suggested solutions to the given task. Thus, the number of combinations produced will be greater than those generated through free association.

Considering the task stated in the previous article of seeking ideas for a novel transportation system, Fig. 2-2 shows how 30 alternative ideas were generated for a mechanical land system considering two variables: type of system and power sources. If the same task is now investigated with the addition of a third variable, funds (which is more often the case), the matrix takes the form of three dimensions in the shape of a cube made of smaller cubes whose number depends on the subdivision of variables. In this case 150 possible combinations of subdivisions exist producing the number of alternative ideas. Many of the ideas are obviously ineffective and will be discarded immediately, but the designer has at hand many to consider that otherwise might have gone unnoticed.

BRAINSTORMING

Brainstorming is a technique of originating new ideas through an organized group's creative collaboration. The term is derived from using the brain to storm

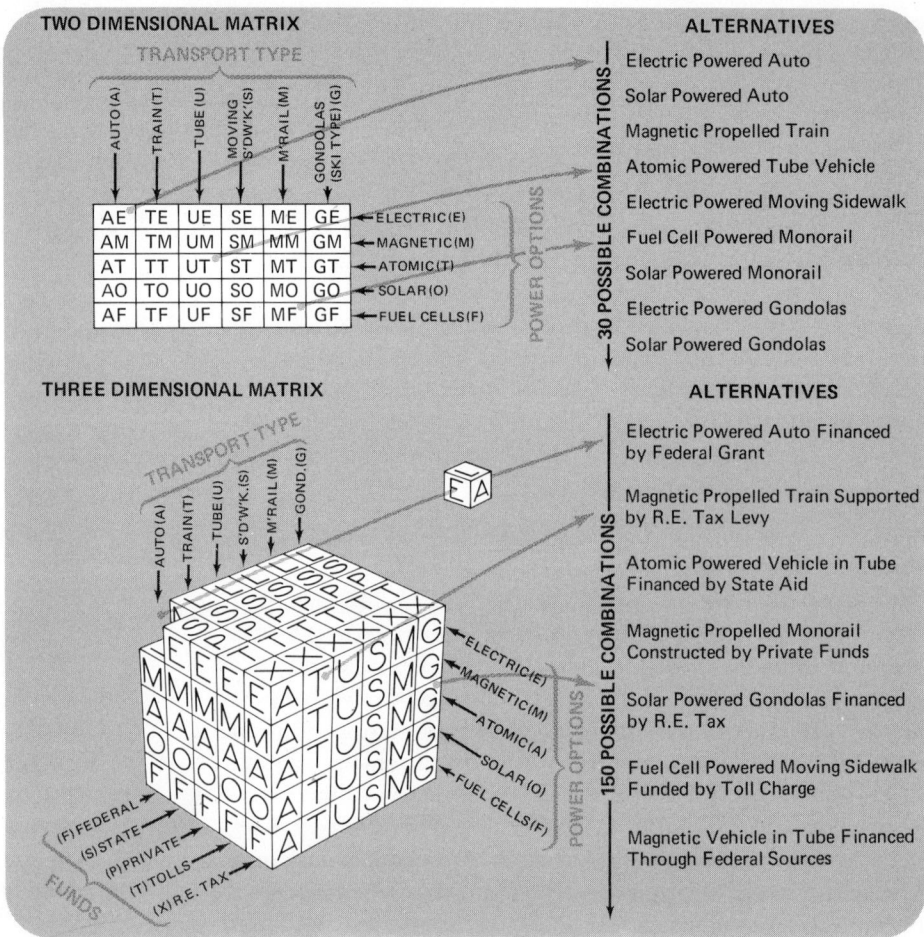

Figure 2-2 **MORPHOLOGICAL IDEA MATRIX**

creative solutions to given problems—and to do so vigorously, with each person focusing his attention on the same objective. This technique was invented in 1939 by Alex F. Osborn, co-founder of Batten, Barton, Durstine, and Osborn Advertising Agency; Vice-Chairman of the University of Buffalo; trustee of several banks; director of four manufacturing corporations; and founder of the Creative Education Institute.

A sizeable number of brainstorming sessions have been held throughout the country spanning a multitude of problem areas and nearly all have been successful in terms of the ideas generated. Leading companies as well as civic groups have found this technique useful. General Motors, General Electric, Westinghouse, federal agencies, state and city governments, the armed forces, and educational institutions are among those using these sessions. If a brainstorming session proves unsuccessful or results in chaos, its failure can usually be traced directly to the chairman. The chairman must under no conditions flout his knowledge of a subject or attempt to dominate the group, for this may result in silence, particularly among the more timid members. It is also important for the leader to suppress any criticism that tends to creep into the proceedings such as "it won't work," "they already sell those," "it won't sell," "whoever heard of doing it that way?" The chairman must insist at the outset that ideas produced during the session not be judged immediately but only after a period of rest following the session. An informal session with all members tossing out ideas about a stated problem area with the chairman acting in the capacity of moderator and sometimes adding stimulus to get things started usually results in a worthwhile and productive session.

Brainstorming is not a bull session but a concentrated group effort to produce creative ideas. Each member of the group must focus his attention on the stated task and not get carried away with opinions and speculations. A group of five to ten seems to work best during a period not to exceed 60 minutes. If the organizer finds he has more than the optimum number interested in participating, these people may split into two or more groups in competition to see who comes up with the greater number of ideas. The session requires a chairman or moderator, a recorder (human or mechanical), and an assigned member who will bring in a few ideas to stimulate initial thought. In some cases, the chairman himself can fill this role. The technique is based on one idea building on another, combining with another, and triggering another, until ideas seem to spill forth with little effort. There must be few if any inhibitions among group members. They must be loose (flexible) in their thinking, and imaginations must be allowed to soar, with people speaking out what is on their minds. Ideas revealed in the session are reviewed later, obviously bad ones are discarded, and a list is prepared in hierarchical form from good to poor for intensive study.

The following rules are considered basic for fruitful sessions:

criticism of ideas is not permitted.

This should be stated early in the session and if violated, the violator should be reprimanded or asked to leave. Criticism often leads to ridicule which stifles creative thinking.

"free-wheeling" or "loose thinking" is welcome.

Alex Osborn states, "The wilder the idea, the better; it is easier to tame down than to think up."

ideas are wanted in quantity.

As stated earlier in this chapter, the mathematical probability of one or more ideas proving truly significant is directly proportional to the number of ideas generated.

transpose thoughts and combine ideas.

Group members should try to improve ideas of others, allow one idea to build on another, and attempt to combine certain ideas into an alternate possibility.

SYNECTICS (like Brainstorming but less ideas)

The synectic approach to idea generation is somewhat similar to brainstorming in that it is based on group effort to suggest possible solutions to a given problem. It differs, however, in that only a few ideas (two or three) are sought and in greater detail, with the chairman playing a dominant role during the discussion.

The problem or task is first explained to the group (chairman, five to ten members, and a recorder) in detail and repeated until all understand it thoroughly. The chairman then begins the session by selecting the method of attack, such as role-playing (described earlier), an investigation of certain minute details of the problem, or presentation of an analogous situation which may or may not have an obvious bearing on the problem. When an interesting idea of possible significance is suggested by someone in the group, the chairman attempts to steer the discussion into an elaboration of and sometimes an analysis of the idea.

The synectic technique might be used, for example, in designing space tools, more specifically, designing a device for drilling a hole in an orbiting platform by an astronaut. The chairman would review the problems related to such a device, such as need for portable power sources, lubrication at extremely low temperatures, zero reaction force devices, and storage of the tool. He may choose the analogy approach by suspending a piece of styrafoam from a length of thread attached to the ceiling and attempting to drill a hole through it with an electric drill without steadying the styrafoam with his free hand. Of course the styraform will sway as drill force is applied and the hole cannot be made. A live and sometimes dramatic demonstration such as this is usually strongly motivating to any observer. The group is now instructed to suggest methods of producing a hole in the styraform and sooner or later someone will suggest burning it with a cigarette. The chairman will then lead the group to zero in on this idea for a more detailed description of a device to achieve the same objective in space.

In general, the technique of synectics is based on the fact that the mind is more productive when dealing with a new or foreign environment. The analogous situation quickly takes one away from the exact problem at hand (with traditional approaches to a solution) and requires him to consider a related one. This has a tendency to make the familiar strange. If a group were considering a novel snow removal system, they might discuss how the soil is filled or fallen leaves are disposed of. Considering the design of an office building, the group might profitably discuss how a beehive is constructed. The tree might be studied when dealing with structural shapes. Novel methods of mowing a lawn might take the form of a detailed analysis on the process of cutting and tearing.

CREATIVE PROCESS

Creative ideas do not usually occur in a flash, as we are led to believe by writers wishing to dramatize the subject. We often read of the inventor who came up with a completely novel idea through a sudden illumination in which a new device was unfolded with all of its details. Such a phenomenon is rare and more often is over-publicized and grossly exaggerated. Creative ideas occur more often to one who has a driving curiosity, something all children have but adults outgrow. Creativity results from attention to troublesome detail that the ordinary mind discards as trivia. It is believed by many that a step-by-step, orderly process will result in the creative solution to a problem or a novel device or idea. It should be remembered, however, that there is no formula for creativity. What works for one may not work for another. The following process, although expressed in step-by-step order, will produce creative ideas when used separately, singly, or in an order preferred by the individual. Its importance lies in understanding the mental processes involved.

step 1: irritation and decision.

Creativity is often triggered as one is confronted with something irritating or disturbing in a situation. This situation raises a problem and the issue of one's decision and will to do something about it. ("Necessity is the mother of invention" but action produces the inventor.)

step 2: preparation.

The preparation stage is a period of conscious, direct mental effort. This vital second step requires mental discipline of the highest order. During this period one explores in great detail all possible solutions and various combinations that may lead to a satisfactory answer. Quite often the problem is solved at this stage. If the problem is not solved to the satisfaction of the designer, at least he is familiar with the task in the most minute detail.

step 3: incubation.

The mind, now fully prepared with all possible variations yet unable to come up with the creative idea, continues to worry about a solution even though one wishes to discard the task and go to another assignment. This period is when the subconscious mind begins to work on the problem. Here the problem is allowed to ferment, to cook, to incubate over a period of time as the subconscious tries different "forgotten" combinations.

step 4: illumination.

When the creative idea or unique solution occurs in a flash, usually during a period of rest or during engagement in an activity completely foreign to the problem, illumination has taken place. Experts feel that the subconscious works harder in probing deeper for ideas when the conscious mind is at rest (often during sleep or relaxation). For example, have you ever searched your memory for the name of a person, place or thing without success, but have the answer flash into your mind while thinking of something else? This is illumination.

step 5: verification. (analysis, experimentation,)

Illumination has occurred and the creative idea is at hand. The idea must now be judged to determine if it is truly a solution. Judgment requires evidence to prove if an idea is actually worthwhile and this can be found through analysis, experimentation, or sometimes the opinion of recognized authorities. The process is known as verification and often requires a great deal of labor. This is the final and most necessary step in the creative process.

Figure 2-3 models graphically the inner workings of the mind during the creative process. Consider the *memory* to be a bottomless box divided into two main compartments, separated by a screen termed the *memory barrier*. The upper level contains *recent* or *hard* facts, such as one's name, address, important dates, facts relevant to one's discipline, such as Newton's three laws, and so forth. The lower level contains *old* or *light* facts that one was once exposed to but has long since forgotten, such as old telephone numbers, the name of one's fifth grade teacher, or what TV show was seen last Tuesday evening. These facts, seemingly

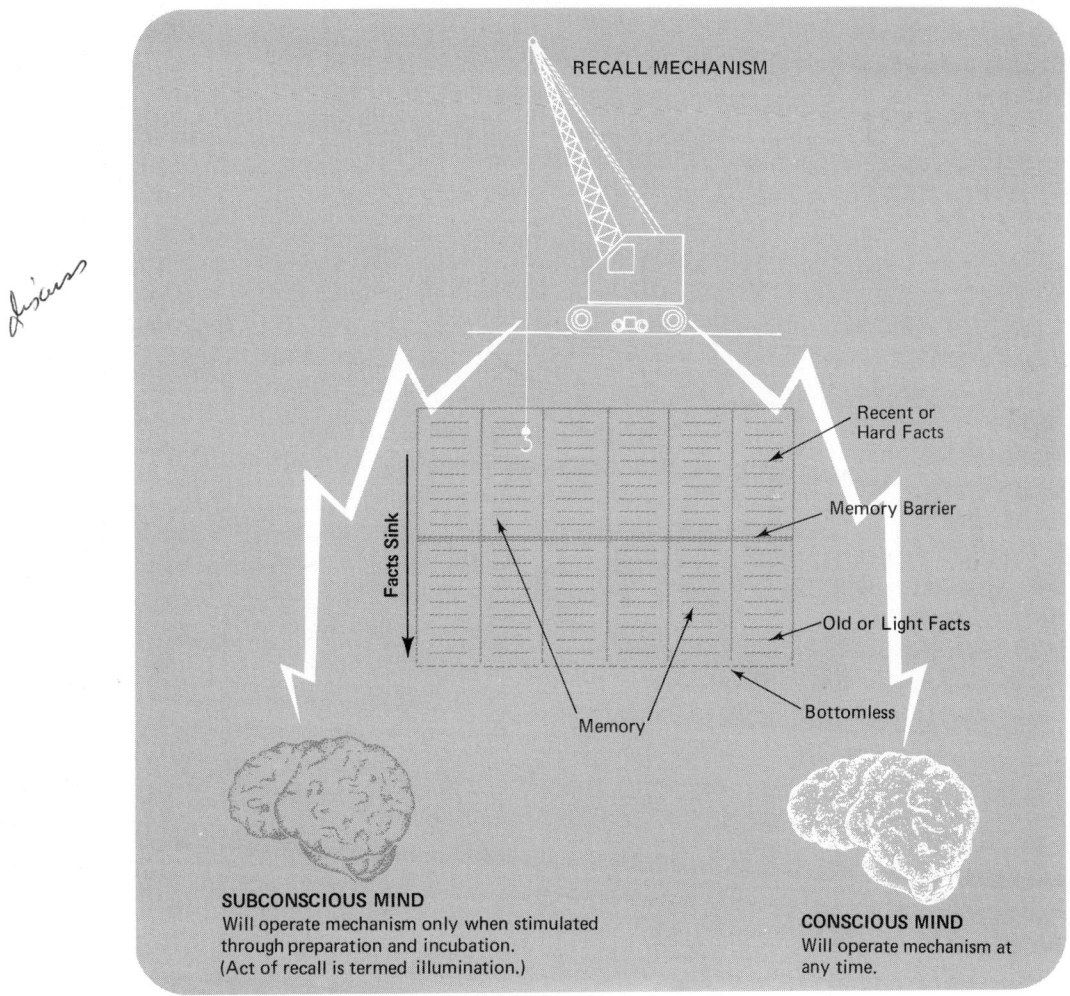

Figure 2-3 **THE CREATIVE PROCESS**

forgotten, are extremely difficult to recall, for they are below the memory barrier. Unfortunately, facts *sink* and the longer one stores a fact and does not recall it at frequent intervals, the lower it will sink below the memory barrier, seemingly never to be recalled.

As stated earlier, a creative thought is not a completely new idea; no one can create something that does not already exist. The creative thinker actually evolves new and untried combinations of ideas that already exist in his mind. The mind is an infinite storehouse for ideas with a great number added every day. The active and curious mind adds many more ideas than a passive one and therefore is able to come up with many more new combinations.

Consider now the act of recalling facts from memory. The *conscious* mind can activate the *recall mechanism* at any time. The slightest signal will cause the mechanism to hook onto a fact and bring it to the attention of the mind. If one wishes to recall a fact lying fairly deep, he must provide a much stronger signal to the recall mechanism to cause the hook to probe deeper. The strength of the signal is directly proportional to the amount of preparation (step 2) the mind undergoes. If this preparation is of a high order and the fact is not recalled, the mind will continue to be bothered and incubation (step 3) will take place. During this period (one of rest of the conscious mind) the subconscious will provide a signal to the recall mechanism, causing it to probe deeper below the memory barrier searching out seemingly forgotten fact. The best time for this activity to take place is during a period of rest, relaxation, or sleep. More often than not the recall mechanism will bring up facts to be recognized by the conscious mind as completely novel and unique (illumination takes place). These facts, ideas, creative thoughts, and inspirations should be immediately recorded for the conscious mind to act on or they will quickly sink below the barrier, never to be recalled. This is why many creative people keep a note pad and pencil on their bedside table, for illumination will often awaken one from a sound sleep.

BARRIERS TO CREATIVITY

In addition to practicing techniques of ideation and the creative process, one must isolate himself from many of the common barriers that prevent full realization of a truly creative approach to design. These barriers may be classified as *personal* and *organizational* in nature. This does not mean there is no overlap between the two, but the division presented here is intended to aid the individual in recognizing certain pitfalls and to show industry how it may provide an environment for creative activity. Some of these barriers, several with brief descriptions, are listed below with examples of how they may be overcome.

PERSONAL BARRIERS	ORGANIZATIONAL BARRIERS
FUNCTIONAL FIXEDNESS	EMPHASIS ON IMMEDIATE FUNCTIONAL UTILITY OF IDEAS.
Familiarity with certain objects or concepts establishes a fixed usage regarding their function, thereby limiting their value. (We think of a paper clip to hold papers together but it may be used as a pipe cleaner, a spring, a small connecting link, a fingernail cleaner, a punch or reamer, and it is probably the best lock pick ever devised.)	Pressure for production usually results in a rush to meet prohibitive deadlines set by management.
	GENERAL DISTRUST OF ORIGINALITY

HABIT TRANSFER

This is a carry-over of past conditioning of thoughts, methods, and actions to new problems. A fixed approach to problem-solution often results in attacking new problems with old methods. (Don't be disturbed if you cannot find a prior art—don't try to fit the solution to an existing mold. Invent techniques that seem to lead to a satisfactory solution. This is especially necessary in the laboratory.)

PRACTICAL-MINDEDNESS

The straight-to-the-point type insists that instead of roaming imaginatively around the problem we get down to the facts immediately, thus approaching the problem too soon. Premature particularization stifles creativity. (It is best when attempting to solve a new problem or design to first think way out—no matter how ridiculous—and then begin to engineer the solution back to reality.)

OVERSPECIALIZATION

Specialization may limit one's horizon to the point where his depth is shallow in the engineering and physical world in general, thereby limiting his interdisciplinary conceptual roamings. (Never discard ideas that seem to be outside your discipline. Learn as much about different areas of knowledge as possible—psychology, medicine, the arts, economics, world affairs. Cultivate a curiosity. Don't dig a trench and find yourself in a rut.)

TENDANCY OF MANAGEMENT TO TELL THE CREATIVE ENGINEER WHAT TO DO AND HOW TO DO IT.

ORGANIZATIONAL SETUP ALONG RIGID AUTHORITY–OBEDIENCE LINES.

REFUSAL OF MANAGEMENT TO DELEGATE RESPONSIBILITY.

LACK OF LONG-RANGE OBJECTIVES

DISAGREEMENT WITHIN MANAGEMENT CONCERNING MAJOR OBJECTIVES.

DISCOURAGEMENT OF EXPERIMENTATION.

FREQUENT CHANGES OF KEY DECISIONS.

LACK OF EFFECTIVE COMMUNICATIONS BETWEEN ENGINEERS AND MANAGEMENT.

HORIZONTAL RATHER THAN VERTICAL FLOW OF NEW IDEAS.

FAILURE OF MANAGEMENT TO RECOGNIZE AND REWARD CREATIVE ABILITY.

DEPENDENCY ON AUTHORITY

Often engineers and students are so impressed by the judgments and approaches of recognized authorities that they immediately accept their leadership and fail to develop leadership qualities of their own. The more quickly one believes all men are ordinary human beings—some are recognized more than others for they seem to meet the challenge—the more creative one will be. Creative ideas need not feed on authoritative stimulus for these ideas are unique and can only be originated through the individual expressing himself.

FEAR OF RIDICULE

The highly creative individual generates nonconventional ideas. The more creative an individual, the more unconventional his ideas. If the designer is continually appraising his ideas to see if others will consider them acceptable, he again stifles creativity. (Almost any proposed idea can be shown to be impracticable, wrong, or insignificant from an immediate logical point of view; and there are many who thrive on idea criticism. Don't be discouraged when you hear, "I've seen it before somewhere," "It won't work," "Someone must have thought of it before," "It will never sell.")

Source: Eugene Raudsepp, Vice President, Princeton Creative Research, Inc., "Removing Barriers to Creativity" *Machine Design*, May 24, 1962.

NEGATIVE ATTITUDE ON THE PART OF MANAGEMENT TOWARD ALL NEW IDEAS.

Donald N. Frey, Ford Motor Company, made the following remark at an Engineering Graduate Seminar at the University of Michigan: "Any manager can burn up talent faster than he can buy it, if he defaults on the opportunity to expose his good men to new opportunities, new experiences, and new challenges . . ."

RELUCTANCE OF MANAGEMENT TO TAKE CHANCES

POOR HANDLING OR OUTRIGHT MISAPPROPRIATION OF CREDIT.

SATISFACTION WITH *STATUS QUO*.

TENDENCY TO OVERDEVELOP ROUTINES AND STANDARD PRACTICES.

FAILURE TO HIRE CREATIVE ENGINEERS BECAUSE "THEY ARE NOT LIKE OUR GROUP."

A wise organization would profit from using this as a check list and honestly practicing a bit of self-interrogation to determine if it is providing the maximum opportunity for creative effort among its personnel at all levels.

Source: Eugene Raudsepp and Editors of Machine Design. "Stimulating Invention." *Machine Design*, April 1, 1965.

EXERCISES

1. An engineer while working on his lawn one Saturday afternoon thought of combining the power lawn mower with the lawn spreader. His idea was to fasten a plastic container to the rear frame of the mower for fertilizing the lawn while mowing it. List as many ideas as you can which combine elements to save time or effort on the part of the user. (Hint:

Think about the kitchen, the bathroom, the automobile, air travel, handyman tools, camping—an area or hobby where you have had some direct experience.)

2. Imagine that you are employed by a soap company in the bar soap division. Your task is to find as many different uses of bar soap as possible other than for washing. The challenge here is to increase sales. List as many different uses as you can think may help sell the product.

3. Visualize an ordinary pair of pliers (two forged parts pivoted by a rivet). List all the ways pliers have been improved to make them more efficient instruments. Now suggest three further improvements.

4. List every possible use that might be made of a toothbrush after it has served its usefulness brushing teeth.

5. Listed below are a number of common machines or appliances. List after each all of the things you think are wrong with them or all improvements that could be made over existing designs. After reviewing each list, identify a need for a new product that you feel is an improvement and seems to have market potential.

    a. Table model radio   c. Lawn mower
    b. Vacuum cleaner   d. Outboard motor.

6. Using the technique of functional visualization, restate the following design problem definitions so the designer will ideate with function foremost in his mind:

    a. Design a novel desk lamp that will compete with the tensor light.
    b. Design an improved snow blower.
    c. Invent a pen or pencil to compete with the ball point pen.
    d. Ideate methods of improving urban transportation systems.

7. Consider that you have been assigned the task of finding a novel way to open cans that will compete with the pull-ring-top. Assign the task three independent variables and construct a three-dimensional idea matrix through a subdivision of each variable. How many possible ideas will your matrix generate?

8. Figure 2-1 illustrated a typical idea diagram for coming up with ideas to propose a novel transportation system. However, the diagram is incomplete. Copy the diagram and complete it as far as your experience will allow. Now, after reviewing the diagram, propose at least one effective transportation system to be used in each classification (land, water, and air.)

9. Operation Paper Clip: Make a collection of the following materials which are common articles found around the average desk. Using only these materials (as many of them as you see fit), create and construct through an organized plan something useful.

    12 paper clips   9 rubber bands
    1 razor blade   2 pieces of poster paper (8½" × 11")
    6 thumb tacks   1 piece of aluminum foil (8½ × 11")
    2 safety pins   4 flexible binding pins
    4 pencils   1 length (2 ft) of insulated wire

10. List as many barriers to creativity (personal and institutional) as you can think affect idea output in your present position.

3 THE DESIGN PROCESS

"Design is regarded as the process of selectively applying the total spectrum of science and technology to the attainment of an end result which serves a valuable purpose."

"It is the segment of engineering which devises and develops new things, in contrast to other segments which emphasize the solving of problems or the generation of engineering information."

"The responsibility of the design engineer is to use the maximum powers of creativity, judgment, technical perception, economic awareness, and analytical logic to devise uniquely useful systems, devices, or processes."

Source: R. J. McCrory, from *The Design Method—A Scientific Approach to Valid Design*, a paper presented at ASME Design Engineering Conference, May 1963.

ENGINEERING DESIGN

To design is to create—to put together something new or arrange existing things in a new way. Design often results in a product that is intended to turn a profit. *Engineering design* is a continuous process whereby scientific and technological information is used to innovate a system, device, or process that will benefit society in some way.

The ability to design is both a *science* and an *art*. As a science, it can be learned through a systematic process (outlined in this chapter), experience, and problem-solving techniques. As an art, it must be practiced with total involvement for one to become proficient. Good design requires both *analysis* and *synthesis*. More often than not analysis is confused with design, for many assume that because a given situation is well analyzed it must be well designed. They neglect to recall where the given situation originated. Analysis, the separation of the situation under consideration into manageable parts and the examination of each part, is performed on the design to verify the goals originally set. Synthesis, which is concerned with the assembling of elements so as to form a whole, is closer to design than any other factor but must be tempered with imagination and creativity so that the assembled elements come out in a unique way. Science and art, analysis and synthesis cannot be separated in the process of design for they rarely occur independently. Usually they are simultaneous activities.

The design of a device, system, or process can be brought about in one of two ways—*evolutionary change* or *innovation*. Evolutionary change is a carry-over from the early days of the Industrial Revolution when product competition was almost unheard of, technological advancement rather slow, and investment risk at a minimum. When a product is allowed to evolve over a period of time with only slight improvements added, the risk of making major errors is minimized and the creative capabilities of the designer hardly excited. More today than ever before, we find innovations in design occurring on a regular basis. The rapid growth of scientific and technological discoveries plus competition among companies for their slice of the market has placed a great deal of emphasis on new products (discussed in Chapter 1) which draws heavily on innovations to produce marketable ideas. Here the creative skills and analytical ability of the design engineer are taxed to the limit, for quite often there is an absence of prior art on which to base assumptions. Therefore, such a practice of designing carries with it a high-risk factor. As with science and art, analysis and synthesis, the designer controls the activities of evolution and innovation to occur somewhat simultaneously. His initial thinking must be along innovative lines in order to conceive a competitive idea that possesses originality. As the design begins to unfold, the designer continually tests his ideas against prior art or state-of-the-art in order to verify them in some way. One can design for the future but he must base results on what is known of the past. Therefore, innovation and evolution are controlled, simultaneous activities of the design engineer.

The individual who has the best chance of success in engineering design activity is one who has acquired the following attributes through education (formal and/or experience), self-discipline, and enjoyment of the creative challenge.

knowledge.

He must understand and be familiar with physical laws and mathematical principles.

skills.

The designer must have developed and continue to improve on analytical techniques, graphical expression, experimental techniques, and oral and written methods of communicating ideas.

attitudes.

It is essential that the designer have a questioning approach, an active curiosity, flexibility of thought, intellectual integrity, responsibility for design, willingness to make decisions, ability to present ideas and to defend them, and a desire to develop judgment through experience.

creativity.

Success also depends on a willingness to explore the nonconventional, to emphasize the unique as opposed to the evolutionary, to continually seek the need for a device or system, and to believe that a creative solution exists.

SCIENTIFIC METHOD

Prior to a discussion of the methodology of design it is of interest to compare the "design method" with the "scientific method" to show the parallel in these intellectual pursuits. Although these activities occur independently, their continued existence depends upon the related success of each. Results obtained through the design method usually finance scientific study which feeds additional information to the designer as fuel for novel designs.

For the purpose of discussion and comparison, the scientific method and design method have been idealized in Fig. 3-1, with only the major stages shown. Each stage may be broken into a number of substages and even further divided when dealing with new scientific ideas as well as when designing new products.

The scientific method begins with a body of *existing knowledge* in the form of scientific laws. As the scientist observes nature in light of these laws, his scientific curiosity requires him to question, research, or explore a new idea from which he conceives a *hypothesis,* explaining the phenomenon or idea. The hypothesis must now be subjected to logical analysis which either confirms or denies the explanation of the idea. Hypothesis and analysis are iterative since analysis often shows a weakness or flaw in the hypothesis which must be altered and then re-analyzed. Once the idea has been confirmed to the satisfaction of the originator, he must have it accepted as proof by fellow scientists. The idea is finally communicated to the scientific community, thereby enlarging the body of existing knowledge, and completing the knowledge loop.

The design method begins with a knowledge of the technical state-of-the-art which includes devices, materials, fabrication techniques, components, processes, the market, and so forth. As society is troubled with problems such as

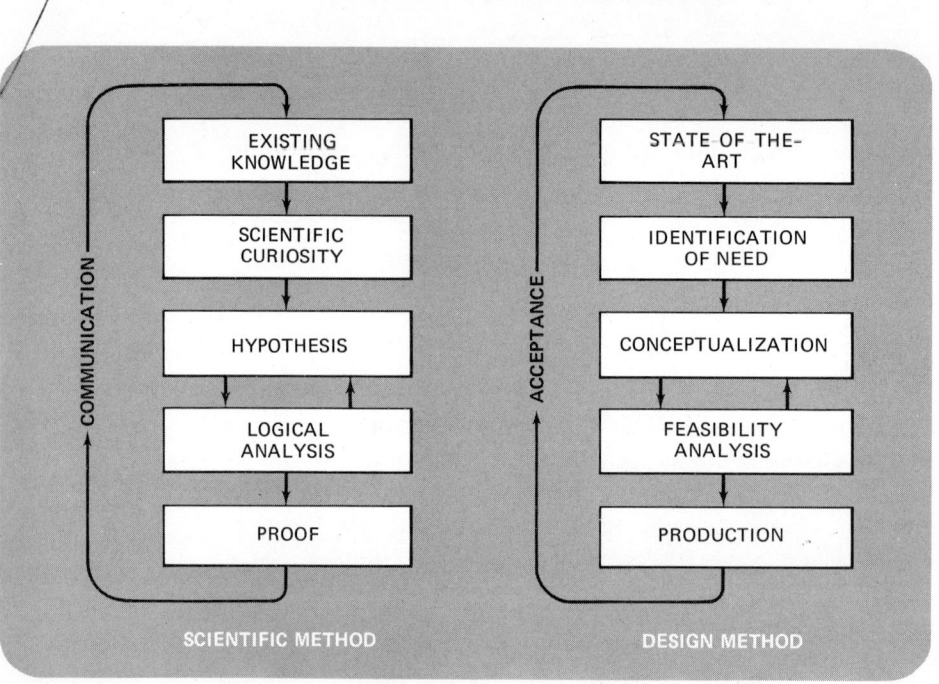

Figure 3-1 **KNOWLEDGE AND PRODUCT LOOPS**

pollution or traffic congestion, and/or the company seeks a new system, process, or device, the design engineer is motivated to identify the need of a system that may solve the problem. The need must now be synthesized into a mathematical or graphical model through conceptualization so that it may be tested against a feasibility analysis to confirm or deny the concept. As in the scientific method, the iteration of analysis and conceptualization will or will not produce a product idea to enter the production stage. The product loop is closed when the device is accepted by technology which enlarges the state-of-the-art and the cycle begins again.

Even though these two methods parallel one another and require similar intellectual activity, they are very different when one considers motivation and the environment in which these activities exist. The scientist is motivated by curiosity, a need for professional acceptance or prestige among his peers and works in an environment in which time and money are not predominant factors. The engineering designer on the other hand must produce something that is useful to society or his place of employment may not survive. He is continually plagued by the time factor, the competition, market analysis, and the spending of capital funds while developing the product. He continually lives in uncertainty, even when the product is accepted, for he must keep it unique and on the market while the competition is trying to "go it one better."

DESIGN PROCESS

The design of a device, system, or process has a number of objectives imposed upon it, including cost, time, screening criteria, feasibility, performance, production, aesthetics, or acceptance. These objectives require that design follow a methodology to insure that something truly useful will result and have a chance of success on the market. Methodology in design is not a formula or even a prescription that will in even the slightest sense guarantee a product, from the awareness of need to the prototype. It should be considered rather as a sequence of events known as the design process within which a design can be caused to unfold logically. The design process can serve as a useful reference in defining where we are, where we ought to be, and the next step in executing a complete design. These steps are illustrated in Fig. 3-2 in the form of a flow diagram of action steps beginning with identification of need and ending with consumption of the product. The steps are iterative and require a series of decisions to move the design along. More often than not a design oscillates back and forth between stages until it reaches a form in which it can occupy the next successive stage. The greatest iteration occurs between conceptualization and analysis, where a general image of the design is tested against the laws of nature, reconceived, and retested to bring it into a reality. Ideation may occur at the position shown in the diagram or earlier, as illustrated by the arrows, depending upon the design situation.

The following list gives a brief definition of each stage in the design process. The article to follow explores each stage in detail and discusses techniques of achieving optimum results.

Identification of need. Imagination supplies the need through the stimulus of being irritated or disturbed by a situation confronting the designer and his decision to do something about it.

Definition of goal. An expression in general terms of one's commitment to a system, device, or process to satisfy the need.

Research. Collection of all available information related to the goal.

Task Specifications. Listing of pertinent data and parameters that will tend to control the design toward the desired goal.

Ideation. The process of originating new ideas.

Conceptualization. Ingenious, innovative, creative, inventive activity in the form of the generation of alternative possible solutions to the required goal. This usually takes the form of free-hand sketches.

Analysis. Testing selective concepts against physical laws.

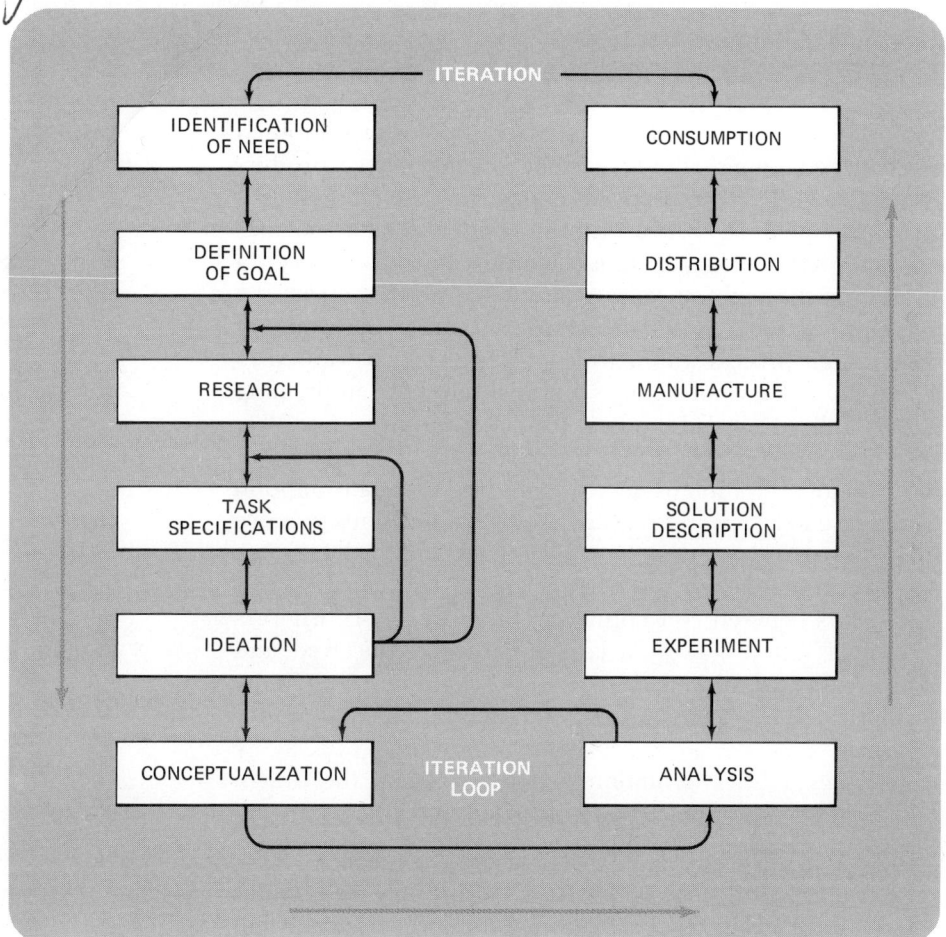

Fig. 3-2   **THE DESIGN PROCESS**

Experiment. Construction of prototype and laboratory testing to determine performance, workability, reliability, durability.

Solution description. Specific information that defines the system, process, or device. It consists of a report containing a description of the devide, drawings, specifications, parts lists, cost estimates.

Manufacture. Consideration of volume of production, shop requirements, fabrication techniques, stock acquisition, automation, scheduling, quality control, inspection.

Distribution. Competitive pricing, advertising, marketing, profit margin.
Consumption. Consumer feedback, repairing, servicing.

STAGES AND TECHNIQUES

Remembering that it takes between 55 and 60 ideas to yield one successful new product (Chapter 1) and that as each idea progresses through the stages of evolution expenditures increase, it takes more than creative thinking to complete the design cycle. Probably the singularly most important use of the design process is to resolve the conflict between creative thinking and logical analysis. Creative thinking is more productive when the imagination is allowed to wander freely among all aspects of the problem; imagination is cut off when tied up in orderly details. On the other hand, analysis works best when a step-by-step procedure is followed and it breaks down when forced to deviate. The design process is a healthy balance between these two activities which are essential to successful design. As discussed earlier, the design process is a series of sequential stages or plateaus which are intended to guide the design. This is certainly not in conflict with analysis. At the same time these sequential stages are broadly defined and can be iterative so they will not restrict creative thinking. Logical analysis and creative thinking can occur as simultaneous activities within this framework. The following discussion of each stage of the design process reveals certain known techniques that may be used in reaching a stage or plateau and effectively moving the design optimally from stage to stage.

IDENTIFICATION OF NEED

The beginning of the design process is an idea. Often the idea is motivated by a company's urgent desire to launch a new product; at other times it may be developed by the individual's desire to solve a social problem. The identification of a need to benefit society in some way requires creative thinking of a high order. One must flex his imagination and continually observe his environment for ideas. When one becomes "tuned in" to need identification through being disturbed by non-ideal situations and makes a decision to do something about them, he will find that idea generation becomes more or less automatic. Once this state is reached, one must pre-screen his ideas (select the best) before continuing to the next stage. Pre-screening amounts to self-interrogation in an effort to determine if the idea will fit company criteria, has a chance of success, and can be accomplished with present state-of-the-art in a reasonable period of time. If these conditions are met, a need (idea) now exists and the design can enter the second logical stage.

DEFINITION OF GOAL

Identification of need is primarily a mental activity resulting in a thought which must be committed to paper as a goal. This expression, in general terms, defines what the designer is seeking in order to resolve the need. It is a broad statement of the wanted end product. Many of the difficulties encountered in design may be traced to poorly stated goals, or goals that were hastily written and resulted in confusion or too much flexibility.

As an example of the design process thus far, assume that an engineer in private practice (consultant) is bothered by the time and trouble of removing fallen leaves from his lawn—that he would rather play golf. Finally provoked by a recent newspaper headline "Autumn Rite of Leaf-Burning Adds to Air Pollution," he decides that there is an urgent need for a device to remove leaves. After careful thought and gathering responses from some neighbors the following goal statement resulted: "Design a device or system for removing autumn leaves from suburban lawns that is safe, simple to use, and will not add to air pollution."

RESEARCH

The collection of all available information related to the goal is essential if the new design-product is to have a chance of acceptance (success in the marketplace). Information sources may include the following: technical and trade journals, abstracts, research reports (government and private), technical libraries, the competition, catalogues of component suppliers, and the U.S. Patent Office (see Chapter 9). This information may reveal an already available design solution, the state-of-the-art, and availability of hardware to accomplish the goal. Oftentimes the results of this information search will require the goal to be altered to accommodate a salable product. Sometimes the goal is abandoned for there is already such a product (or near-product) available. Certainly this knowledge will save a designer as well as his company a lot of embarrassment, time, and money (and usually the designer his job).

TASK SPECIFICATIONS

This stage requires the designer to list all pertinent data and parameters tending to control the design and guide it toward the desired goal. The list includes factors that must be considered, takes into account the industrial environment in which the design is to be produced, and deals with cost and maintenance. It is important to point out here that task specifications are not intended to be a set of rules to which the designer must adhere but instead a reminder to help channel effort toward the specified goal. Since this stage requires imaginative thinking involving the overall design plan, the designer often questions the goal, may wish to revise it, and may abandon the project if it is proven risky.

Referring to the earlier example expressed as a goal "to design a device or system for removing autumn leaves from suburban lawns . . . ," the following partial list is intended as a sample of task specifications.

1. Device must be safe to operate, especially when used near small children.
2. Controls must be such that the average housewife will have little to no difficulty in learning to operate.
3. Device may collect and remove leaves in one of the following ways:
    a. Shredded and bagged.
    b. Shredded and deposited as mulch (already available).
    c. Compressed into blocks to be burned in fireplace during winter.
    d. Shredded and compressed into a type of peat for planting and covering.
    e. ………………………………
    f. ………………………………
4. Types of power to be considered:
    a. Storage battery
    b. Electricity (110v)
    c. I. C. engine (2 cycle or 4 cycle)
    d. Manual power through gearing
    e. ………………………………
    f. ………………………………
5. Device should be easily stored in garage or basement.
6. Possibility of combining lawn tasks such as:
    a. Collecting leaves and mowing lawn.
    b. Collecting leaves and fertilizing.
    c. Collecting leaves and spreading lime.
    d. Collecting leaves, mowing lawn, and spreading lime.
    e. ………………………………
    f. ………………………………
7. Device should use presently available hardware and require only sheet metal and tube forming work, or possibly fabricated plastic or vacuum-forming.
8. Material should be selected on the basis of workability, strength, cost, appearance, and ability to withstand varying weather conditions.

9. Device must be reliable, require little to no maintenance, and sell at a price below the average power lawn mower.
10. ....
11. ....
12. ....

IDEATION

Ideation, the process of coming up with new ideas, although shown in the design process as a single event, often bridges many stages and frequently occurs out of order. This all-important stage precedes and provides the fuel for conceptualization. Ideation deals with initial ideas, while conception transforms those ideas into realities. Several techniques that may be effectively used during the ideation stage are discussed in Chapter 2.

CONCEPTUALIZATION

The process of generating alternative solutions to the stated goal in the form of concepts requires creative ability of a high order. Here the designer must consider the research phase and continually review task specifications as he engages in ingenious, innovative, creative, and inventive activity focused on the end product. This activity usually takes the form of free-hand sketches in producing a series of alternative solutions. In this way, the designer begins to realize his thoughts on paper. The alternatives need not be worked out in detail but instead are recorded as possibilities to be tested against selected criteria to determine which has the best chance of success. The process of conceptualization is the same whether in the design of an educational toy, an aspirin-size hearing aid, or a missile-launching system. Considering the leaf-removing device once more. Figure 3-3 shows examples of conceptual work.

DECISION MATRIX

Once a number of concepts have been generated in sufficient detail, a decision must be made as to which one or ones will enter the next and most expensive stages of the design process. An excellent technique to guide the designer in making the best decision regarding alternatives is a scoring matrix which forces a more penetrating study of each alternative against specified criteria. The use of the matrix can best be illustrated within the framework of an example, so let us consider the six alternative concepts developed for the leaf remover weighed against design criteria.

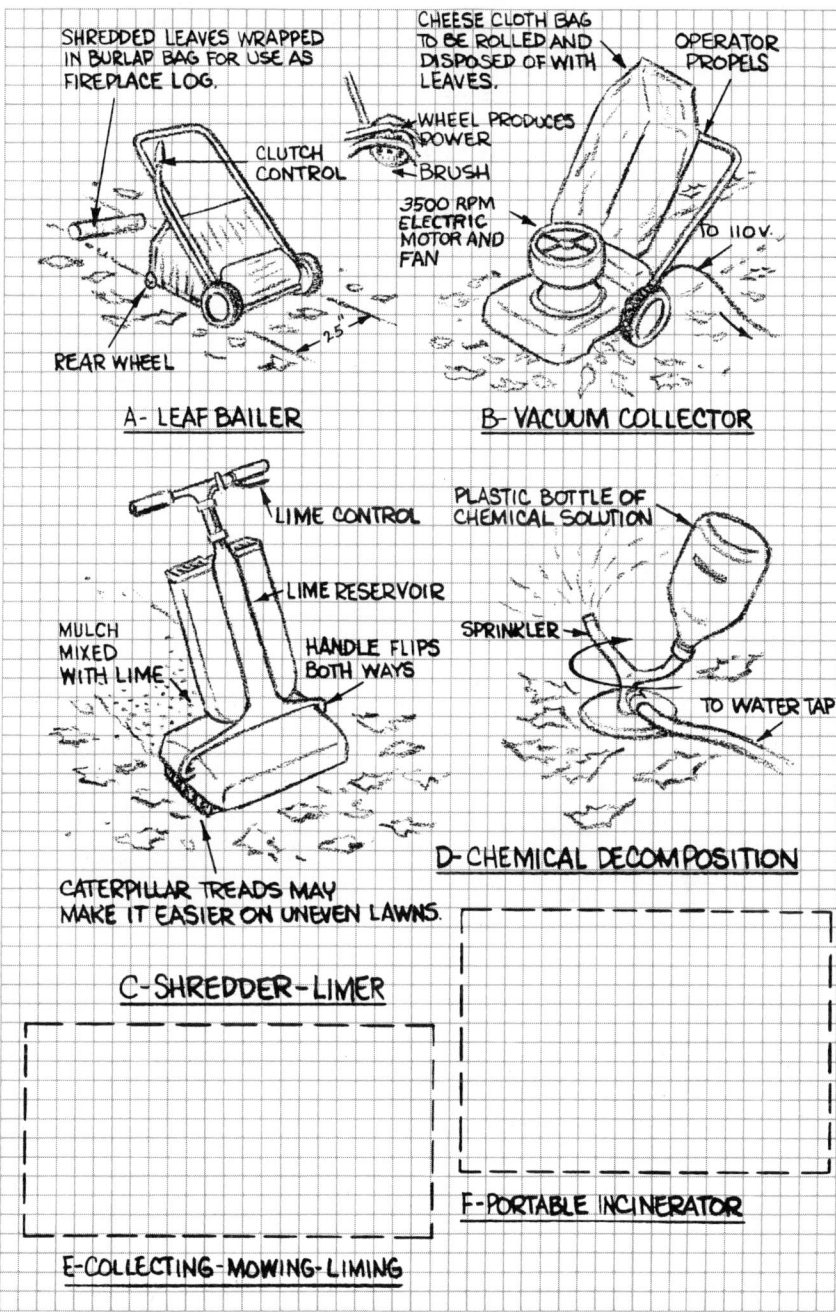

Figure 3-3 **CONCEPTUAL SKETCHES**

The decision matrix in Fig. 3-4 shows the six alternative concepts listed at the left of the chart and design criteria listed across the top. Each criteria is now given a *weighting factor* from 0 to 1.0, based on the measure of relative importance of each criteria in supporting the design goal. The sum of all weighting factors must total 1.0. If we are very familiar with these criteria from experience, we may be able to assign the numbers immediately. Otherwise we divide the number 1 by the number of criteria ($1/C_n = 1/9 = 0.11$), and then vary the resulting number 0.11 as the relative importance of each criteria varies as to its influence on the total design, keeping the sum equal to 1.0. This sum (shown by ①) will give an indication of commercial feasibility to be explained shortly. Each alternative is now rated (given a *rating factor*) on the basis of 0 to 10 against design criteria. It is important to note that when dealing with cost, the weighting and rating factors are scored at the low end of respective scales to indicate high cost. Once each alternative is rated, rating factors are multiplied by weighting factors and a sum is taken to indicate which alternative best suits the goal. The example of Fig. 3-4 shows sums of concepts B and D to be most appropriate to enter the next stage of the design process (indicated by ② and ③). On the basis of a perfect score of 10, considering that all judgments are perfect (although they never are), concept D has an 8.18 to 10 chance of success, while B has 8.14 to 10. If the total falls below 5, there is serious doubt that any of the con-

Fill in a decision matrix

DESIGN CRITERIA / ALTERNATIVES	USE OF ST'D PARTS	SAFE	SIMPLICITY AND MAINT.	DURA-BILITY	PUBLIC ACCEPT.	RELIA-BILITY	COST TO DEVELOP	COST TO BUYER	PERFORM-ANCE	SUM	
WEIGHTING FACTOR	0.08	0.12	0.10	0.10	0.18	0.20	0.03	0.04	0.15	1.0	①
A) Leaf Bailer	3 / 0.24	5 / 0.60	2 / 0.20	4 / 0.40	9 / 1.62	6 / 1.20	1 / 0.03	1 / 0.04	3 / 0.45	4.78	
B) Vacuum Collector	9 / 0.72	10 / 1.20	10 / 1.00	8 / 0.80	6 / 1.08	7 / 1.40	10 / 0.30	10 / 0.40	8 / 1.24	8.14	②
C) Shredder	5 / 0.40	6 / 0.72	7 / 0.70	7 / 0.70	8 / 1.44	6 / 1.20	3 / 0.09	4 / 0.16	5 / 0.75	6.16	
D) Chemical Decomposer	8 / 0.64	10 / 1.20	9 / 0.90	8 / 0.80	9 / 1.62	7 / 1.40	2 / 0.06	8 / 0.32	8 / 1.24	8.18	③
E) Collector-Mower-Lime	6 / 0.48	8 / 0.96	3 / 0.30	2 / 0.20	6 / 1.08	4 / 0.80	2 / 0.06	2 / 0.08	6 / 0.90	4.86	
F) Portable Incinerator	5 / 0.40	4 / 0.48	8 / 0.80	6 / 0.60	5 / 0.90	10 / 2.00	3 / 0.09	4 / 0.16	9 / 1.35	6.78	

Weighting Factor (W.F.) = Measure of Relative Importance (0 to 1.0 Σ = 1.0)

Rating Factor (R.F.) = Measured Value of Alternatives Against Design Criteria (0 to 10).

ANALYZE THEN RUN SUB-MATRIX

R.F. / W.F. X R.F.

Figure 3-4   **DECISION MATRIX**

cepts are worthy of consideration. Since it usually is too costly to construct two different prototypes for testing and experimentation, it would be wise in this case to run a submatrix of these two alternatives after the analysis phase, in hopes of selecting only one to enter the final stages of the design. Needless to say, the assignment of values to weighting and rating factors is of the utmost importance in the accuracy of data on which the ultimate decision will be based. Therefore one must base these values on the greatest possible amount of information, a full understanding of the problem, and judgment of the highest degree.

ANALYSIS

Once a concept has been chosen that accurately defines a possible solution to the stated goal, it must be tested against physical laws. This verification with the laws of nature is known as analysis and forms an iteration loop with conceptualization. Often analysis requires a concept to be altered or redefined, then reanalyzed and reconceived, so that the design is constantly shifting between analysis and concept until it begins to take on some meaningful physical significance. Analysis takes the form of two primary activities: estimation, followed by order-of-magnitude calculations. Estimation is judgment based on experience; it is not a guess. It would not be difficult for an engineer to estimate the inside dimensions of a one- or two-car garage to house the modern passenger automobile, but it would be purely a guess if he attempted to predict auto size over the next ten years unless he had some active experience in this area; and then it would be an estimate, not a guess. An estimate, when possible, gives the designer a general feeling for the concept and helps him to concretize certain details. Following the estimate is an order-of-magnitude analysis. This is considered to be approximately a factor of ten. For example, there are 9 options (factors) lying between 10^{-4} and 10^5, each being an order-of-magnitude. Since analysis is expensive, tedious, and often time-consuming, one of the responsibilities of the designer is to decide when a first-order-of-analysis (rough calculation) or a higher order is required to produce the needed information. The use of high-speed computation devices, once a program is available, makes possible a higher order of analysis than ever before at a reasonable cost.

EXPERIMENT

In the classical sense, an experiment is a set of observations carried out under controlled conditions. Classical statistics require that the environment in which the experiment occurs be held constant with controls. In design, the imposing of controls to insure a fixed environment frequently converts what was intended to be a real-life situation into a nonrepresentative system. This results in data that is incomplete or inaccurate when studied in the real-life context. It has been proven that the real test of an automobile is on the proving ground, not in the laboratory.

The experiment phase of engineering design requires that a piece of hardware or software be constructed and tested to verify the concept and analysis of the design as to workability, durability, and performance characteristics. Here the design on paper is transformed into a physical reality. The three techniques of construction available to the designer are the *mock-up*, *model*, or *prototype*. The mock-up is generally constructed to scale from plastics, wood, cardboard, and so forth to give the designer a "feel" for his design. It is often used to check such things as clearances, assembly techniques, manufacturing considerations, and appearance, and it is a valuable aid in selling the design idea to management or the client. It is the least expensive technique and provides the least amount of information but it is quick and relatively easy to produce. The chemical industry produces to-scale, nonworking breadboard-type mock-ups of refining processes to aid in visualization of the full-size system. A more expensive method of collecting experimental information is through a model representing the physical system and related to it through the mathematics of dimensional similitude. The four types of models used to predict behavior of the real system are *true, adequate, distorted,* and *dissimilar*. A true model, as the name implies, is an exact geometric reproduction of the real system, built to scale and satisfying all restrictions imposed by design parameters. An adequate model is so constructed to test specific characteristics of the design and is not intended to yield information concerning the total design. A distorted model purposely violates one or more design conditions. This violation is often required when it is difficult or impossible to satisfy the specified conditions due to time, material, or physical characteristics, and it is felt that reliable information can be had through the distortion. Dissimilar models bear no apparent resemblance to the real system, but through appropriate analogies give accurate information on behavioral characteristics. An example of this type of model is the increased use of an analog computer to study a complete system.

The most expensive experimental technique and the one producing the greatest amount of useful information is the *prototype*. The prototype is the constructed, full-scale working physical system. Here the designer can see his idea come to life, learn about such things as appropriate construction techniques, assembly procedures, workability, durability, and performance under actual environmental conditions. The greatest difficulty in prototype construction is the time and expense involved and one's reluctance to change or improve the design as the finished product confronts him.

As a general rule, when entering the experimental stage of the design process, to allow beneficial iteration with concept and analysis, one should first deal with the mock-up, then the model, and finally the prototype, after the mock-up and model have proven the real worth of the design.

SOLUTION DESCRIPTION

This stage requires specific information that defines the device, system, or process. Here the designer is required to put his thoughts regarding the design on paper for purpose of communicating with others. Communication is involved in selling the idea to management or the client, directing the shop on how to construct the design, and serving management in the initial stages of commercialization. The solution description should take the form of a report containing a detailed description of the device; how it satisfies the need and how it works; detail and assembly drawings; specifications for construction; list of standard parts; cost breakdown; and any other information that will insure that the design will be understood and constructed exactly as the designer intended. Examples of such a description may be seen in Chapter 4 under Design Case Histories.

MANUFACTURE

The designer is usually assisted by a specialist in the stage of manufacture. Attention must be paid here to available shop facilities, training of personnel, volume of production, stock availability, scheduling of work and production of end product, quality control, inspection, and so on. A successful designer must have a knowledge of manufacturing capabilities within his or his client's company and should work closely with the manufacturing engineer throughout the design process.

DISTRIBUTION

Distribution of the product falls under the classification of commercialization, discussed in Chapter 1. Unfortunately, present-day practice excludes the designer from this stage, for it is usually handled by specialists. This stage is responsible for placing the product on the market and consists of competitive pricing, establishing the best product-release time, advertising, promotional literature, market testing, profit studies, and so forth. Management would be well advised to consult with the designer during this stage, for he knows the product better than anyone else and often will have ideas that can affect a favorable distribution.

CONSUMPTION

Accurate records and studies made during the consumption stage can greatly aid a model change in present products and be very useful in the successful design of new products. Consumption deals with such things as who buys the product, where it is used, consumer feedback, competition's reaction, repairing, and servicing. Much of this information can be gathered by the sales force which usually has the confidence of the customer or distributor.

Even though the design process terminates with consumption (probably the most unglamorous stage), it is here that the entire process is focused and where the real test of a design lies, both in company reputation and in profit.

MORPHOLOGICAL APPROACH

Morphological may be defined as "pertaining to the science, structure, and shape of an organized body or system." Scholars of the science of design chose this term to describe an approach to the logical organization of ideas in designing as opposed to traditional methods of intuition and experience. When the morphological approach is applied to the design process it offers a precise and somewhat rigorous mathematical method of directing and combining stages to arrive at the stated goal. It is of particular benefit to the inexperienced designer in getting off to an early start on a problem and as an aid in organizing and guiding work along fruitful paths. Figure 3-5 shows how the morphological approach is applied to stages in the design process. It can be seen here that the main thrust of this approach encompasses stages from research to conceptualization. It will be shown later that although the morphological approach includes the research stage, research must still be conducted independently and then compared to the results of steps 1 through 5.

All techniques dealing with the design process are based on the asking and answering of questions related to the goal and methods of making decisions regarding these answers. The morphological approach does just this and can be examined through the five steps that follow.

step 1: lay out the field of investigation.

This consists of combining task specifications, ideation, and conceptualization in the form of *parameters* and *possibilities*, listed on a morphological chart as illustrated in Fig. 3-6. The parameters describe in a general way the features and functions of the design (what the design must be or have). The possibilities describe in general the ways or means of achieving the required parameters. Such a chart will indicate all possible solutions to the problem within the bound-

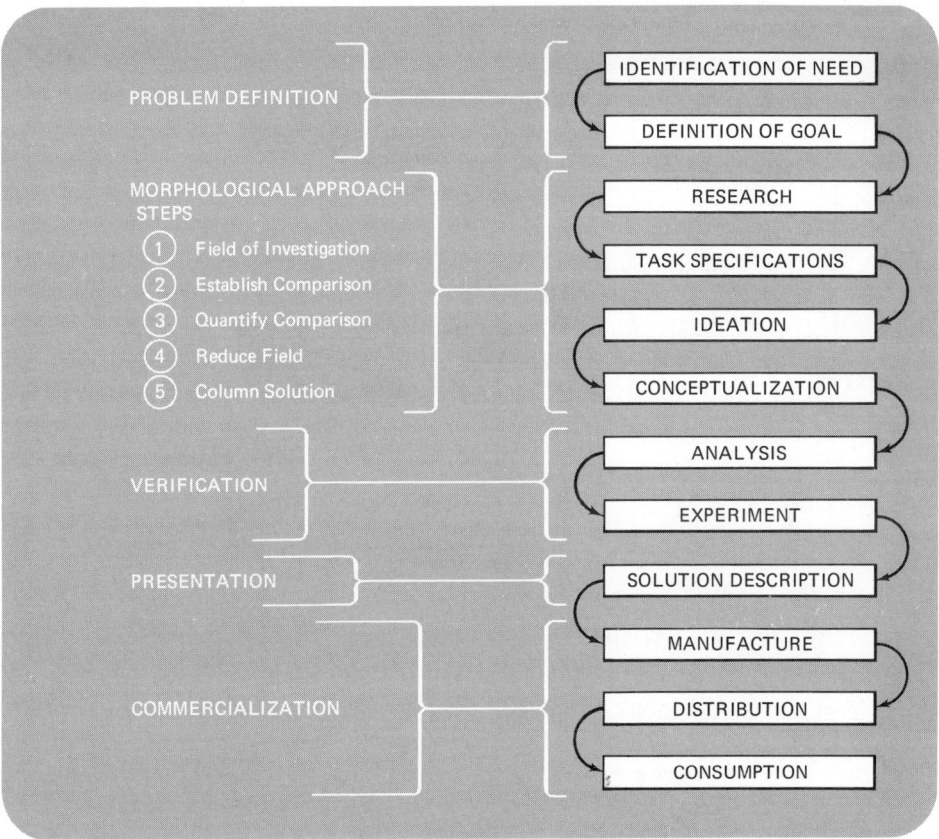

Figure 3-5 **THE DESIGN PROCESS WITH MORPHOLOGICAL APPROACH**

ary of the field, with the best solution listed somewhere therein. Research for background material is very necessary to both the listing of parameters and possibilities. If research brings commercially available solutions to the stated goal, they should already be listed on the chart. Such a technique forces analysis and synthesis to occur simultaneously in preparing and reviewing the chart. Parameters and possiblities require analysis, while synthesis occurs in the column solutions explained in step 5.

step 2: establish basis (or bases) of comparison.

Since step 1 will generate all of the possible solutions to the design within stated boundaries, the designer must establish the basis or bases of comparing solutions

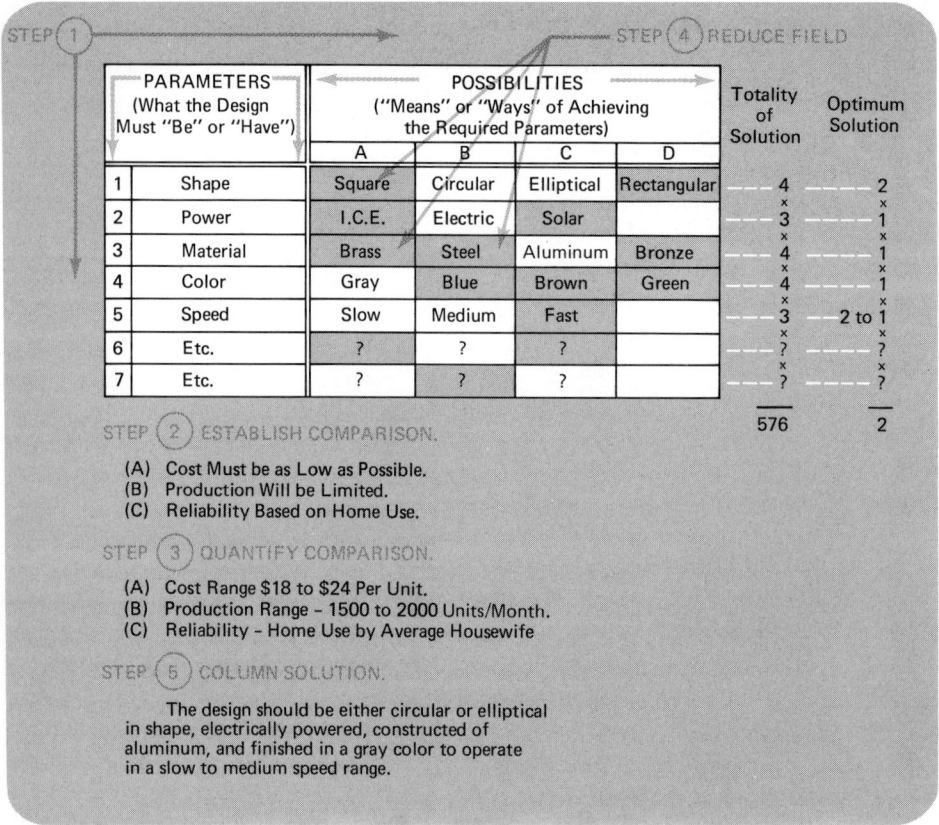

Figure 3-6 **MORPHOLOGICAL CHART**

in order to select the best one. A basis might be cost, simplicity, durability, or weight. If a number of criteria are established as a basis, the designer should rank them in order of importance.

step 3: quantify basis (or bases) of comparison.

This step requires rather specific limits of acceptability to be placed on the basis or bases of comparisons. In other words the basis or bases of comparisons must be made as specific as possible either through research, knowledge of manufacturing capabilities, the market, assumptions, or consultation with the client.

step 4: reduce field of investigation.

Through the process of elimination using the basis or bases of comparisons, the field of investigation is reduced to a number of selective optimum or best solutions. This reduction can be carried out through an analysis of rows or an analysis of columns. Since the number of rows is the same as the number of parameters, each row can be analyzed separately. Those that seem less appropriate can be crossed off, thereby allowing the optimum to remain, as shown in Fig. 3-6. Once several rows have been eliminated, an analysis by columns is in order. This is more detailed, requiring judgment of the first order, and should reveal the solution most likely to succeed.

step 5: column solution.

By now matching the remaining possibilities appearing in each column against parameters, a number of solutions will emerge. These solutions are now compared to each other, and the best one is selected for analysis and experiment to serve as the ultimate solution to satisfy the stated goal.

OPTIMIZATION

Optimization is an important activity in the design of a product, system, or process. It is the essence of good design. Optimum design exists when one has achieved, for example, the best appearance at the lowest cost, the most power with the least weight, or the best quality with the least material. Designers have been optimizing since the first patent was granted and probably since the invention of the wheel, which is a good example of optimum design in itself. The process of arriving at the best design, maximizing performance while minimizing cost has in recent years been given the name optimization or optimum design and has received attention from scholars of design. It will someday become an engineering science if it is not already accepted as such.

Many important contributions to technological knowledge came about through optimization measures. Pre-stressed concrete was the result of a search for lighter pavements to carry greater loads on aircraft runways in Germany during World War II. The Japanese Zero aircraft of World War II is good examples of high performance, light but strong, easily constructed airframe that outperformed other aircraft of its time. The process of sanforizing fabrics was discovered because of the need to minimize shrinkage. Dural (aluminum–steel alloy) was introduced to maximize strength while minimizing weight. The ball point pen is a good example of maximizing ink supply, convenience, and performance while minimizing cost. Each of us optimizes every day of our lives while seeking bargains when purchasing goods, finding the shortest route to work, taking advantage of opportunities, and the like.

To optimize the design itself is not enough; the entire process from inspiration to finished product must be optimized to produce the best possible result. This is basically what this book deals with—methods of achieving the best possible design from a management, engineering, and commercial point of view. It is a good idea to optimize the design time, the research phase, planning new product management, and market analysis, but not too much should be expected as far as the design itself during the highly creative stages. Optimum design as a technique plays its most important role during the analysis, experimental, and manufacturing stages. The following procedures may be of some help to the engineer in dealing with optimum design activity.

subjective decisions.

Optimization must be a state-of-mind and exercised by the engineer as an overview of design. Experience builds this overview with ingredients of cost, material, production procedures, customer likes and dislikes, awareness of new developments, knowledge of the competition, and so on. It adds up to making the correct intuitive decision about each minute detail of the design so the result will be the best of all possibilities.

general principles.

Through both formal education and experience, the engineer learns certain facts that he will not allow to be violated and others commonly known as "rules-of-thumb" that he automatically applies to the design situation. Even though these general principles are sometimes applied subconsciously, they are nonetheless an accepted technique of optimization. For example, the greater the number of pistons in an engine the easier it is to achieve dynamic balance; members of a truss or structure should be geometrically arranged to distribute the load evenly; sharp corners, notches, and small radius on stressed parts should be avoided for this usually results in stress concentration; an electric motor should not be required to start under full-load conditions; bending stress may be reduced by increasing the moment of inertia of the section, and so forth.

graphical and analytical methods.

There are both graphical and analytical methods of optimizing the design. The graphical method requires a scale drawing to test the design against specified criteria and this method is based on the fact that it is cheaper to draw the product than to make it. The drawing can be altered many times until the designer achieves what he is seeking, while it is usually too costly to change the physical system. Examples of graphical optimization could include sizing the windshield of an automobile for maximum operator visibility, designing a windshield wiper to sweep the largest possible area, or calculating the smallest rear-view mirror to cover viewing area of the rear window. Analytical techniques of optimization are based on writing an equation dictated by imposed criteria and consisting of terms to be optimized, such as cost, weight, dimensions, volume, and speed. Once an equation is written, a cursory inspection of terms will often indicate how certain variables change when one manipulates (increases or decreases) physical characteristics. If the equation lends itself, the process of differentiation may be used as well as curve-plotting and nomography.

trade-off.

Unfortunately, the physical world does not aid the designer in his pursuit of the optimum. We know that one cannot have the strongest and lightest structure at the same time, the fewest parts to do the most work, the smallest surface area to contain the greatest volume. The engineer must continually make decisions as to what must be traded off to achieve one or more aspects of the optimum. Examples might include: to achieve greatest strength, weight might have to be traded as well as cost; to accommodate the 95th percentile male operator, smooth lines might have to be traded for a boxy shape; and for maximum television or radio reception, concealed antennae might have to be traded for the exposed type.

RELIABILITY

[handwritten: definition of reliability & calculate it.]

The ultimate goal of the design engineer is to create a device or system that will optimally satisfy the stated need, conform to task specifications, and perform its function satisfactorily over the required time period under predicted environmental conditions. To do this means that the design must not malfunction (it must not burn out, overheat, overdeflect, or overload) over a pre-defined period. This element of design has been included as a design parameter since its introduction in the early 1950s as a formalized method known as *reliability engineering*. The reliability of a design is the probability that it will perform its function during the period of time specified under known operating (environmental) conditions. One can readily see the difficulty faced by the automotive manufacturer who must produce a standard passenger vehicle to operate satisfactorily in the state of Maine as well as in Florida or California; whose paint must withstand desert sun, winter road salt, industrial smog, and the ocean salt spray; which will be driven by the average male or female, the hot rod sport, and the little old lady. Fortunately reliability has an exact meaning. Not only can it be defined; it can be calculated, evaluated, measured, tested, predicted, and designed into a piece of equipment. When human life depends upon the proper functioning of a component in a system, such as the environmental pressure hardware in a jet aircraft, the steering mechanism in an automobile, or an articifial heart during surgery, reliability plays its most important role.

The manufacturer as well as the consumer would like to have a satisfactory level of reliability in everything that is produced. It is quite possible within the present state-of-the-art to achieve this goal. All required is that the designer increase the factor of safety and over-design each component of a system, and thereby over-design the entire system. This was the general rule in the past before mass production, when material was plentiful, skilled labor readily available, competition nil, and the rate of technological progress slow. Today, however, the device or system must be designed to perform its function at the lowest cost, with minimum weight and size, optimum use of materials, and within an acceptable reliability range.

The determination of a numerical value for reliability is based on the theory of probability and chance. This can be illustrated through the example shown in Fig. 3-7. Table (A) lists test data for a component that was designed for a 50-hour operating life. When 500 components were tested, it was found that 85 failed during the first hour, 43 the second, 24 the third, and so forth. The number of failures began to level off at about 8 hours of operation, and the last survivors operated for 99 hours. This data follows a normal or Gaussian probability distribution with approximately 250 survivors (one-half) at the 50-hour design time that continued to operate until the last component failed at 99 hours. Considering the reliability of these components to be the percentage of time each is to operate, it may be expressed numerically as follows:

$$R = \text{reliability} = \%\text{ of time a part is to operate} = \frac{\#\text{ survivors}}{\text{population}}$$

$$R @ 5 \text{ hr} = \frac{330}{500} = 0.66; \quad R @ 12 \text{ hr} = \frac{295}{500} = 0.59; \quad R @ 97 \text{ hr} = \frac{4}{500} = 0.008$$

Reliability values were calculated for each hour of the test data and plotted against operating time producing the *standard reliability curve* at (B) Fig. 3-7. This curve, R = f(t), is of the exponential form and when fitted to an equation comes very close to the solid line curve. This may be expressed as $R = e^{-Ft} = e^{-t/M}$, where F = failure rate and M = 1/F = mean time between failures. This equation is known as the *exponential reliability (probability) law*. The two curves do not match because $R = e^{-Ft}$ is based on a constant failure rate F. This is usually the case when the customer receives the product, for the manufacturer will have discarded defects during the initial test period.

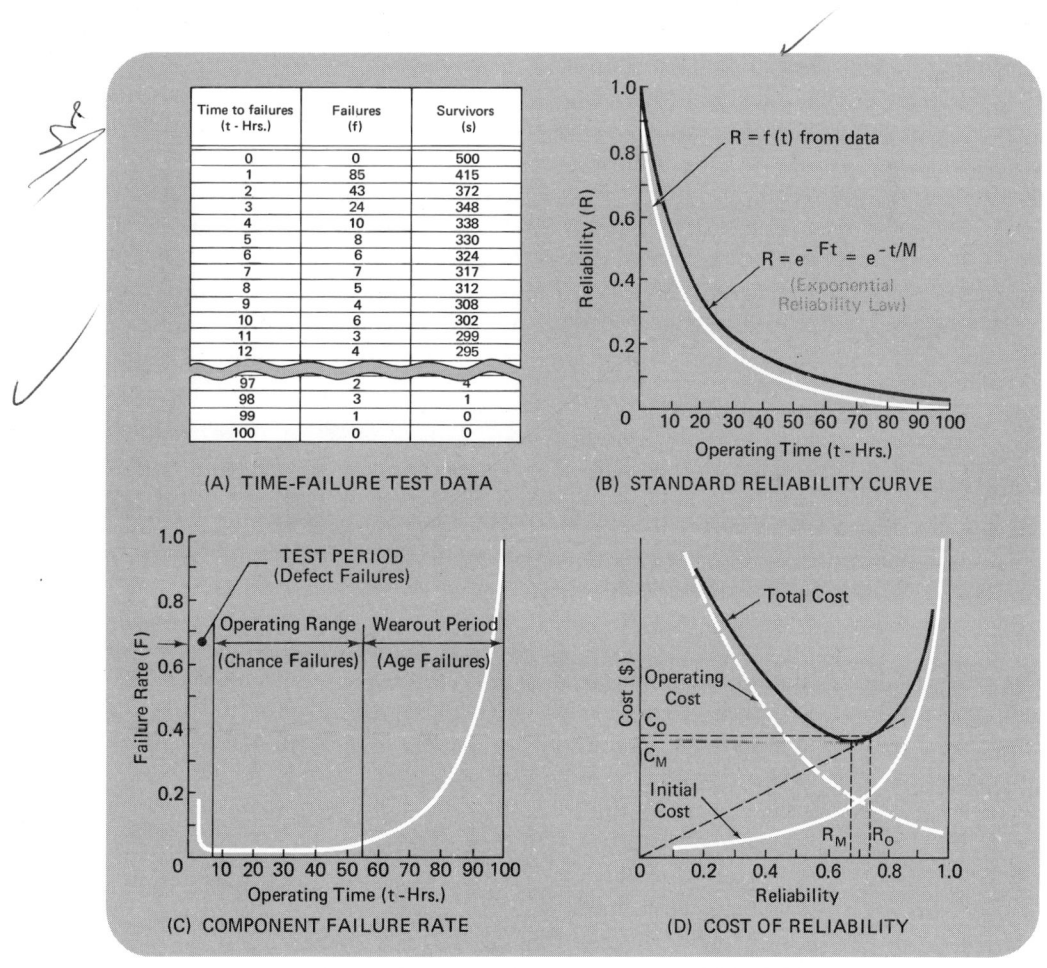

Figure 3-7 **RELIABILITY AND PROBABILITY**

The failure rate is now calculated from the data at (A):

F = failure rate = # failures per unit of time as % of population
N = average population during time interval.

$$N @ 5 \text{ hr} = \frac{338 + 330}{2} = 334; \quad N @ 12 \text{ hr} = \frac{299 + 295}{2} = 297; \quad N @ 97 \text{ hr} = \frac{4 + 4}{2} = 4$$

$$F @ 5 \text{ hr} = \frac{8}{334} = 0.024; \quad F @ 12 \text{ hr} = \frac{4}{297} = 0.013; \quad F @ 97 \text{ hr} = \frac{2}{4} = 0.50$$

These values were plotted against operating time, resulting in the curve shown at (C), the *component failure rate*. This curve shows a decreasing failure rate during the manufacturer's test period as a result of defect failures. Failure rate is low and constant during the operating range, when components fail by chance, up to their design life which in this example is 50 hours. About one-half of the components can be expected to function during the *wear-out period* with a steady increase in failure rate. Considering now the plotted data in total, one can easily generalize that it is essential for the manufacturer to run initial tests on products to detect defects. These defects may be kept to a minimum through appropriate quality control measures. If the consumer is dealing with light bulbs, he may as well leave them on during the wear-out period until they eventually fail; but if a television transmitter is involved, he should replace components near the end of their designed operating range. In this way, failure of equipment (mechanical, chemical, or electrical) will be a rare occurrence, and theoretically a system should operate forever through planned maintenance and component replacement.

Figure 3-7(D) shows some interesting curves on the cost of reliability. As one may expect, initial cost increases as reliability is increased and grows to a prohibitive amount as a reliability of 1.0 is attempted. At the same time operating cost of equipment decreases as individual component reliability is increased. This is achieved through savings in maintenance, shut-down time, and start-up time caused by frequent failures. Total cost is arrived at by simply adding the ordinates of the first two curves. The minimum total cost C_M is found at reliability R_M through a horizontal tangent to the curve. The optimum ratio of reliability R_o to total cost C_o is found by drawing a tangent to the curve from the origin o. Data of the type plotted in Fig. 3-7 collected on products will greatly aid the design engineer in fixing the reliability of related products.

Another value of interest to the design engineer as well as the consumer is the mean time between failures which may be calculated as follows: M = mean time between failures = 1/F

Considering the operating range of Fig. 3-7(C) where the failure rate F is about 0.015, the mean time between failures is M = 1/0.015 = 66.7 hr.

In addition to improving quality control and increasing factors of safety, the design engineer may improve the reliability of a system through *redundancy*. This consists of duplicating elements or components (adding to the system) in such a way that the duplicate is not required for successful operation of the system under normal conditions, but functions normally at time of system failure. A crude example of this is the fifth wheel (spare tire) on an automobile which is not required for successful operation of the car until there is a flat. This wheel is considered redundant to the system.

Another example of redundancy is illustrated in Fig. 3-8. Let us assume that the switch at (A) has a reliability of $R = 0.8$. This reliability is based on the ability of the switch to close on demand. Part (B) shows the elements of a system in which two identical switches must be wired in series and both must function properly to close the circuit. The reliability of this system is $R^2 = (0.8)^2 = 0.64$. This is much lower than for the single-switch system due to dependency of one switch on the other for proper functioning. Since the design specification requires that two switches must operate in series, the designer may introduce redundancy by wiring a third switch in parallel with switch S_2 so that S_2 and S_3 are closed at the same time as S_1, as shown at (C). The reliability of this system is $R^2(2 - R) = (0.8)^2(2 - 0.8) = (0.64)(1.2) = 0.768$. This is an improvement on system (B) and approaches the reliability of a single-switch system.

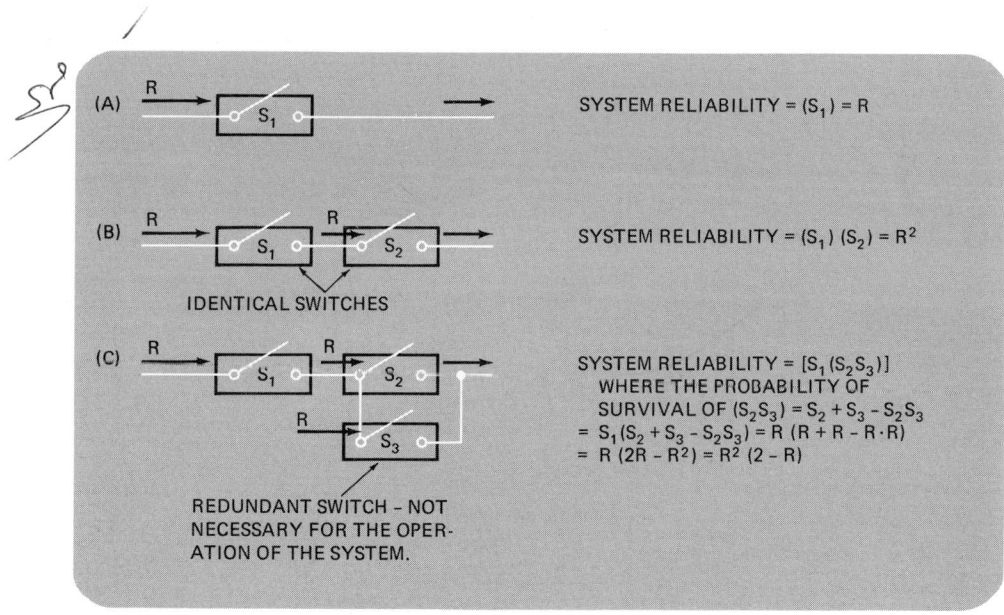

Figure 3-8 **REDUNDANT SYSTEMS**

The addition of redundancy to a system is expensive in terms of cost, weight, and size and should be considered carefully beforehand. The addition of redundancy depends wholly upon the circumstances of the design. Certainly it is not required when normal maintenance (replacement of failures) does not seriously interrupt the operation of the system, but it is required when even the shortest possible shut-down or turn-off will endanger the result of the system.

The discussion of reliability here is intended to give a brief introduction to the subject and to describe the way it fits into the scheme of design. A thorough mathematical and theoretical treatment of reliability and probability is beyond the scope of this book. References listed under Bibliography with Abstracts will present to the more serious or interested reader material enabling him to gain a thorough working knowledge of the subject.

EXERCISES

1. The product with the biggest payoff is a unique device, whereas a product which offers only slight improvements on what is already available is often doomed to failure. Name products that fall into each of these categories that were/are available in recent years.
2. Write a set of task specifications for the design of a device that will automatically dispense a single prescription drug in the form of a pill three times a day.
3. Show three conceptual sketches as alternative solutions to the automatic pill dispenser of question two.
4. The decision matrix is used not only for making decisions on alternative design schemes, but it may also be used to make a decision whenever a choice is available. Assume that you are planning to purchase a new automobile and have narrowed your choice to five makes or models. Construct a decision matrix, assign values, and make calculations to show which car best suits your needs.
5. Show through means of a chart how the creative process of Chapter 2 compares with the design process. (Hint: Are these processes in conflict or do they complement each other?)
6. Give an example of the following types of experimental mediums that you have read about or seen:
    a. Mock-up
    b. Models (true, adequate, distorted, and dissimilar)
    c. Prototype
7. Construct a morphological chart for the leaf-removing device presented as an example earlier in this chapter. What is the totality of solutions? After reducing the field of investigation, how many optimum solutions are there? How do they compare to the results of the decision matrix chart explained earlier?

8. Give an example of a product that was optimized through the trade-off principle.

9. Consider the reliability of a standard automotive tire to be 0.85 based on probability of a relatively new tire going flat during an initial 12,000-mile period. What is the reliability of a four-tire system (auto without spare tire)? What is the reliability of a five-tire system (auto with spare tire)?

10. Construct a diagram involving a double redundancy of the switches shown in Fig. 3-8 in accordance with the stated series specification. Calculate the reliability of this system.

4 CASE HISTORIES

"... and everything in the world must have design or the human mind rejects it. But in addition it must have purpose or the human conscience shies away from it..."

John Steinbeck
from Travels with Charley

It is not enough to just read about the design process as discussed in Chapter 3. One must experience it to truly understand the role of each phase and the iteration that causes interaction between phases. This chapter attempts to provide such an experience through three carefully chosen case histories. Case 001 is presented in step-by-step detail to show the efforts of the designer in idea development, conceptualization, and decision-making through to the ultimate solution in the form of a proposed new product. Case 002 is presented in less detail and is intended to show the design of a device that was arrived at through a slightly different approach. Case 003 is presented in even less detail and more as a design report to management. These cases were chosen to give the reader insight into the design process and the thinking of designers when involved with three different design situations.

CASE HISTORY 001

NEED

DESIGN ENGINEERING INC,
1460 Boston Avenue
MEDFORD, MASSACHUSETTS 02155

INTER-OFFICE CORRESPONDENCE

Date: 3/6/ 19 - -

ACTION COPY TO: CHIEF PROJECT ENGINEER

ATTENTION: ADVANCED DESIGN GROUP

SUBJECT: Design of new flashlight product line for the Brawley Manufacturing Company, Brawley, Vermont.

    Our company has recently signed a contract with the Brawley Manufacturing Company to submit designs of a new flashlight product line for their consideration on April 3, 19--. The Advanced Design Group is being assigned this task due to its reputation in design innovation and competence in getting a job done on time. We expect to review proposals and design layout drawings with Brawley's top management on this date so that a selection may be made of the products that best fit into the company's manufacturing facilities. We would also like to present working models of new product ideas at this time.

    I am enclosing a brief history of the Brawley Company, a description of their manufacturing facilities, and an outline of their immediate problem as we see it.

James P. Doe
Vice-President

enclosure
JPD/ch

NEED (cont'd)

CASE HISTORY 001 (CONT.)

BRAWLEY MANUFACTURING COMPANY
Brawley, Vermont

HISTORY

Albert Mancuso, father of Thomas, founded the Brawley Manufacturing Company in 1930. The company is located in Brawley, Vermont, a town of 1,500 people; the surrounding area is mostly farm lands. Starting with 7 employees, the company grew rapidly during World War II and currently employs 60 production workers. Thomas Mancuso became president of the company in 1961 after the death of his father. Thomas had been around the shop from the time he was a small boy, and by the time he was eighteen he could operate practically any machine in the plant. After acquiring an undergraduate engineering degree, he continued his education in the graduate business school of a large eastern university. Following his graduation in June, 1952 he assumed a full-time job in his father's business.

THE PRODUCT LINE

The Brawley Company has been engaged in the manufacture of outdoor and light sporting equipment (flashlights, alcohol camping stoves, waterproof containers for matches, field compasses, fresh water bait and spin casting fiberglass fishing rods, etc.) since its founding. These quality items marketed at a competitive price have made a good reputation for Brawley in the northeast and the company has enjoyed a reasonable sales position for many years. A number of quality sporting items are sold through L. L. Bean of Freeport, Maine and Sears stores under their trademark. Other items are sold through sporting goods stores in New England and upstate New York and northern Pennsylvannia.

EMPLOYEES

The employees of the company are all local residents; 30 skilled machine operators, 15 non-skilled personnel, and 15 women who perform hand assembly operations. The shop is non-union and the working force is easily trained for different operations.

THE PLANT

The company is located on two acres of land just outside the town of Brawley. The single one-story building of about 25,000 sq. ft. floor area consists of two small offices, stock room, and shop. Machinery presently owned by the company is as follows:

DRILLING
Small drilling
Reaming Head

SCREW MACHINES
Hand screw machine
0 B & S automatic
  screw machine

NEED (cont'd)

BRAWLEY MANUFACTURING COMPANY
Brawley, Vermont

PROFILE MILLING
Profile mill small parts
Automatic mill small parts
Vertical mill

GRINDING
Hand cylindrical grinding
Internal grinding
Centerless grinding

LATHES
Engine Lathe
Automatic Lathe

CAM MILLING
Fellows gear shaper
Barber Coleman hobbing
Thread milling

HEAT TREATING
Annealing
Brazing

FINISHING
Ball burnishing
Tumbling
Polishing

RAW STOCK
Band saw
Power Hack saw
Abrasive cut-off

BENCH
Power riveting
Punch press
Small oil grooving
Air press work
Breaching machine
Bench lathe

THE PROBLEM

Thomas Mancuso, president of Brawley, at a recent meeting of the stockholders, announced that the company has experienced a steady drop in sales during the past two years. This drop is due to the stream of competitive items supplied to this area through foreign manufacturers. The quality of these items is slightly below that of Brawley, but the price to the wholesaler is 20% lower than U.S. companies.

Brawley will continue to sell to many of its long standing customers, but is unable to aquire new dealers and has had several stores drop its line in the past year in favor of the imported items. The largest loss in sales is in the flashlight line. This is undoubtedly due to the fact that Brawley has not changed its flashlight design in the past six years while the competition has introduced many novel ideas. If the present situation were allowed to continue, Mr. Mancuso feels the company would be required to liquidate or merge with a larger manufacturer who would be in a financial position to absorb these losses over a period of time. Such a merger or liquidation would in all probability have a determinal effect on the Brawley community at large.

The only hope for the Brawley company is to come up with a new product line that could be produced in its shop with a minimum of capital investment for new tooling. The new product must be unique in nature, useful to the point of commanding a reasonable market potential, and produced from materials with which the company has had some experience.

CASE HISTORY 001 (CONT.)   65

GOAL

DESIGN A NEW PRODUCT FLASHLIGHT FOR BRAWLEY MANUFACTURING COMPANY TO FIT IN WITH THEIR SPORTING EQUIPMENT LINE AND SO CONSTRUCTED TO BE PRODUCED IN PRESENT SHOP FACILITIES. DESIGN IS TO BE PRESENTED TO BRAWLEY MANAGEMENT ON APRIL 3, 19 - -.

RESEARCH

POPULAR FLASHLIGHTS PRESENTLY AVAILABLE

TYPE	SIZE (IN.)	BATTERIES	SPECIAL FEATURES	PRICE RETAIL	SALES RATE
BIG BEAM CYLINDRICAL LIGHT	3x12	3-D Cells	Long Beam	$2.80	7
LARGE REFLECTOR CYLIN. LIGHT	3x8	2-D Cells	Flood Beam	2.60	6
COMMON CYLINDRICAL LIGHT	2x8	2-D Cells	Impact Case	1.75	10
MAGNETIC MODEL COMMON LIGHT	2x8	2-D Cells	Magnet Clip	1.30	8
LANTERN WITH RED FLASHER	2x3x4	2-D Cells	Compact Lant.	2.00	6
LANTERN WITH RED FLASHER	10x8x5	6V Battery	3500 Ft. Beam	9.75	5
PLASTIC LANTERN FOR BOATS	5x5x8	6V Battery	Impact–Floats	2.95	3
PLASTIC LANTERN-IMPACT CASE	5x5x8	4-D Cells	Impact Case	2.25	8
SKIN DIVERS LIGHT	3x3x4	8-D Cells	Pressure to 150'	9.95	1/8
PURSE LIGHT	1½x¾	Jap. Cell	Cosmetic Size	1.00	1
PEN LIGHT	½x4	2-AAA Cells	Pencil Size	1.00	1
KEY CHAIN LIGHT	½x2	1-AA Cell	Miniature Lt.	1.00	1
OUTDOORSMAN'S LANTERN	5x7	Ni.–Cad.	Wtrprf. Sld. Bm.	18.95	2
CAMPER'S EMERGENCY LIGHT	5x7	Ni.–Cad.	Read. Rchgble.	12.00	2
LARGER REFLECTOR SELF CHARGE	3x5¾	Ni.–Cad.	Impact–Rchgble.	9.75	2
POCKET LIGHT SELF CHARGE	2x4¾	Ni.–Cad.	250 yd. Beam	5.75	3
SAFETY BOATING FLASHER	4x16	Ni.–Cad.	Starts–water	35.00	1/4

** Note: All Dry Cell Lights without Batteries.

STANDARD BATTERIES

COMMON DRY CELL - Carbon, zinc, ammonium chloride. First of the dry cell batteries. Metal clad to prevent harm to flashlight casing. Short life. Poor discharge characteristics. Low cost.

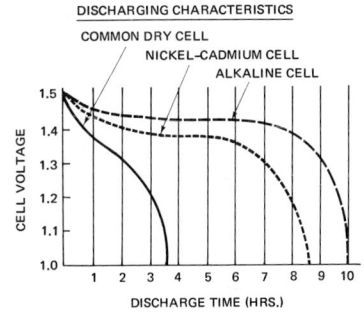

DISCHARGING CHARACTERISTICS
COMMON DRY CELL
NICKEL-CADMIUM CELL
ALKALINE CELL

RESEARCH (CONT'D)

CASE HISTORY 001 (CONT.)

ALKALINE DRY CELL — Ten times more service then regular dry cells. Remains at about peak power for life of battery. Most suitable for frequently used items. Medium cost.

NICKEL-CADMIUM CELL — About same service life as alkaline dry cell and performance characteristics. Main advantage ability to be recharged. High cost.

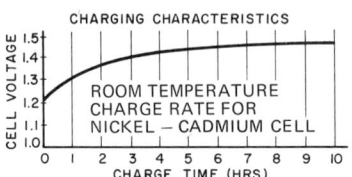

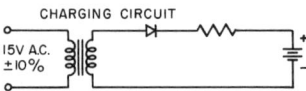

DESIGN DATA AVAILABLE FROM NICAD BATT. DIV., GOULD NATIONAL BATTERIES, INC., ST PAUL, MINN.

PHYSICAL CHARACTERISTICS (1.5V BATTERIES)

TYPE	DIMENSIONS				WT (OZ)	AMP. HRS	RETAIL PRICE ($)		
	A"	B"	C"	D"			DRY CELL	ALK. CELL	NI. CAD.
AAA	1.625	1.75	0.375	0.0094	0.5	1.0	0.15	0.30	1.50
AA	1.875	2.00	0.50	0.156	1.0		0.20	0.45	2.50
C	1.875	2.00	1.00	0.25	2.0		0.20	0.55	4.00
D	2.250	2.375	1.25	0.25	4.0	5.3	0.20	0.70	4.00

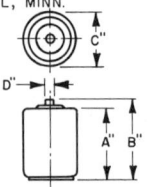

STANDARD 1.0 TO 2.5V BULBS (G.E. EVEREADY, WESTINGHOUSE)

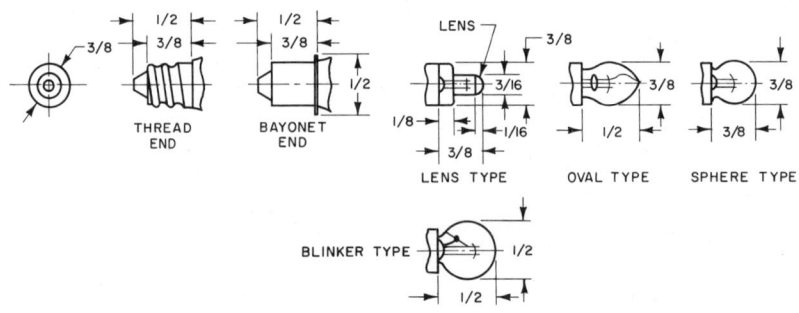

STANDARD SUBMINIATURE SWITCHES

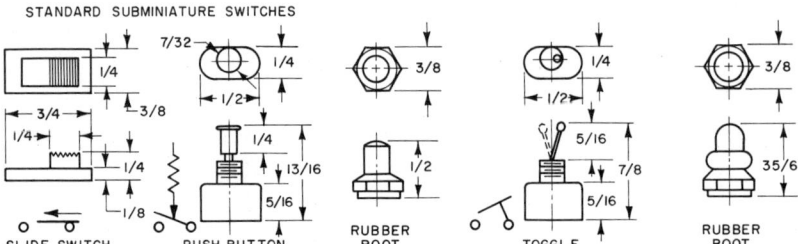

TASK Specifications

CASE HISTORY 001 (CONT.)

1. Flashlight case must be waterproof and impactproof.

2. Case must be able to float.

3. Flashlight must be dependable, light in weight, and unbreakable.

4. On-off switch should be easily found in the dark (feel).

5. Possible sources of power.
   a) Common dry cell.
   b) Alkaline dry cell.
   c) Nickel-Cadmium cell.

6. Size of flashlight to be determined by standard batteries. See research phase.

7. Flashlight to be designed for sporting use.
   a) Fisherman
   b) Hunter
   c) Boating
      (1) Sail
      (2) Power - outboard
      (3) Power - cruising
   d) Auto
   e) Etc.

8. Provide ease of portability

9. Set costs so that price will compete ($1.25 to $3.00).

10. Use as many standard parts in the design as possible.

11. Case should be shaped so that it will not roll.
    a) Hexagon
    b) Elliptical
    c) Square
    d) Weighted

12. Flashlight must be easily maintained by average woman and parts available at most hardware stores.

13. Reliability must be relatively high (say, 0.85).

14. Flashlight must have a solid feel, one that gives the sense of "can be depended upon."

15. Flashlight design must be unique in nature to command a market with little or no competition. (An innovation in flashlights or application of present lights.)

16. Design must be such that it fits in with and complements company's present product lines.

68 CASE HISTORY 001 (CONT.)

IDEATION

1. Provide for zoom lens (flood to distance).
2. Nickel-Cadmium cell to be recharged through solar energy (solar cells).
3. Hooks (fishing) and matches stored in handle.
4. Safety pin on back of case.
➡ 5. Case to be integral with hat.
6. **Could** be worn in a sheath like a hunting knife.
7. Cork section for storage of fishing **flies** and lures.
➡ 8. Scale (weight) and measuring tape built into handle.
➡ 9. Flashlight integral with fishing rod.
➡ 10. His and her light (two flashlights in one).
11. Contains spare bulb and batteries.
12. Resistant heater as a hand warmer.
➡ 13. Cigarette lighter at bottom of handle.
14. Red and green lens cover for use in boat.
15. Warning buzzer for emergencies (when lost in woods).
16. Map and hunting and fishing laws storage.
17. Add thermometer and possibly barometer to case.
➡ 18. Storage of cigarette package.
19. Small radio receiver or possibly walkie-talkie.
20. Small SOS transmitter to Coast Guard.
➡ 21. Compass at back of handle.
22. Storage of first-aid kit.
23. Flashlight should indicate (visual or audio) when batteries must be replaced.
24. Combine **flashlight** with spotter telescope.

CONCEPTS

CASE HISTORY 001 (CONT.)   69

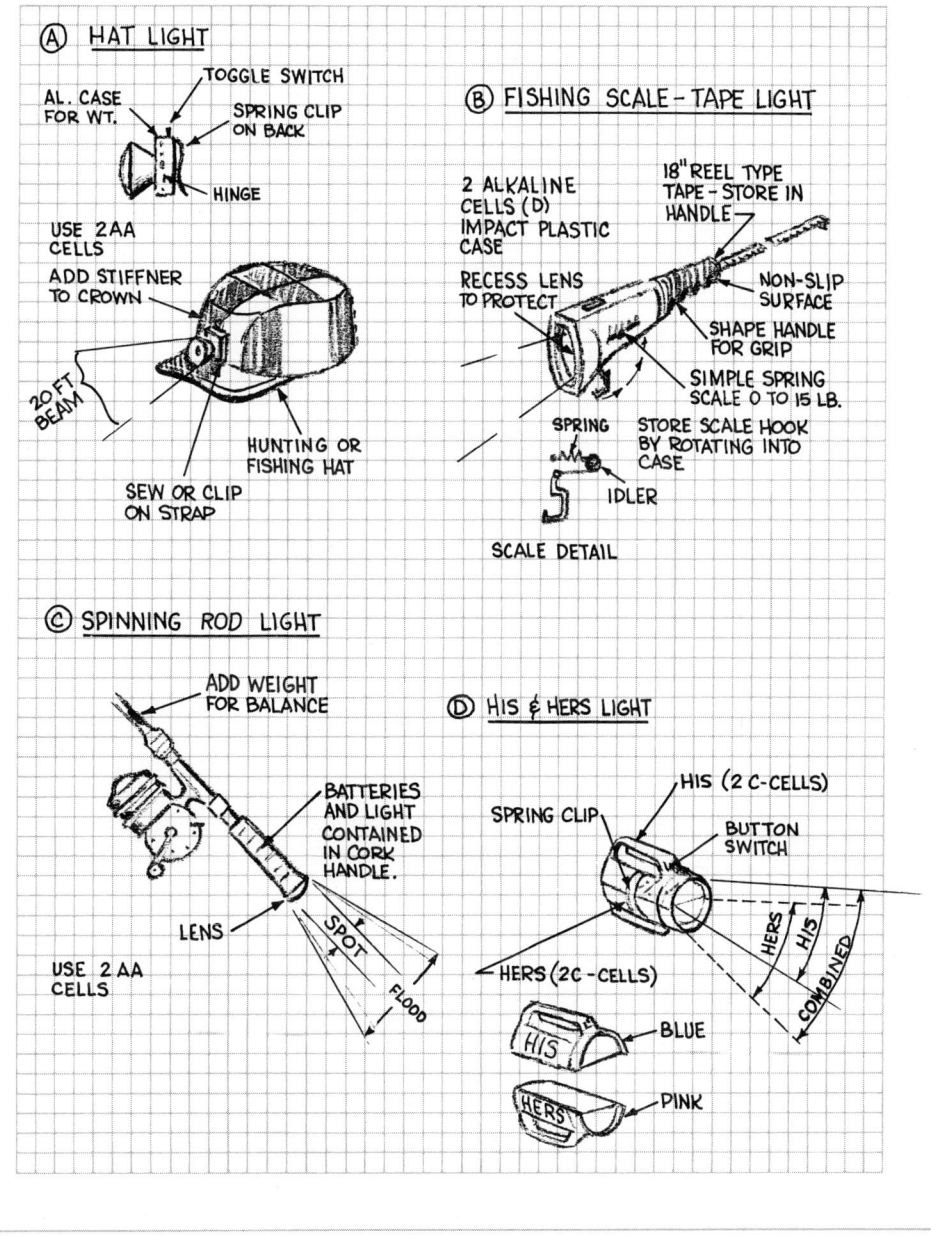

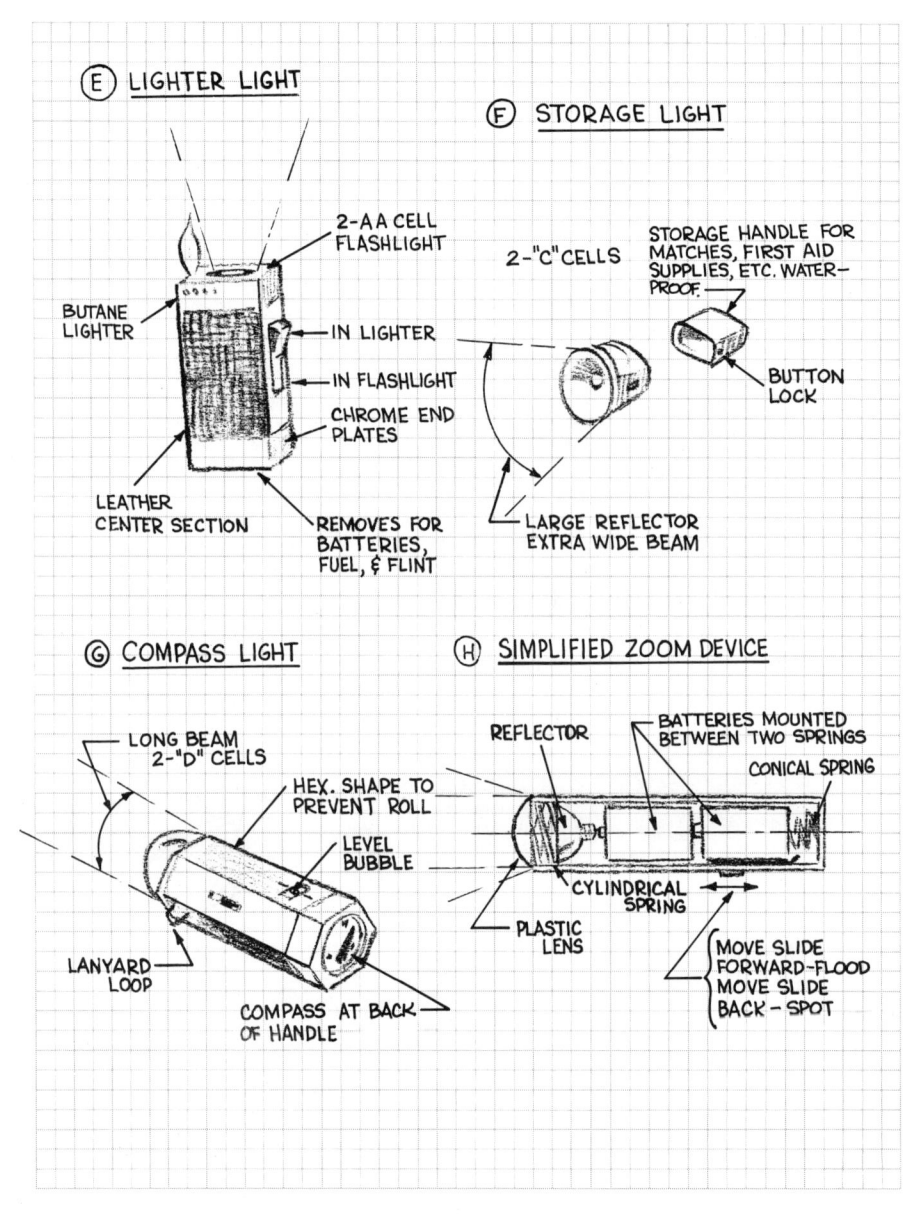

CASE HISTORY 001 (CONT.)   71

Alternatives \ Design Criteria (Weighting Factor)	USE OF STD. PARTS (0.1)	DURABILITY (0.1)	RELIABILITY (0.1)	SALES APPEAL (0.2)	SIMPLICITY (0.1)	COMPLEMENTS CO. LINE (0.2)	COST TO DEVELOP (0.1)	PRICE TO BUYER (0.1)	SUM (1.0)
A HAT LIGHT	7 / 0.7	6 / 0.6	8 / 0.8	6 / 1.2	8 / 0.8	3 / 0.6	6 / 0.6	7 / 0.7	6.0
B SCALE-TAPE LT.	6 / 0.6	7 / 0.7	7 / 0.7	8 / 1.6	3 / 0.3	9 / 1.8	6 / 0.6	3 / 0.3	6.6
C ROD LIGHT	9 / 0.9	8 / 0.8	8 / 0.8	9 / 1.8	6 / 0.6	9 / 1.8	6 / 0.6	7 / 0.7	8.0
D HIS & HERS	5 / 0.5	8 / 0.8	8 / 0.8	7 / 1.4	2 / 0.2	3 / 0.6	2 / 0.2	5 / 0.5	5.0
E LIGHTER LIGHT	7 / 0.7	7 / 0.7	7 / 0.7	6 / 1.2	7 / 0.7	3 / 0.6	6 / 0.6	7 / 0.7	5.9
F STORAGE LIGHT	9 / 0.9	8 / 0.8	8 / 0.8	8 / 1.6	8 / 0.8	4 / 0.8	6 / 0.6	8 / 0.8	7.1
G COMPASS LIGHT	9 / 0.9	7 / 0.7	8 / 0.8	6 / 1.2	8 / 0.8	8 / 1.6	6 / 0.6	8 / 0.8	6.4

CONCEPTS (DECISION MATRIX)

72 CASE HISTORY 001 (CONT.)

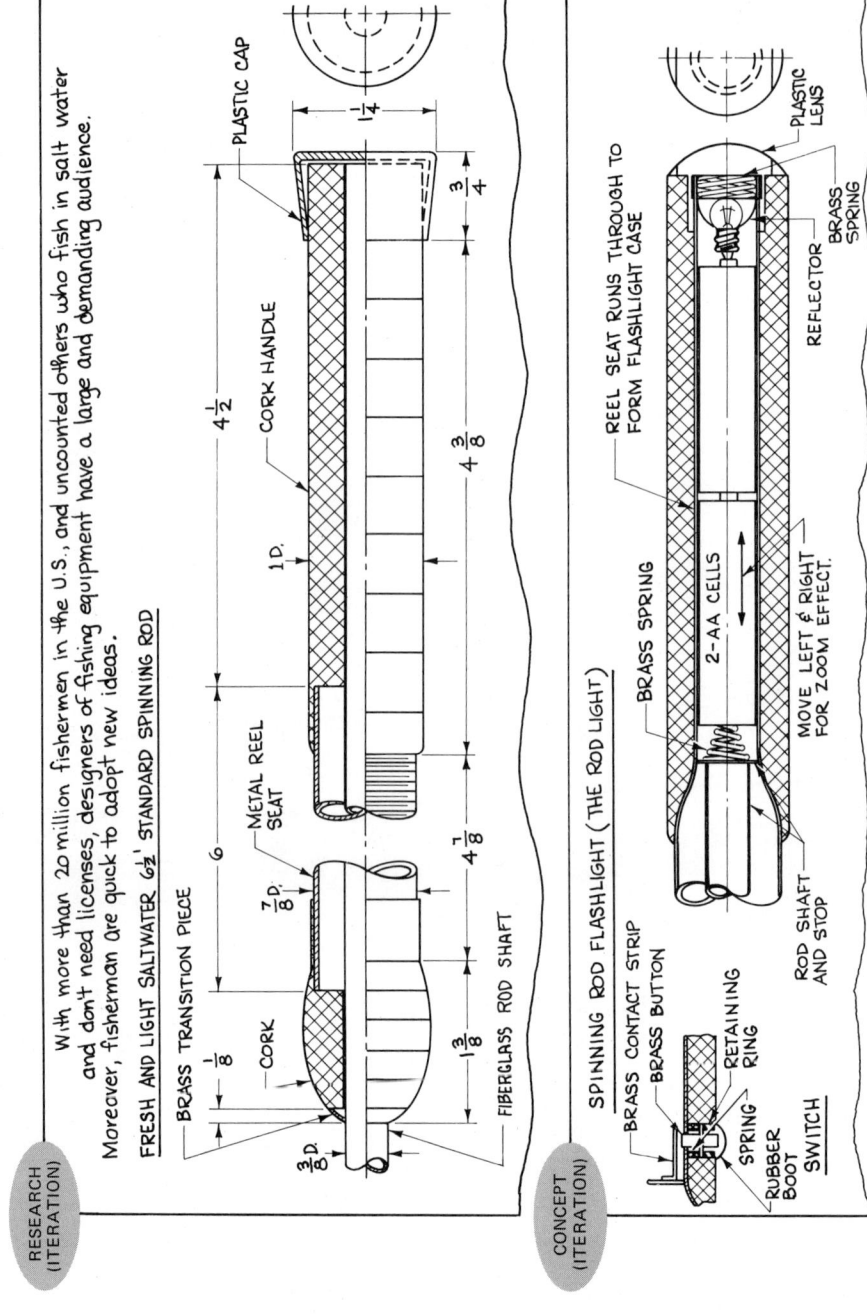

With more than 20 million fishermen in the U.S., and uncounted others who fish in salt water and don't need licenses, designers of fishing equipment have a large and demanding audience. Moreover, fisherman are quick to adopt new ideas.

FRESH AND LIGHT SALTWATER 6½' STANDARD SPINNING ROD

RESEARCH (ITERATION)

SPINNING ROD FLASHLIGHT (THE ROD LIGHT)

CONCEPT (ITERATION)

CASE HISTORY 001 (CONT.)   73

ADDITION OF WEIGHT TO BALANCE ROD

$\Sigma M_{CG} = 0$
$6.5(2.7) - 4.5 W = 0$
$4.5 W = 17.55$
$W = 3.9 \text{ oz.}$

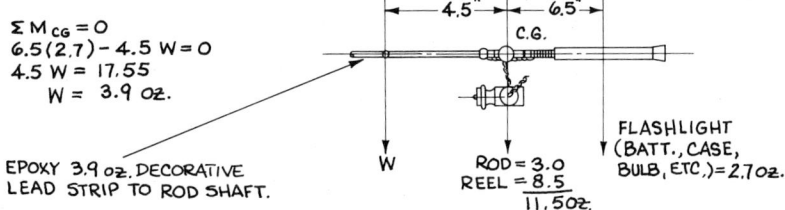

EPOXY 3.9 oz. DECORATIVE LEAD STRIP TO ROD SHAFT.

ROD = 3.0
REEL = 8.5
—————
11.5 oz.

FLASHLIGHT (BATT., CASE, BULB, ETC.) = 2.7 oz.

MATERIAL REQ'D. TO FLOAT ROD & REEL WITH LIGHT

$$(W_{RR} - V_{RR} \cdot d_{H_2O}) = (V_F d_F - V_F d_{H_2O})$$

VOLUME OF ROD & REEL
ROD = $\pi (0.094)^2 (6.5)(12) = 2.16$
REEL SEAT = $\pi (0.44)^2 (6.5) = 3.96$
HANDLE = $\pi (0.5)^2 (5.25) = 4.13$
REEL = $\pi (1)^2 (1) = 3.14$
$\pi (1.25)^2 (1) = 4.92$
—————
$V_{RR} = 18.31 \text{ cu. in.}$

WHERE:—
W_{RR} = WT. OF ROD & REEL
   = 3.9 + 11.5 + 2.7 = 18.1/16 LB.
V_{RR} = VOL. OF R&R (CU. IN.)
d_{H_2O} = DENSITY OF H_2O
   = 62.43/1728 #/CU. IN.
V_F = VOLUME OF FLOTATION
d_F = DENSITY OF FLOTATION
   COMMON CORK = 0.00722
   STYROFOAM = 0.00288

BASE CALCULATIONS ON 15% OF ROD & REEL FLOATING OUT OF WATER OR 85% SUBMERGED.

$(W_{RR} - 0.85 V_{RR} \cdot d_{H_2O}) = (V = d_F - V_F d_{H_2O})$

$\left[\dfrac{18.1}{16} - 0.85(18.31)\left(\dfrac{62.43}{1728}\right)\right] = \left[0.00722 V_F - \dfrac{62.43}{1728} V_F\right]$

$0.02888 V_F = 0.568$
$V_F = 19.7$ CU. IN. REQD. IF CORK

$0.03322 V_F = 0.568$
$V_F = 17.1$ CU. IN. REQD. IF STYROFOAM

LENGTH OF STYROFOAM HANDLE REQD. TO FLOAT ROD & REEL.

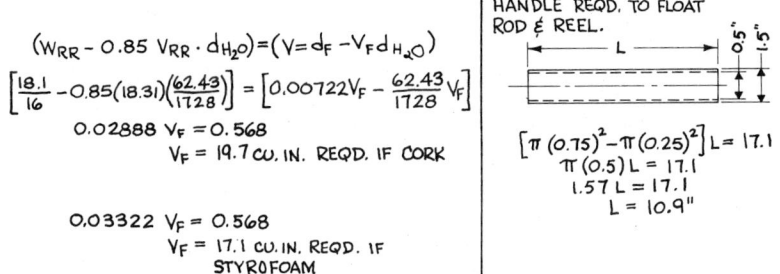

$\left[\pi(0.75)^2 - \pi(0.25)^2\right] L = 17.1$
$\pi(0.5) L = 17.1$
$1.57 L = 17.1$
$L = 10.9"$

IMPRACTICAL TO MAKE ROD & REEL FLOAT — TASK SPEC. #2 WILL BE VIOLATED.

74 CASE HISTORY 001 (CONT.)

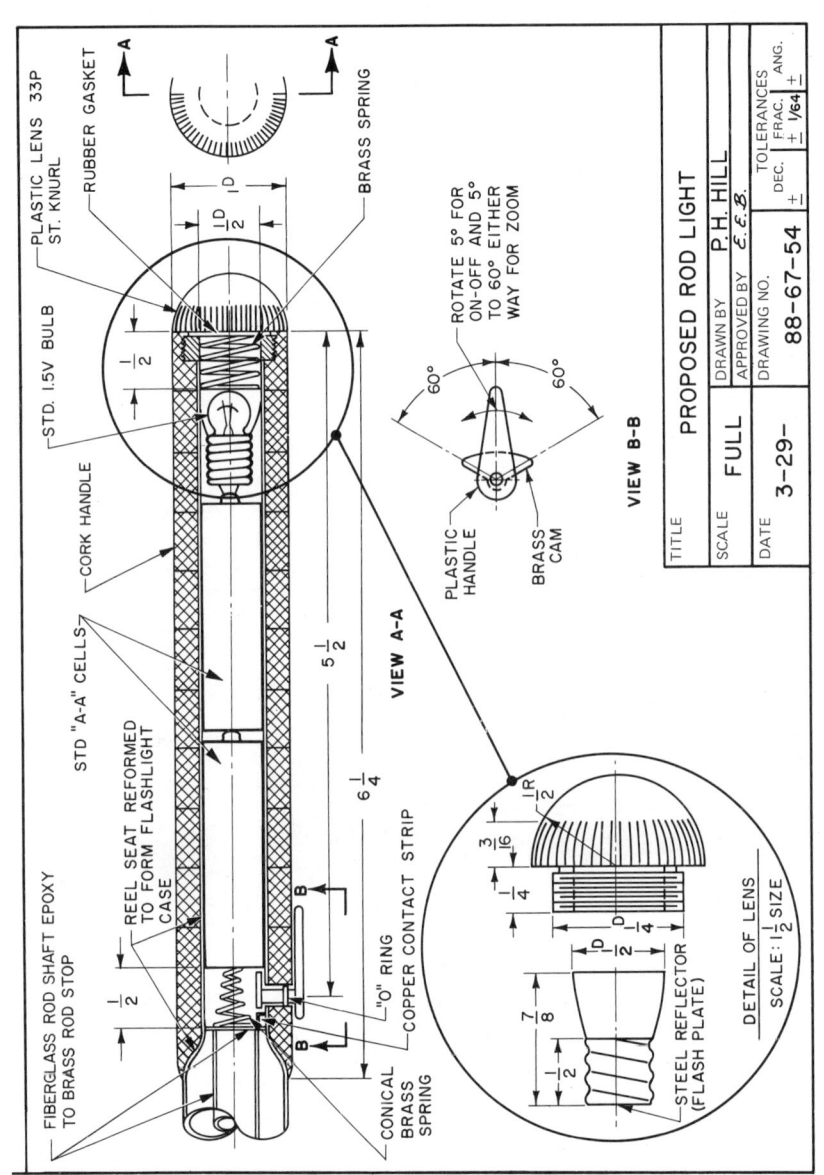

MANUFACTURE

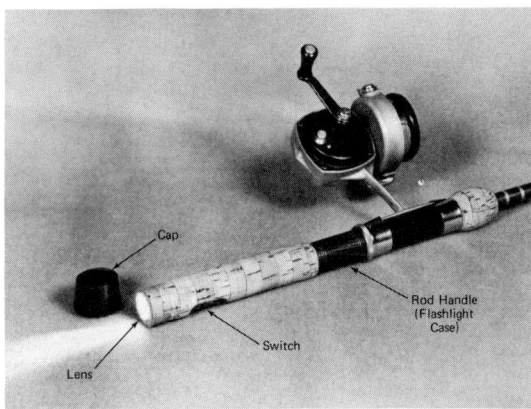

PROTOTYPE: FISHERMAN'S ROD LIGHT

The prototype shown in the photograph above was constructed from a standard fresh-water light spinning rod. The cork handle was drilled out and reinforced to receive an inexpensive "pen light" type flashlight containing two "A-A" cells. This model was planned to duplicate balance and weight as close as possible to the design shown in the drawing on the preceding page.

This prototype was used during the summer months of July and August by ten fishermen, five of which are considered experienced and the remaining as novices. All ten praised the design innovation and expressed a desire to own such a rod light. No fisherman found an appreciable difference in balance between the rod light and a standard model when casting for distance or accuracy. There was concern, however, among the five experienced fishermen as to the increased weight of the rod.

The light was used in the following ways while fishing or returning from fishing after sunset:
    (a) To inspect contents of tackle box when changing lures.
    (b) As a beacon to warn oncoming boats.
    (c) To pick out one's footing when returning along a dark path.
    (d) To check the time on wrist or pocket watch.
    (e) To remove hooks from a boated fish and place fish on a tow chain.

Results from this feasibility test with the prototype has proven to this designer the value of the ROD LIGHT as a marketable item.

MANUFACTURE

CASE HISTORY 001 (CONT.)

COST BREAKDOWN

1. REFORMING OF REEL SEAT TO FORM FLASHLIGHT CASE------------- $0.18
2. FABRICATION OF COPPER CONTACT SHIP------------------------- 0.02
3. FABRICATION OF BRASS ROD STOP----------------------------- 0.06
4. PLASTIC LENS (PRESSURE MOLDING)--------------------------- 0.15
5. ON-OFF AND ZOOM CAM AND HANDLE (ASSEMBLED)---------------- 0.20
6. SPRINGS (2)--- 0.10
7. REFLECTOR--- 0.10
8. STANDARD 1.5V BULB-- 0.10
9. RUBBER GASKET--- 0.03
10. LABOR AND TOOLING-- 0.75
11. ASSEMBLY LABOR--- 0.50

            SUB TOTAL $2.19

OVERHEAD (management, plant, packaging, promotion, etc.) 75%- 1.64

              TOTAL $3.83

PROFIT (25% to Distributor)------------------------------ .95

            NET PRICE $4.78

PROPOSE THAT PRICE OF ROD BE INCREASED BY $5.00 TO INCLUDE LIGHT FEATURE. BASED ON A RUN OF 1500 UNITS.

SALES POTENTIAL

- FLASHLIGHT IS INTEGRAL WITH ROD - WILL ALWAYS BE WITH FISHERMAN

- WATERPROOF AND NEARLY UNBREAKABLE

- RELIABILITY AND DURABILITY RENDERED THROUGH SIMPLICITY OF DESIGN

- COST IS LOW - WILL ENHANCE SALES OF ROD

- ZOOM LIGHT HAS SALES APPEAL

- WILL APPEAL TO THE GADGET MINDED INDIVIDUAL

- WILL APPEAL TO A SENSE OF SAFETY

- UNIQUE DESIGN SINCE IT IS ONE-OF-A-KIND

- SHOULD BE ABLE TO CAPTURE 1/5 OF THE ROD BUYING MARKET

➤ PROPOSAL - NO MARKET SURVEY REQUIRED.
LAUNCH PRODUCT IN EARLY MARCH
SECURE ENDORSEMENT OF "OUTDOOR LIFE" AND "FIELD & STREAM".

1) insurance for many years (10 years)
2) light
3) good light (wide visibility)

CASE HISTORY 002

CENTRAL DYNAMICS, INC.

MEMORANDUM

March 10, 19--

To: Advanced Design Group

From: Chief Project Engineer

Subject: Design of Underwater Work Tools under NAVY Contract.

    Our company has been fortunate in being awarded a pilot design contract by the U.S. Navy (Special Projects Office COMSUBPAC) to develop initial designs and prototypes of UNDERWATER WORK TOOLS to be used for deep diving operations. Historically, tools used by divers have been adaptations of similar implements used for land operations. The increased activity of industrial and Navy divers have demanded tools more appropriately designed for undersea use. Also the human factors of undersea work (diver strength limitations, buoyancy, ability to carry tools from surface vessel to work area, etc.) have not really been considered in conjunction with deep sea work.

    Tools required for undersea exploration involve both light and heavy duty, automatic and manual, as well as specialized devices. Items suggested by the Navy include hammers, drills, and wrenches as well as an observer's recording board, underwater crane for heavy objects, transit, stakes for instrument mounts, and an acoustic marker and receiving homing system.

    Our plan of operation is to have each designer submit a proposal on or before March 15 stating his request to design a specific tool and the reasons for this design. Upon written approval from the Chief Engineer the designer is to proceed with the design and proportion his time so that a reasonable solution may be presented to U.S. Navy officials on April 30, 19--. All experimental work may be carried out at our swimming pool facilities on College Avenue.

    Needless to say, designs of work tools that are accepted on April 30, will go a long way in securing future contracts with the Navy as well as private industry who are becoming more and more interested in this area. (See chart of Undersea Research Vehicles).

-1-

BACKGROUND MATERIAL

The following background material was collected as initial research or resource information for management to negotiate the pilot design contract with the Navy and is included here as a starting point for the designer.

> "Do not expect, however, for rare sea animals to be found in nets, nor seek for too many clues by dredging the bottom, or by drilling for cores. All these endeavors will help, but in the final analysis man must be present, using his own eyes, his own ears, and his other senses, combining with instruments derived especially for the enviornment of the sea and the depths".
> Captain E. John Long, U.S.N.

OCEAN STATISTICS:

Free ocean surface is 70% of total earth's surface or 141×10^6 square miles.

Estimated weight of all ocean water is 1.6×10^{18} tons.

Weight of the annual catch of all fish is 45×10^6 tons.

Dissolved minerals include magnesium, bromine, boron, uranium, copper, manganese, gold, and silver. All of the magnesium processed in the U.S. comes from the sea as well as a large percentage of bromine yet it costs $50,000 to extract $20,000 worth of gold from sea water.

The ocean under 80% of the area is deeper than 9000 ft. (almost 2 mi.) and under 50% is deeper than 12,000 ft. (2.3 mi.). Only a very small part of the area lies over depths less than 1000 ft.

One of the most troublesome environmental conditions is the magnitude of the hydrostatic pressure of great depths. It may be considered linear to range from 1.0 atm at sea level to 1090 atm or 16,000 psi at 36,000 ft.

The speed of sound in sea water is 4978 fps near the surface at a water temperature of $74°$ F. It increases with temperature rise at the rate of 5 fps per deg. F. It increases with depth (at constant temp.) at the rate of 1.8 fps per 100 ft. of depth.

The temperature of ocean waters varies with location. The surface temperature averages $30°$ F. near the poles and $80°$ F at the equator. At altitude $45°$ N the variations are as high as $40°$ F near Japan and as low as $6°$ F. on the California coast. All variations disappear at great depths and stabilize near $32°$ F. It is estimated that the average temperature of all the ocean water is $39°$ F whereas the surface average is $63°$ F.

-2-

Another important factor to the designer of undersea equipment is the absence of light even at relatively shallow depths. Red light is absorbed much more rapidly than ultraviolet, which penetrates to about 250 ft. In addition to cutting down the level of illumination, the apparent color of objects changes as the residual color of the light changes.

AQUACULTURE

Aquaculture (farming the sea) activity is picking up. Last year, Congress passed the Marine Resources and Engineering Development Act, establishing a Cabinet-level Council on Marine Resources, headed by the Vice-President. Several high-priority projects are already in the works.

Leading the list is a program to produce fish-protein concentrate; two pilot production plants will be built in the U.S. and three more in foreign countries to be selected. Next highest on the list is the concept of "sea-grant" colleges, patterned after the land-grant colleges of the last century which did so much to develop agricultural technology.

The most promising starting points for the development of aquaculture are bays and estuaries, where the waters are fairly well fenced in and are easily accessible from land. Today, in the U.S., brackish near-shore areas are usually regarded as waste space, and are neglected and often polluted. By contrast, in the Gulf of Taranto, Italy, the brackish zone is used to produce an average of 108,000 pounds of meat per acre per year.

INDUSTRIAL OPPORTUNITIES

A recent Harvard Business School survey entitled "The Businessman's Guide to Oceanography" observes that the full exploration of the sea depends on the development of new ocean technology, particularly in the fields of instrumentation and offshore construction, and that the most important task for successful exploitation is the scientific exploration of the ocean.

Before minerals can be mined from the sea floor, for example, the floor must be surveyed and charted. To conduct this exploration, oceanographic scientists need a great variety of instruments whose manufacture therefore appears as a prime commercial area for the ocean industry, especially for smaller firms. In fact, the instrumentation market, it is predicted, will be the first major boom in oceanography.

ADVANCED DESIGN GROUP MEMORANDUM

COMSUBPAC CONTRACT Pg.1 of 3

Subject: COMSUBPAC - Special Projects Proposal

To: Chief Project Engineer

From: R. Cox

Date: 12 March 19--

  In response to COMSUBPAC-RFP it appears that an area which will be of great benefit to both COMSUBPAC and this organization is that of support cable stabilization.

  The stabilization of support cables is a pressing problem not only to the military but to any and all systems which require surface support whether it be for power and instrumentation or for life support. There is also an obvious extension of this area into the commerical fishing industry as a power source for the manipulation and maneuvering of large undersea fishing nets. This latter area being a second generation of the techniques learned from the prime area of development. It is expected that a complete proposal package would be available for submission to COMSUBPAC on or before 30 April 19--. To accomplish this task the following services and funding are requested by the Advanced Design Group:

1. Personnel
   - (a) One staff engineer/project manager (full time)
   - (b) One senior administrator (part time)
   - (c) One senior planner (part time)
   - (d) One senior design engineer (full time)
   - (e) One Secretary (full time)
   - (f) One model maker (part time)

2. Office Space
   Office space is requested for the above personnel in order to provide a single area within which this project may be carried out. Appropriate equipment for these personnel is also requested.

3. Funding
   Funds are requested for the following:

(a)	Salaries for personnel	$5,000.00
(b)	Office Supplies	500.00
(c)	Publications	1,000.00
(d)	Machine Shop	500.00
(e)	Raw Materials	1,000.00
(f)	Transportation and Telephone	500.00
(g)	Misc. Petty Cash	500.00
	Total Funds	$9,000.00

REC/al

ADVANCED DESIGN GROUP MEMORANDUM

COMSUBPAC CONTRACT Pg.2 of 3

12 March 19--

INTRODUCTION:

The need to provide cable stabilization between surface vessels and underwater equipment has been an area of concern to many for a number of years. To date, stabilization is used only on the most demanding underwater projects but the general utilization of these concepts will produce beneficial results for any underwater system which requires support from surface vessels.

This proposal is for a preliminary investigation into the area of cable stabilization between surface ships and underwater work platforms. It is intended to point out the basic problems with such a concept and to offer a practical solution through the preliminary design concept as shown in this proposal. If approved, Advanced Design Group proposes to carry out a program for development of the cable stabilization unit as follow:

PHASE 1 - STUDY

Upon receipt of work approval, Advanced Design Group proposes to investigate problem areas considered important to proper control of the cable stabilization unit. In this phase detailed engineering efforts will be utilized to provide the systems analysis and unit design. Where necessary breadboard systems will be constructed to demonstrate the feasibility of certain operating principals. During this phase detailed drawings will be prepared which will be suitable for the manufacture of prototype equipment.

ADVANCED DESIGN GROUP MEMORANDUM

COMSUBPAC CONTRACT Pg.3 of 3

12 March 19--

PHASE II - LABORATORY PROTOTYPE

Upon completion of Phase I, and upon receipt of notice to proceed from COMSUBPAC, Advanced Design Group will construct and evaluate a laboratory prototype of the cable stabilization unit. The prototype unit will be exposed to extensive testing to determine unit's ability to comply with the requirements. Once satisfactory performance is achieved in the laboratory, actual field testing will be conducted to test ability to perform under actual operating conditions. This phase will conclude with a full report on the systems capability to performances required.

PHASE III - FIELD PROTOTYPE UNITS

Upon approval of the final design drawings and upon receipt of notice to proceed, Advanced Design Group shall fabricate three prototype cable stabilization units. The three units shall be delivered to COMSUBPAC together with three sets of calibration equipment and instruction manuals for proper operation. Advanced Design Group shall also furnish a reproducible set and two copies of all engineering, shop and fabrication drawings in sufficient detail to permit solicitation of bids for procurement of additional quantities on a competitive basis.

CENTRAL DYNAMICS, INC.

MEMORANDUM

Date: March 15, 19 --

To: R. Cox

From: Chief Project Engineer

Subject: COMSUBPAC CONTRACT

    Your proposal dated 12 March 19-- regarding a CABLE STABILIZING SYSTEM as the design study related to the COMSUBPAC contract has been reviewed by this office. We fully agree that such a system will be required in the future as exploration as well as technology involving the ocean depths becomes more commonplace. Although your proposal deviates from items suggested in my memorandum of March 10, 19--, we find it to be of sufficient interest to the Navy to approve of the request.

    It should be remembered that as this system is designed primarily for military use, your design group should review and adhere closely to standards set by the U.S. Government. We also request that you keep this office informed of design progress on a weekly basis. To insure that material will be ready for review by Navy officials on April 30, 19--, the preparation of a P.E.R.T. network is strongly recommended.

Ships at sea tending underwater work stations presently have a great deal of trouble holding position accurately above these stations even in calm weather. The problems are magnified when the sea is running or if the weather is other than ideal. This means that the ships have to put out half again as much cable as is required to service a given station so that they do not sever the connecting cable while trying to maneuver above the work station. Part of the problem is not the surface vessel's ability to hold position but how the sea is running below the surface (between the surface and the work station). Putting these problems together, one can now appreciate why cable stabilization is an important part of providing a connecting link between surface vessel and underwater work station.

GENERAL SPECIFICATIONS:

1) To support cables between surface ships and underwater work stations.

2) To maintain cables between underwater work stations and the surface within acceptable coordinate location above the work station.

3) To maintain cables at some predetermined tension level separate from that of the surface vessel. Surface vessel can now maintain a slack cable.

4) To maintain cable position by an internal sensor system and not rely on signals sent from auxilary sounding buoys.

SPECIFICATION STANDARDS

The cable stabilization unit shall comply with naval military specifications for equipment which is intended for underwater service. Other specifications which this design will comply with are as follows:

- A. Workmanship Standards - Advanced Design Group
- B. Safety Standards - Advanced Design Group
- C. Human Engineering - Advanced Design Group
- D. US Navy MATERIALS SELECTOR Specification
- E. US Navy Underwater propulsion specification
- F. US Navy Underwater power sources specification

DETAILED SPECIFICATIONS

To set some ground rules from which to base the design the following information must be considered:

- a) The unit should be of portable design
- b) The unit should be capable of cable attachment by simple mechanism by underwater swimmers
- c) The unit should be capable of being used alone or in series with other units
- d) The unit should be safe to be handled by underwater personnel
- e) The unit shall be capable of extended service within sea water
- f) The unit shall have three sigma reliability factor for all systems
- g) Power source may be self-contained or provided by surface vessels
- h) Unit shall be capable of holding position within required cone angle from work station

SPECIFICATIONS (continued)

    Using these ground rules the unit to be designed shall be as compact as possible and shall contain only the essential equipment for its thrusters and controls. Power equipment will be maintained by the surface vessel and shall be supplied to the cable stabilization unit by separate cable which is to be loosely attached to the cable being supported.

    Because thruster power is an important parameter in the cable stabilization unit the selection of a drive motor type is not an easy choice to make since the design is to be predicted on the use of a hydraulic drive system using sea water as the operating fluid. Systems analysis will be concerned with a study of all motor types.

    Because hydraulics are provided for this design, actuation of all control surfaces will also be hydraulic with inputs obtained from a fluidic control system which will use sea water as its operating fluid. Modulation of the control system will be provided by mechanical force sensor systems. All of these control systems have been developed as considered here; however, modification of existing equipment will be a prime consideration and will be closely evaluated as a part of systems evaluation.

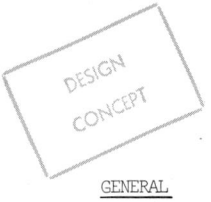

GENERAL

This design is the evolution of a series of ideas and this report shows the culmination of these ideas into the proposed configuration. The cable stabilization unit is described by reference to Figures 1, 2, 3, and 4. Some of the features of this design are:

- a) The compact unit
- b) The ease with which the unit may be cable attached
- c) The considerations of personnel attaching the unit while underwater
- d) The method of control stabilization
- e) The method of coordinate sensing.
- f) The arrangement of component parts
- g) The simplicity of construction

This design will utilize the best available materials where required to produce a highly reliable rugged unit.

METHOD OF OPERATION

The cable stabilization unit will be attached to support cable at specified intervals depending upon cable length and weight per linear foot. Once attached to the support cable and power from the surface vessel is available, the unit will operate as follows:

A. The inertial sensor system will sense acceleration forces and through displacement of a proof mass will vary the output signal of a fluid control system. This output will be amplified and transmitted to the control surface actuator mechanisms which will operate in a closed loop and provide corrective thrust to bring the unit back to its original position. The inertial system will be specified to have positional sensitivity of $\pm$ 2 meru* with a dynamic range of $\pm$ 10 G's.

B. Upon receiving a corrective error input the control surface flaps to direct unit thrust in a direction so as to compensate for side forces acting on the thruster unit. Continued rectilinear acceleration inputs to the system over a determined time period will cause the control unit to activate a rotation mechanism which will tilt the cable stabilization unit and direct greater side thrust in the event such thrust is necessary.

C. The cable stabilization unit motor for thruster power is hydraulic of the constant displacement type. This motor operates at a speed depending upon operating pressures supplied. The operating pressure to the motor is set at some minimum level and is varied upward from this level by the fluidic control unit with inputs of the inertial sensor systems. The maximum upper limit of pressure to the motor

* meru = milli-earth-radian-unit = 66.48 $^{\circ}$/hr. (angular measure of gyrascopic drift)

is predetermined so that the cable stabilization unit will under no condition exceed the safe cable tension.

UNIT ADVANTAGE:

The cable stabilization unit has the following advantages:

a) The unit is of small size

b) The unit uses sea water as its operating fluid not requiring special operating fluids nor return lines

c) The unit is easy to attach to present support cable systems

d) The unit relieves surface vessels to maintain a taut support cable system

e) The unit relieves surface vessels to maneuver in high seas

f) The unit does not expose swimmers to hazardous equipment

g) The unit is of simple construction

h) The unit provides its own inertial sensor system

UNIT DISADVANTAGES:

Because of the unit's simplicity and lack of internal power supply, the unit is limited in the following ways:

a) The unit requires power to be provided by surface vessel

b) The unit is not an absolute sensing system and as a result over an extended time period the inertial sensor system develops error drift which may decrease its ability to remain on station over the work platform. Realignment may be achieved by using pressure pulse modulation to the control system but this technique is untried as yet.

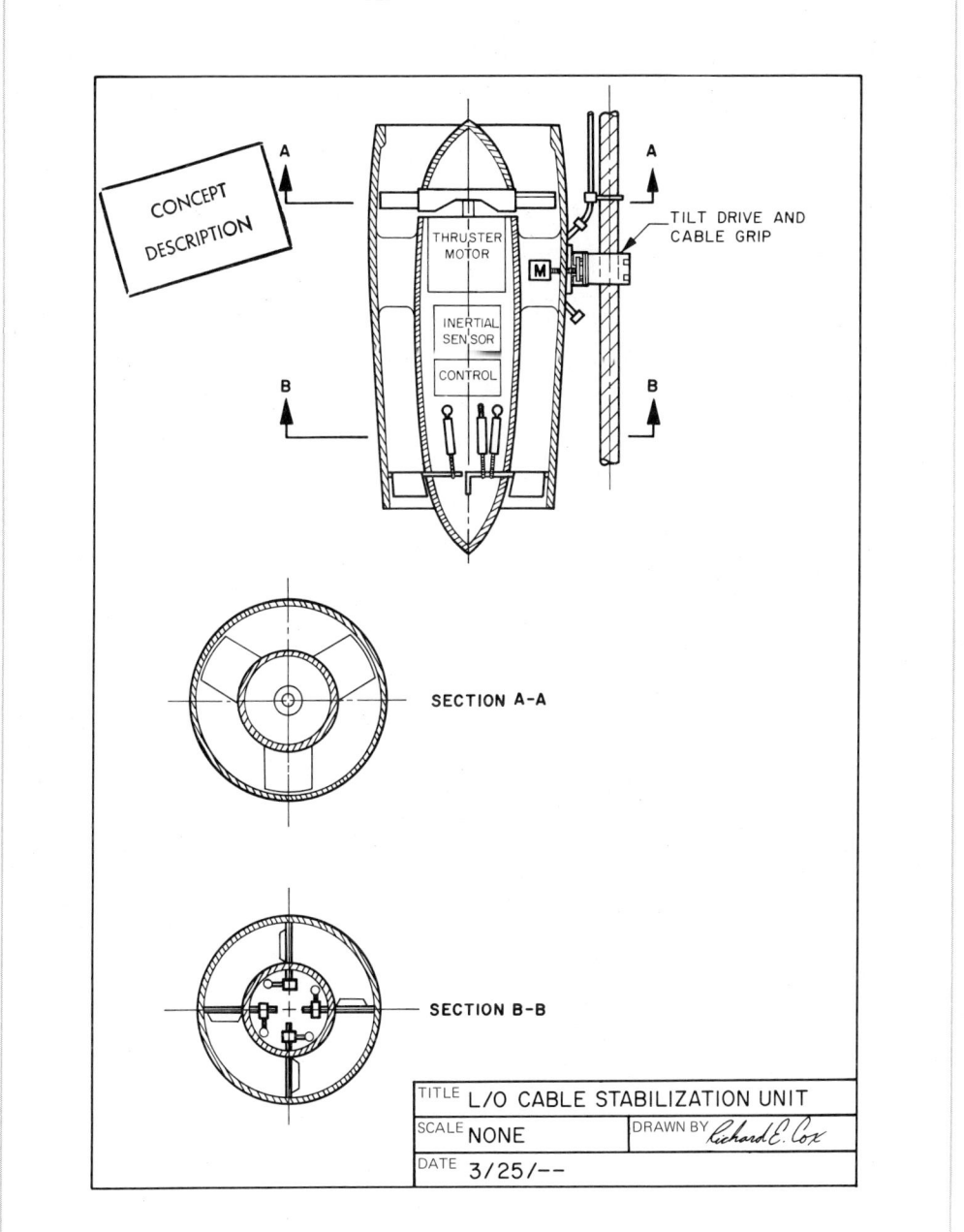

92 CASE HISTORY 002 (CONT.)

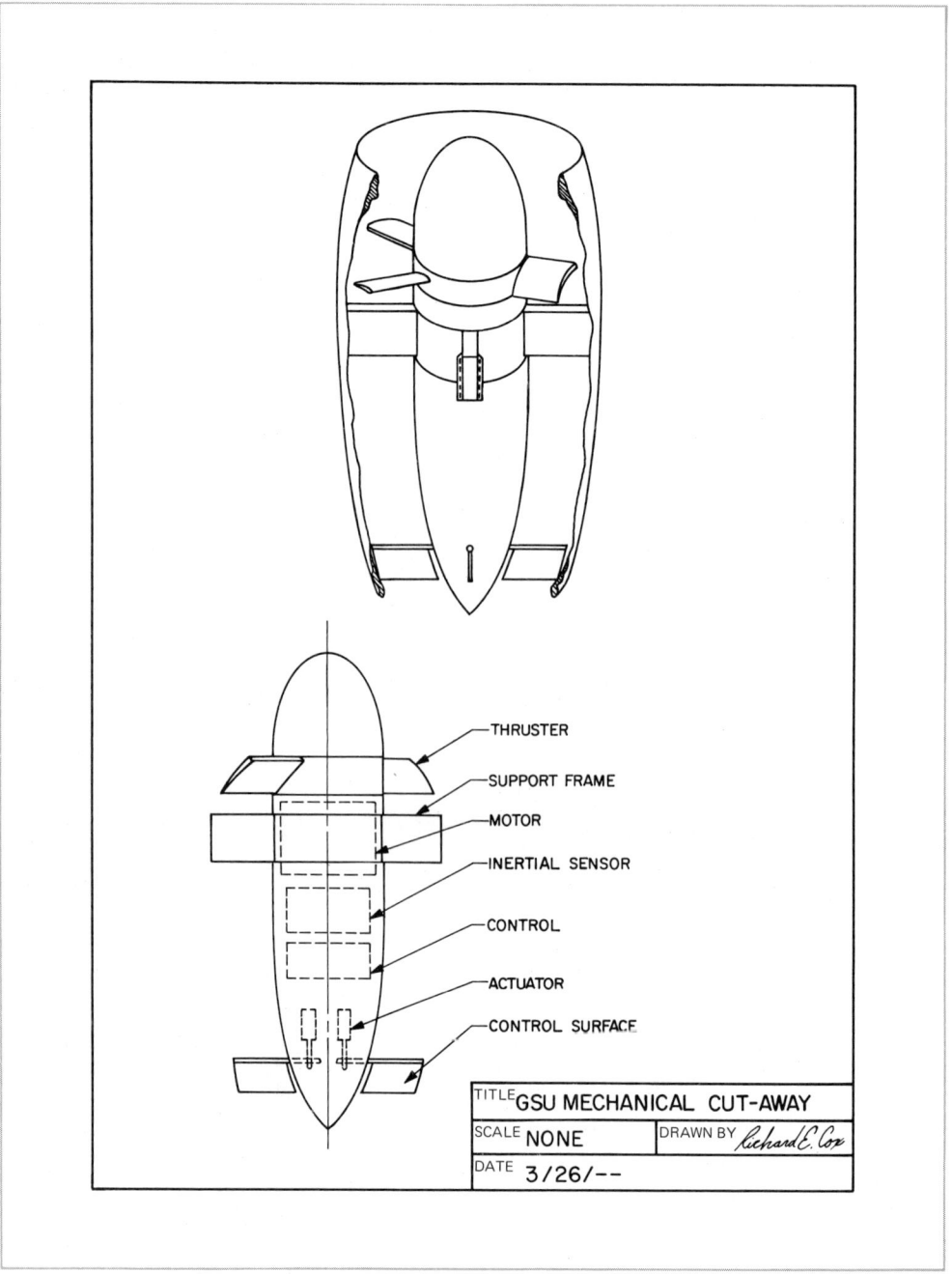

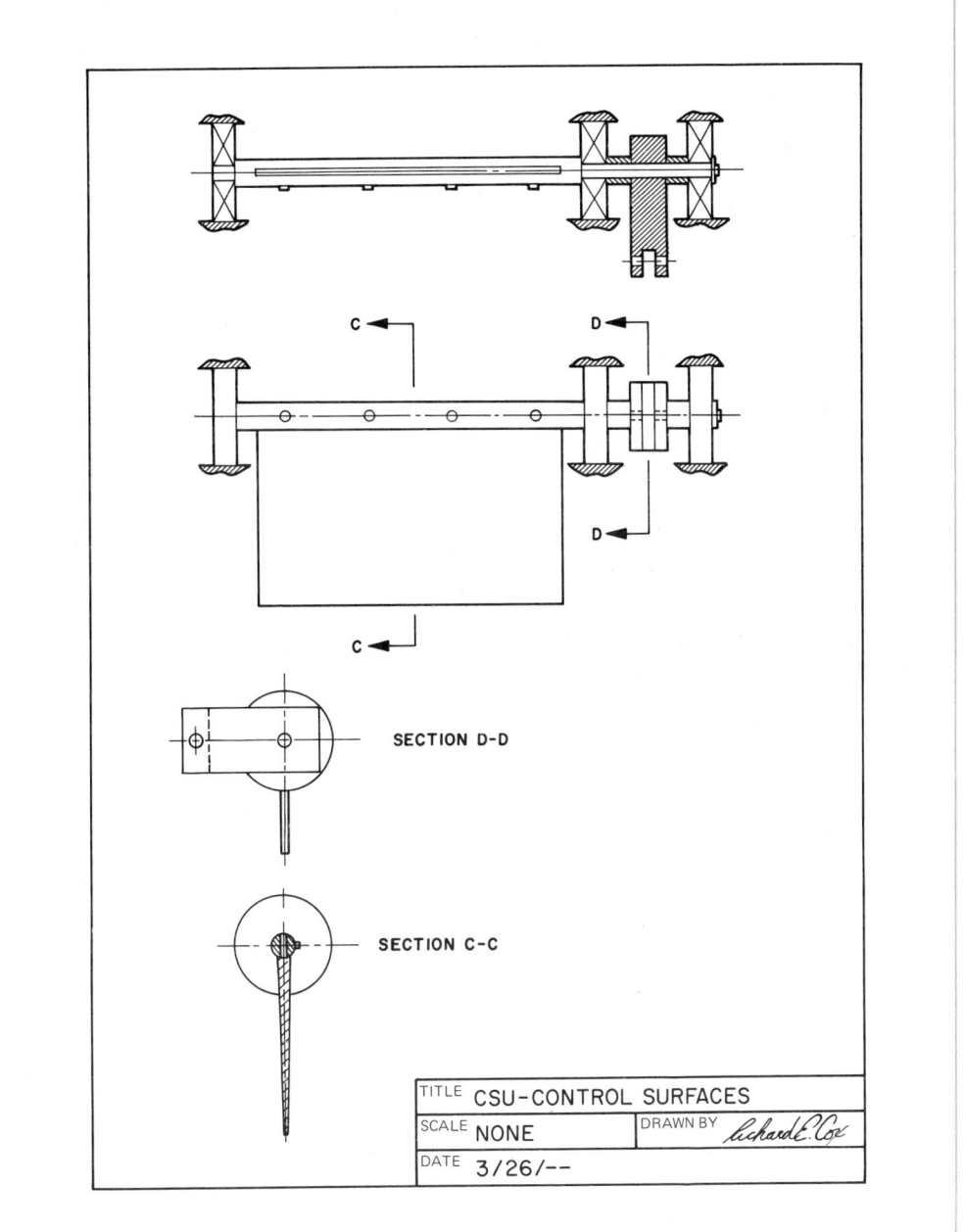

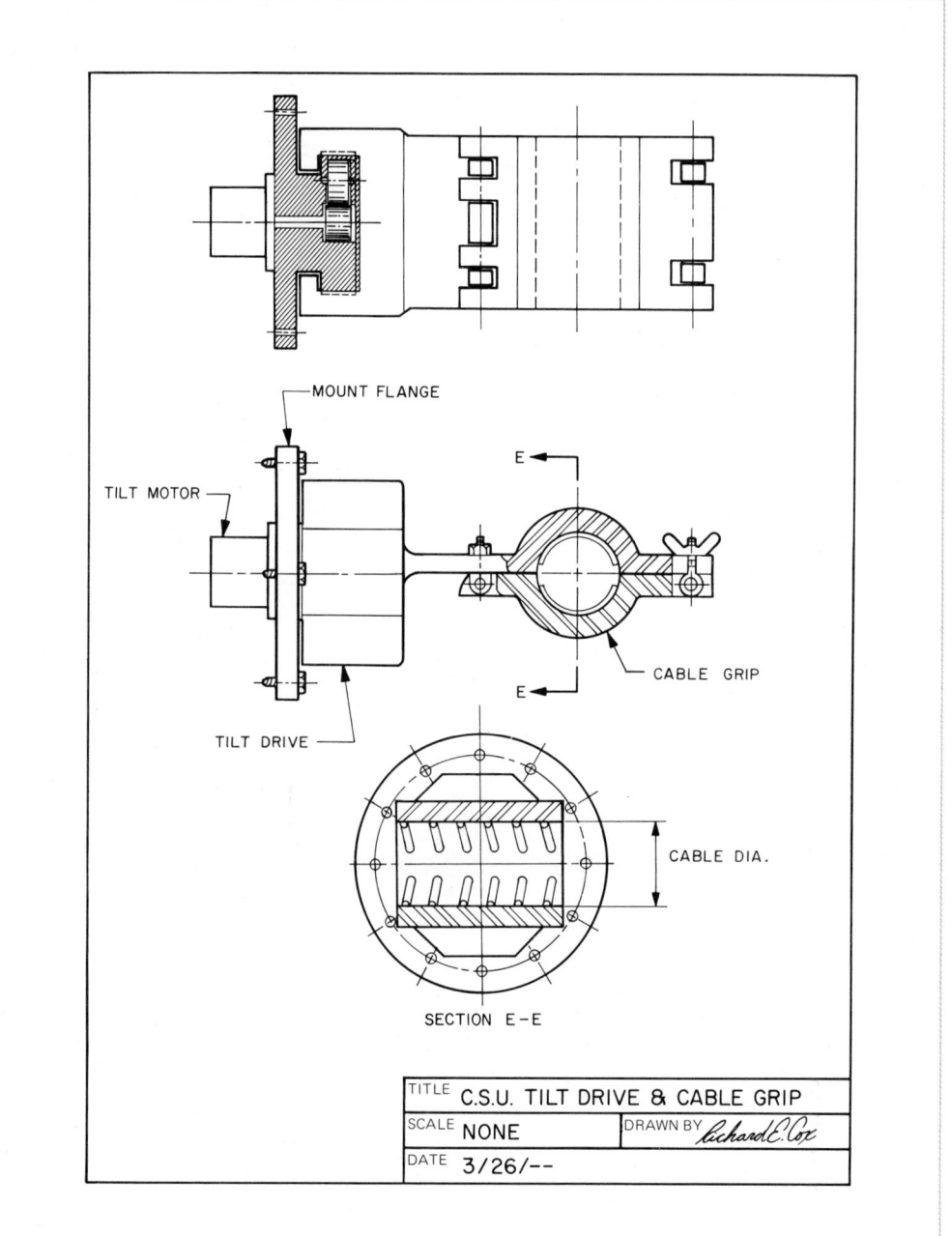

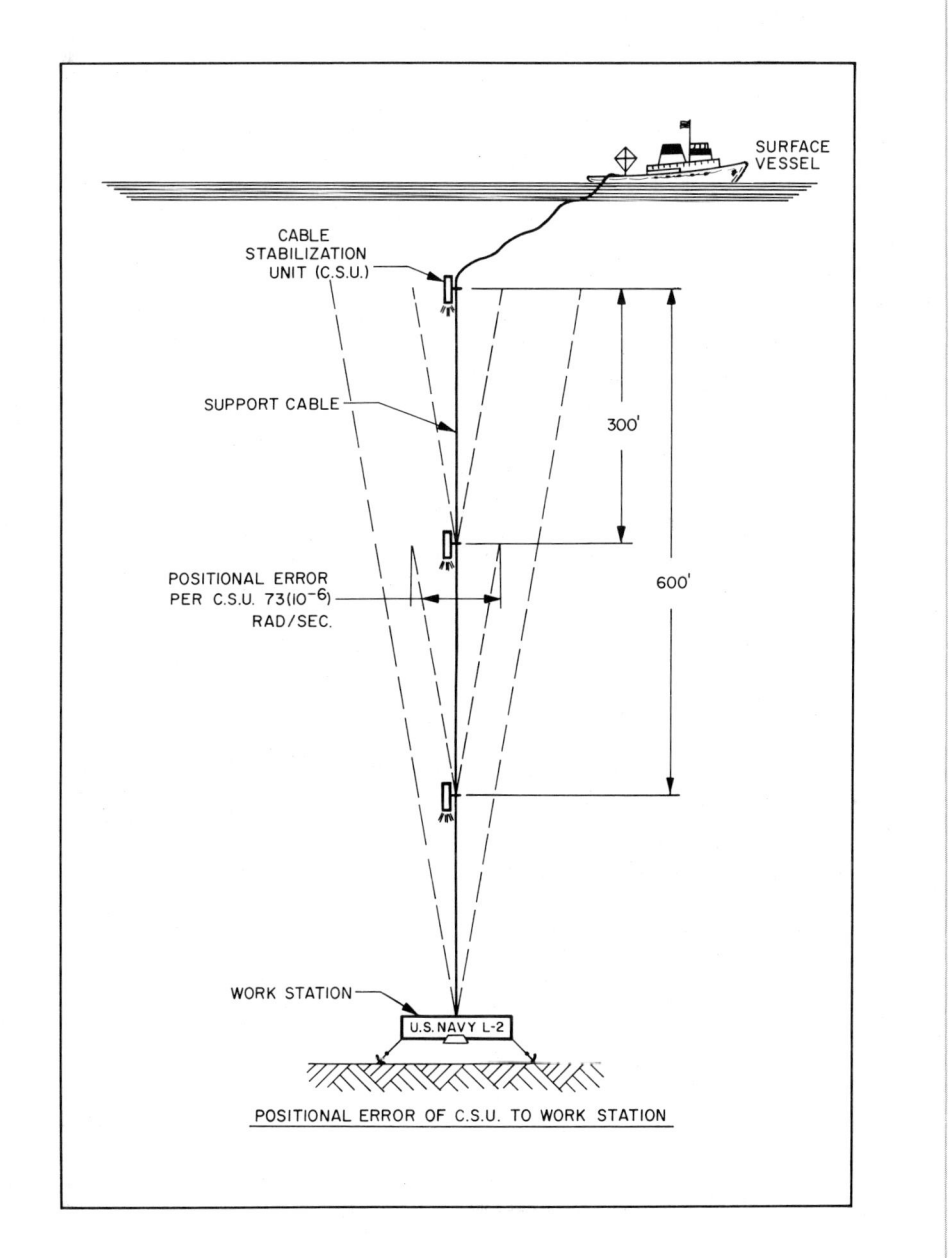

ANALYSIS

The cable stabilization unit has been engineered for function and basic power level only. The inertial control unit and the control system in general will require detailed engineering and will be conducted in Phase 1 of the Work Statement. The following is a preliminary concept of unit control.

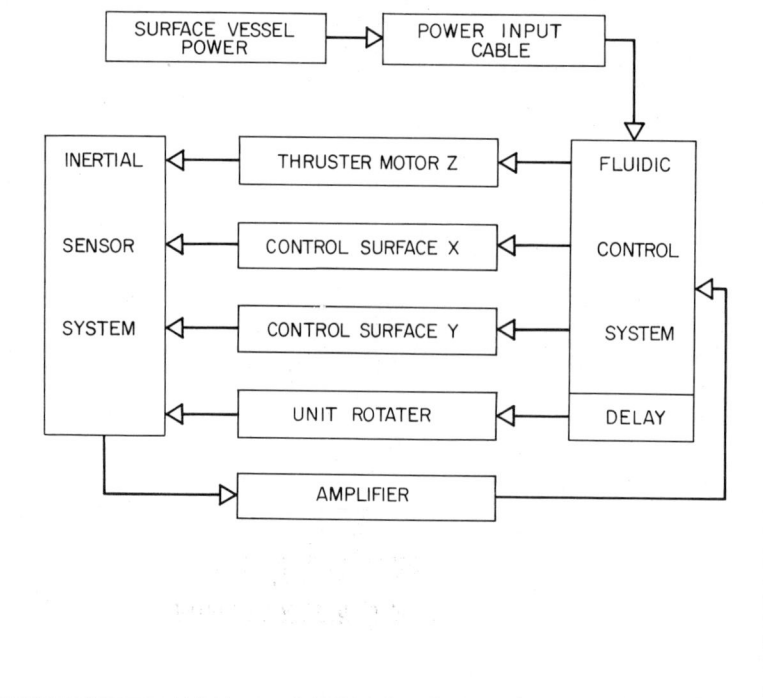

Where surface vessel power is supplied to the fluidic control unit and distributes to the thruster motor and control surfaces, the accelerator output of the unit is then sensed by the inertial sensor system and the error is used to modify the fluidic control system and drive the cable stabilization unit back to its true position.

THRUSTER MOTOR

The thruster motor size and power required is dependent only upon the necessary thrust required by the cable stabilization unit. The thruster is obtained by driving an impeller unit within the cable stabilization unit.

Considering that the thruster has to be able to lift a cable which weighs one pound per linear foot with normal spacing 300 foot increments the thruster must develop 300 pounds of thrust. Designing for a 200 per cent overload the thruster must develop 900 pounds output thrust. This percentage increase in thrust is required to allow for various operating conditions and operations which require considerable side thrust correction.

Based on an impeller which is 2 foot outside diameter with a one foot inside diameter the following shaft power is required to deliver 900 pounds output thrust:

O.D. = 24"   I.D. = 12"   No. of Blades - 3/8"
Blade $5°$ Rotational speed = 1000 rpm

Because thruster is located within a ducted cylinder we can apply the following momemtum theory to determine the thrust and shaft torque:

$$\text{Thrust} = T = \rho V_o A (V_s - V_o) \quad (1)$$

$$\text{Power} = TV = \rho V_o^2 A (V_s - V_o) \quad (2)$$

in these equations nonuniformity in velocity and stepstream are neglected

$$\text{Power input} = 1/2 \rho A V_o (V_s^2 - V_o^2) \quad (3)$$

where

T = Thrust in pounds

ρ = Density in slugs/ ft.^{3}

A = Cross section area ft.^{2}

V_o = Initial flow velocity ft/sec.

V_s = Final flow velocity ft/sec.

Substituting into equations 1, 2, and 3 where V_s and V_o are related by thruster slip factor for ducted fans.

$$V_o/V_s = 0.6$$

therefore

Thrust = 900 = (2.06)(2.36)(0.6)(V_s)(V_s - 0.6V_s)

V_s = 27.8 ft/sec

Power = TV_o = (900)(0.6)(27.8) = 15,000#ft./sec.

$$\text{Power input} = (2.06)(2.36)(0.6)[(27.8)^2 - (16.7)^2]/2$$
$$= 20{,}000 \text{ \#ft/sec.} = 36.4 \text{ H.P.}$$

This power level is not uncommon for hydraulic motors and motor envelope will be approximately 9" outside diameter by 10" long.

Using a radial piston motor with eight power pistons and a supply pressure of 1500 psi the flow rate will be in the order of 20 gallons of fluid per minute. This flow rate will allow motor capability of approximately 60 H.P.

CONTROL SURFACE ACTUATORS

The design of control surface actuators will be based upon system engineering calculations to be conducted in Phase 1 of Work Statement. Such factors as response time position accuracy etc. will have an important bearing on equipment design.

CABLE STABILIZATION UNIT-COST

Phase 1- Study

Program Administration	@ 30*mm	$77,400
Engineering		
Mechanical	@ 12 mm	31,200
Controls	@ 7 mm	18,200
Aeronautical	@ 7 mm	15,600
Systems	@ 20 mm	52,000
Drafting		
Designers	@ 36 mm	62,400
Draftsman	@ 43 mm	74,500
Detailers	@ 20 mm	34,600
Checkers	@ 10 mm	17,300
Model Shop & Fabrication		20,000
(fixed price)		
Total Services Costs (w/overhead)		$403,200
Profit 10%		40,320
Total Costs Phase I		$443,520

* mm = man months

PHASE II - PROTOYTPE UNIT - LABORATORY

Item	Unit Name	Cost
1	CSU Housing	$ 1000
2	Support Frame	700
3	Inner Housings	600
4	Thruster Motor	2000
5	Thruster	500
6	Inertial Sensor System	5000
7	Control System (Fluidic)	1000
8	Control Actuators	200
9	Control Surfaces	100
10	Misc. Plumbing	100
11	Tilt Mechanism & Cable Holder	200
12	Fluid Power Hose 500'	100
13	Misc. Parts	100

Total Material & Fabrication Costs	$11,600
Engineering Services	15,000
General and Administrative	9,000
Overhead and Profit 115%	40,940
Total Costs Phase II	$76,540

SUMMARY

Advanced Design Group feels that this problem of Cable Stabilization for underwater work platforms is an important area of interest to COMSUBPAC Special Projects Office and to the National Need.

Advanced Design Group has experienced personnel on its staff which are familiar with the various problems of designing operational undersea equipment and its systems engineering group is presently evaluating major systems of a similar type for COMSUBPAC command.

CASE HISTORY 003

STONEHAM BABY DIES IN FIERY RTE. 93 CRASH

"Hair singed, hands and feet burned, a young Stoneham mother pulled her four-year-old son from the burning wreckage of their station wagon Friday afternoon on Route 93 in Medford.

Then she looked for her five-year-old girl and one-year-old boy. The baby was dead.

One minute earlier Mrs. ----, was heading home with her children, ----, 5, ---- 4, and the baby, ----, 1.

Mrs. ----, who is six months pregnant, was driving in the passing lane when she was involved in a collision with a car driven by Mr. ---- of Windham, N.H. He was following the ---- car.

Both cars burst into flames and skidded onto a grass median strip. Air Force Capt. ----, stationed at Pease Air Force Base, Portsmouth, N.H., was one of the first to reach the scene.

"I saw the cars hit about 100 feet up the road," he said. "They just seemed to explode into flames. I pulled over to the right and ran through heavy black smoke to the cars. First, I came upon a little girl. I moved her away from the flames and went back. The mother was helping the boy get out of the station wagon. She yelled, "My other boy is still in there." But I could see the little boy on the grass. I picked up the little boy and brushed some dirt off his face. Then they took him away in the ambulance...."

Rush-hour traffic northbound was stopped and southbound was slowed for almost an hour by the fire and police vehicles that responded to the accident....

One witness said the ---- station wagon was moving slowly at the time of the accident...."

Case printed by permission of Ronald A. Feeny

THINK, INC.
123 WASHINGTON AVENUE
SOMERVILLE, MASSACHUSETTS 02155

March 5, 19--.

Mr. James Q. Smith, President
Automotive Safety Industries, Inc.
1162 East Main Street
Cleveland, Ohio 05188

Dear Mr. Smith:

Please find enclosed a report covering the basic design of an AUTOMATIC ELECTRICAL SHUTOFF. This device is designed to shut off all electrical power in an automotive vehicle upon impact (sudden deceleration) or roll over.

Our company undertook this design in light of recent accidents that resulted in flash fires. It is apparent to us that these fires were caused by arcing from the vehicle's electrical system which ignited the high octane gasoline used in todays cars.

We are covered by U.S. patents 2-814-976 and 977 on this device and are presently investigating the possibility of leasing the rights to manufacture and market the design as a product. Since your company is a known leader in the automotive safety field, our advisors have suggested we contact you first in regard to this item.

After you and your engineers have had an opportunity to review the design, please advise me of your interest. If you wish additional information or would like to witness the demonstration of a prototype, please contact me.

Sincerely yours,

Ronald A. Feeny
Chief Engineer

RAF:h
Enclosure: Design Report 003

LETTER OF TRANSMITTAL

INTRODUCTION

Much attention has been paid in recent times to the design of devices and systems to prevent or lessen injury during the second collision (human body and vehicle interior) in the event of an automobile accident. However, little to no effort has been directed to either the extinguishing or preventing of flash fires in accidents caused by the ignition of the high octane gasoline used in today's cars. It is possible to escape injury in even the most severe accidents, but it is highly unlikely that one can escape serious burns when involved in an automotive flash fire. The possibility of extinguishing a fire was investigated but due to the complexity and inability of a system to instantaneously lower the temperature or remove the oxygen, it was abandoned in favor of a device to decrease the possibility if a flash fire. This report presents the design of a device that will instantaneously switch off the electrical system in an automobile upon sudden deceleration, thereby preventing the cause of flash fires through arcing of parted wires in the system.

SPECIFICATION

1. Device must be such that it can be built into the existing framework of todays automobiles.

2. Hardware and installation must not raise the overall vehicle cost.

3. The device must shut off all electrical current from the battery upon impact or roll over.

4. It must be located so that it cannot be made un-operative due to an accident.

5. The device must provide the ability to be reset.

6. It must be simple in operation and relatively maintenance free.

7. It must be easily installed on older cars and easily built in new ones that conform to a mass production process.

8. The device must have the highest reliability factor (say 0.90 to 1.00).

DESCRIPTION

The device described here is an electrical disconnect unit for any type of motor vehicle. This unit is built around a mercury capsule through which a circuit is completed under normal driving conditons. As soon as the motor vehicle becomes involved in any type of collision, the mercury in the capsule is forced up and away from the adjustable contact, thus breaking the circuit and shutting off all electrical power in the vehicle. For a complete description of the circuitry involved, see the following page.

The unit is relatively inexpensive, easily fabricated and assembled and accomplishes the following safety functions when installed in the electric system of an automobile.

1) Automatically disrupts all electric power upon impact of crash, in any condition of travel or park, irrespective of the ignition key position.

2) When ignition key is in "off" position and removed from lock, all electric power is shut-off. This eliminates the possibility of leaving the car with the lights, etc. "on".

3) Burglar proofs car, because all power is removed and cannot be restored by jumpers. The only source of reset voltage to restore the power is a crush resistant armored cable which runs from the switch case to the safety control and is inaccessible.

4) Once the safety device is actuated and electric power is disabled, it cannot be restored until the key is inserted and turned to either the start position for running or the reset position for accessory use while parked.

5) The lock is designed so that the key may be removed when in accessory position in case lights etc., need to be used while parked.

The unit pictured in this report, is designed primarily for new automobiles, as it includes the starter solenoid inside the unit. However, it could also be manufactured and sold for use in older vehicles, if the box excluding the starter solenoid were redesigned.

A disadvantage comes forth when the unit is installed in a vehicle with power steering and power brakes. If, for example, a car with the unit installed should sideswipe another vehicle, the impact would break the circuit through the capsule shutting off all power. The driver would then experience some difficulty in steering and braking. However, we feel that a person in this type of accident often panics and loses control anyway so we do not think this disadvantage is detrimental to the sale of the unit.

CIRCUITRY

    Beginning at the battery (lower right hand corner of the wiring diagram) and following the battery leads, it can be seen that no current is flowing to the fuse block or to the starter solenoid (RY 1). The circuit is complete however, through the mercury capsule to point A on the ignition switch. Turning the key to the "on" position completes the circuit through line 4 (which joins line 2) to RY 3. At this point the circuit is still open. By turning the key to position 5 (a momentary contact with a spring return to position 4) lines 4 and 5 are shorted together. Current then flows through line 5 to RY 3 energizing that relay and closing it. As soon as RY 3 is closed the current in line 4 flows through the relay contacts to RY 2. As soon as the current from line 4 flows through RY 3 it automatically holds RY 3 closed. Upon reaching RY 2 the current closes RY 2.

    The battery current can now flow to the fuse block so that the lights, horn, radio, etc. operate. Current also flows to point B on the ignition switch through line 5 and activates the starter solenoid (RY 1) and through line 4 to the ignition. When the key is returned to position 4 the starter solenoid opens and the circuit through line 5 from A to RY 3 opens. The accessory position 2 operates the same way (position 1 has a momentary contact and a spring return to position 2) except the ignition and starter solenoid are not activated. Position 2 is necessary so that one may leave the car with its parking lights on or listen to the radio, etc. The key may be withdrawn in the off or accessory position.

    Upon impact or rollover the circuit through the mercury capsule is broken. This breaks the circuit through line 4 or 2 of the ignition switch at point A. Relay RY 3 opens which in turn opens relay RY 2 which breaks the circuit from the battery to all accessories and the ignition. Even _if_ the circuit to point A is completed again, RY 3 and RY 2 will _not_ close. These relays can be reset only by turning the key to position 1 or 5. The only wire in the vehicle that could potentially cause a spark is an armored cable thus removing all possibilities of fire from short circuits.

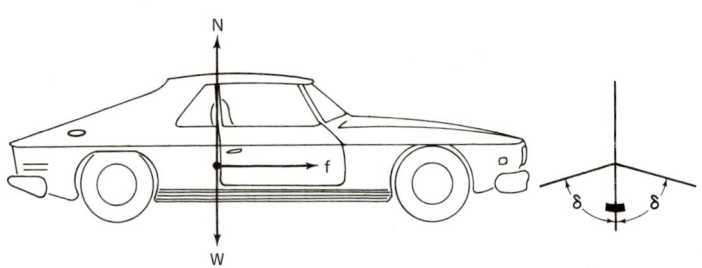

REASONING FOR 70° CRITICAL MERCURY ANGLE

Consider an automobile moving along a straight horizontal road with a speed V_o. If the coefficient of static friction between the tires and the road is μ, what is the acceleration of the car?

To determine a, apply the second law of motion to the "X" component of motion $f = ma = \left(\dfrac{W}{g}\right) a$ or $a = -g\left(\dfrac{f}{W}\right)$

From the "Y" components is obtained
$N - W = 0, \quad \therefore N = W$
$\mu = \dfrac{f}{N} = \dfrac{f}{W}$
$\therefore \boxed{a = -\mu g}$

From the Mark's M.E. Handbook it is found that the μ of rubber on concrete = 0.6 TO 0.9

$\therefore$ Max. deceleration = $0.9 (32.2) = 28.8$ ft./sec.^{2}

The acceleration along the spherical bowl at any point $a = g \sin \alpha$

where α = the angle between the vertical and the point in question.

At the critical point $a = 28.8$ ft./sec.^{2}

$\therefore 28.8 = 32.2 \sin \alpha \qquad \alpha = 69°$

By adding 1° to the 69° angle it is now sure that contact will not be broken even if the vehicle is put into a full brake under ideal conditions, i.e. $\mu = 0.9$. A more accurate value of μ is 0.65 to 0.75.

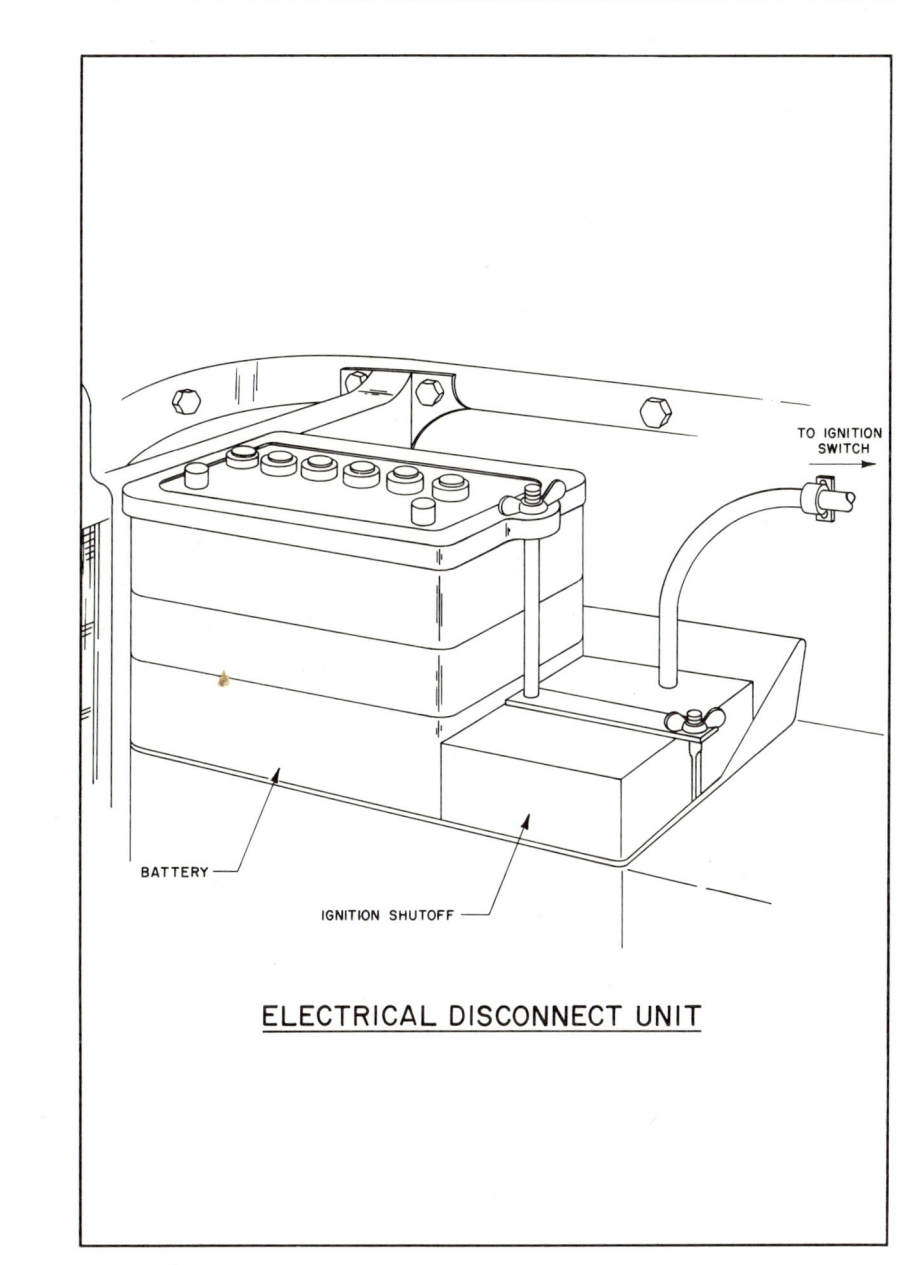

ELECTRICAL DISCONNECT UNIT

CASE HISTORY 003 (CONT.)

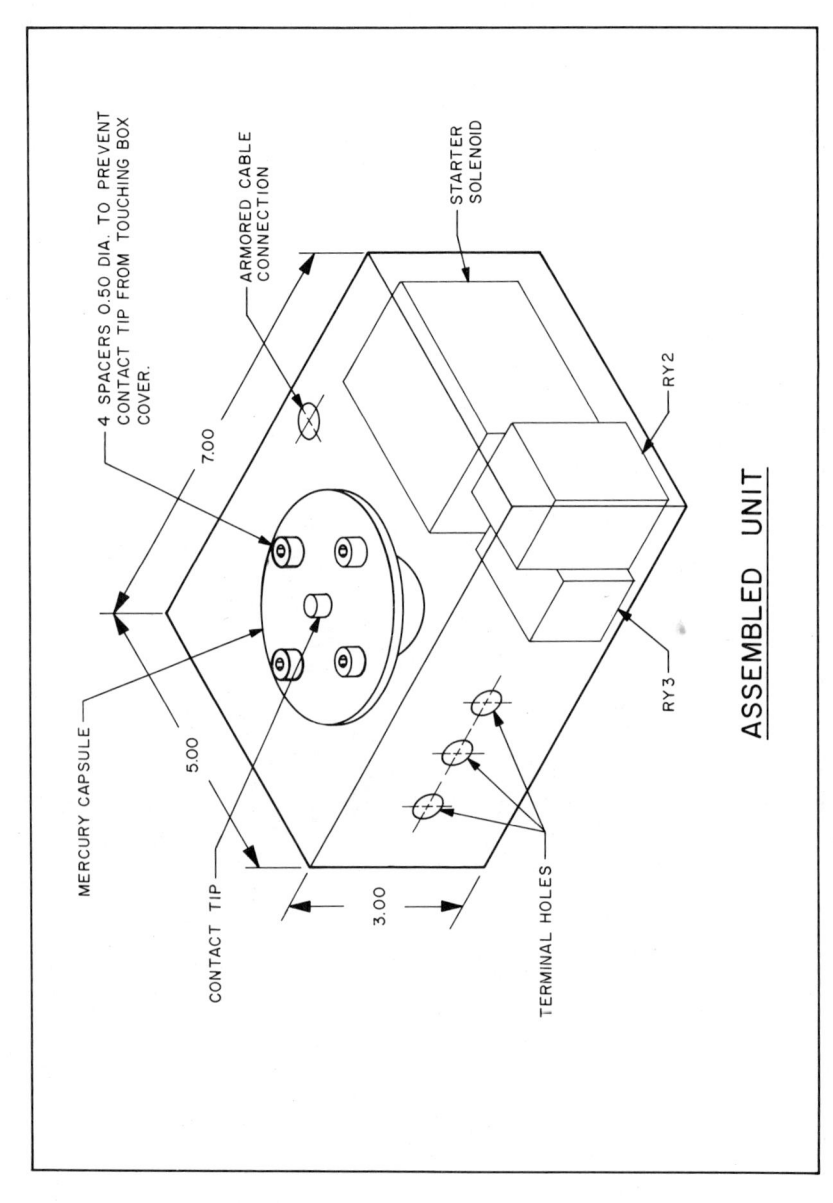

CASE HISTORY 003 (CONT.) 111

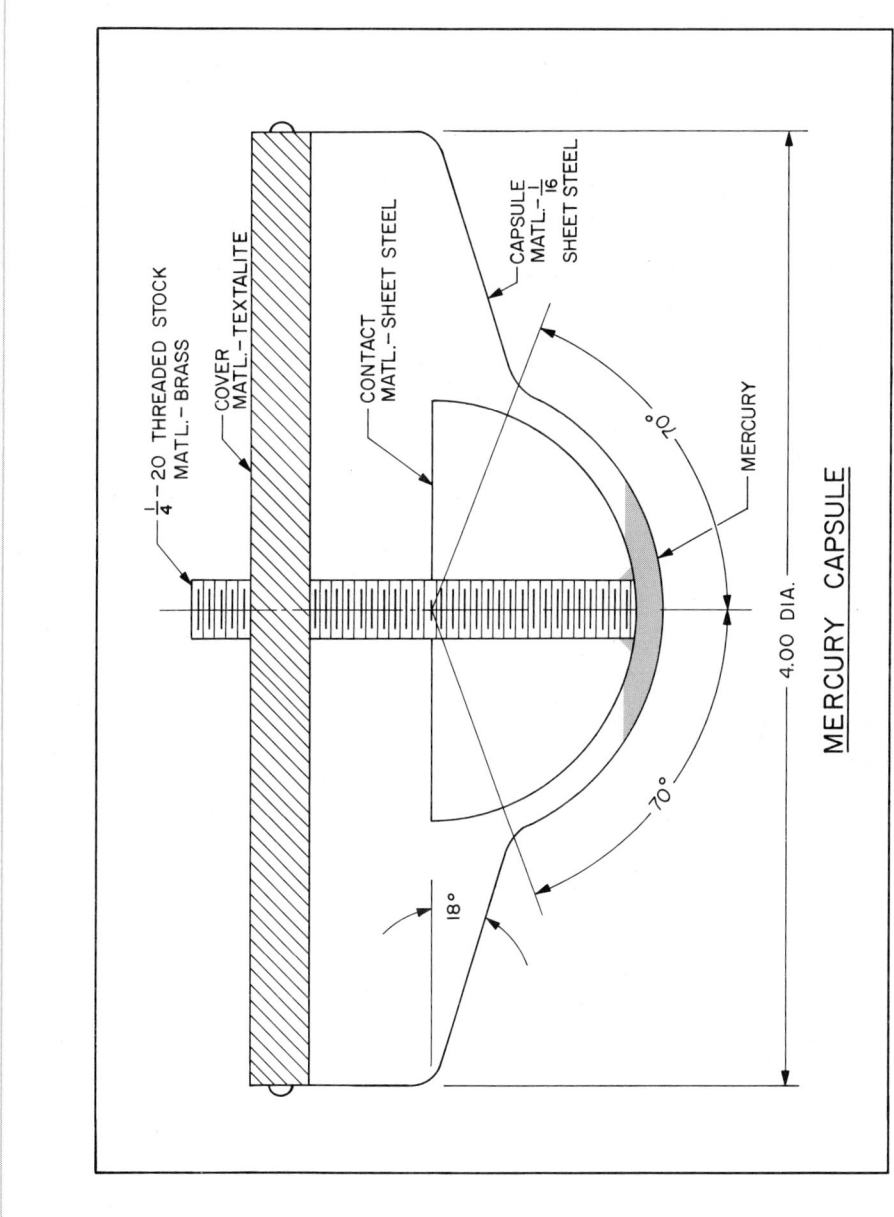

MERCURY CAPSULE

GRAPHICAL DESCRIPTION

112   CASE HISTORY 003 (CONT.)

GRAPHICAL DISCRIPTION

KEYLOCK SWITCH POSITIONS

POS. 3 - OFF
POS. 4 - ON
POS. 3 - START (MOMENTARY) (SPRING RETURN TO 4)
POS. 1 - RESET (MOMENTARY) (SPRING RETURN TO 2)

(KEY MAY BE WITHDRAWN IN OFF AND ACC. POSITION).

RY1 - STARTER SOLENOID
RY2 - D.C. CONTACTOR 12V D.C. COIL 1.5 WATTS 50 AMP D.C. CONTACTS (MAGNECRAFT W11DX39)
RY3 - RELAY 12V D.C. COIL 1.5 WATTS 5 AMP D.C. CONTACTS (MAGNECRAFT W88X-7)

SAFETY CONTROL UNIT SEALED ENCLOSURE MOUNT NEAR BATTERY.

WIRING DIAGRAM

COST ANALYSIS

PART	NO. REQ'D	OPERATION	$ LABOR	$ MATERIAL
Mercury Capsule	1	Punch & Draw	0.01	0.01
Capsule Cover	1	Turn–Drill–Tap	0.30	0.20
Box	1	Punch-Draw-Punch	0.03	0.02
W11D X 39 Relay	1			2.84
W88X - 7 Relay	1			4.36
Spacers	4	Turn & Drill	0.20	0.08
¼-20 Mach. Screws	8			0.16
10-32 Mach. Screws	4			0.08
Brass Threaded Stock	1			0.20
Adjustable Contact	1		0.01	0.01
Electrical Connection	1			0.08
Electrical Connection	3			0.18
Mercury	1 Oz.			0.25
		SUB TOTAL	0.55	8.47

Assembly time – 20 min. = $1.00

Engineering, Overhead, Management, etc. @ 100% = $10.02

Price to Distributor – $25.00

COST ESTIMATE

EXERCISES

1. Do you feel that the rod light fits the Brawley Manufacturing Company as a new-product line? Why?

2. Add three more specifications to the 15 listed for the flashlight design proposal in Case 001.

3. Add three more product ideas to the 24 listed in the ideation phase of the design of the sport flashlight.

4. What additional features can you add to the list of factors that suggest sales potential in presenting the rod light to management?

5. Considering the memorandum from Central Dynamics, Inc. and background material in Case 002, list at least five ideas that are needed and fulfill the Chief Project Engineer's request.

6. Add two more specifications to the list of eight detailed specifications for the cable stabilization unit.

7. How would you improve on the graphical concept description of the cable stabilization unit in order to better convey its physical characteristics to the U. S. Navy?

8. Considering the need shown in Case 003, would you have arrived at the same design goal as Mr. Feeny did, as explained in the INTRODUCTION? Why?

9. Explain why the electrical system in Case 003 does not return to its normal "on" state when the vehicle comes to rest after an accident.

10. Provide a sales potential list for the automatic electrical shut-off device to be presented to Mr. James Q. Smith in order to convince him to manufacture and market this product.

5 GUIDE TO MATERIAL SELECTION

> *"Like a Yankee farmer, the engineer must do. Husbanding materials and energy, he tries to get the most for the least. He conserves rare substances, such as the platinum catalysts needed for oil refining, by using them over and over again. He exploits the readiest sources, diligently squeezing from the atmosphere and oceans useful elements like nitrogen and magnesium."*
>
> Source: Life Science Library, *The Engineer*

There is a continuous spectrum of materials development ranging from minor improvements in existing materials to radical innovations. This spectrum of materials development has been brought about largely through demands imposed by science and technology on the metallurgist to provide materials for hostile environments: the environments of high-speed transportation, undersea exploration, and space. The materials specialist has met these demands through an ever-increasing supply of new and improved materials, from metals to plastics and ceramics to unique coating agents.

The experienced design engineer who is concerned with the broad range of design from phonograph needles to motor cars, from medical isotopes to nuclear power stations, from outboard motors to space rockets, and from coat hangers to suspension bridges chooses from experience and histories of prior-art the materials to produce these products. This is not the concern of this chapter. Instead, it is written for the inexperienced engineering designer and for the manager who must understand the role materials play in design situations. This chapter will assist them in making an intelligent choice of the optimum material(s) to serve their design needs. The theoretical aspects of material science, adequately treated in other available literature, will not be discussed

PROPERTIES OF METALS

The following definitions are fundamental to an understanding of the mechanical properties of metals. They are included here as a reminder and for ready reference. Since fundamentals are so easily forgotten or misunderstood, a continued review of the following list will assist the designer in the selection of a functional material.

SAE METAL ALLOY CODING SYSTEM.

So that the designer might understand the composition of metal alloys, the following numbering system is considered standard by the SAE (Society of Automotive Engineers):

The first digit denotes characteristic material as follows:
1—Plain carbon steel
2—Nickel
3—Chromium-nickel
4—Molybdenum
5—Chromium
6—Chromium-vanadium

The second digit denotes the approximate contents of characteristic alloying elements in hundredths of one percent.

The third and fourth digits denote approximate carbon content in hundredths of one percent.

Examples:
   Plain steel 1010
   Free-cutting 1112, 1340
   Nickel 2315, 2340
   Nickel-chrome 3120, 3140, 3312
   Molybdenum 4140

TENSILE TEST

A metal sample of specific shape is subjected to a tension load (F) until it breaks.

As the load increases, the sample elongates, at first in proportion to the stress (elastic range), later at a greater rate than indicated by the stress (plastic range). From this test a number of useful properties can be obtained which are directly applicable to engineering design with metals. These properties are illustrated in Figure 5-1 and are defined as follows.

1. Elastic limit (proportional limit)—The greatest stress a metal can withstand without permanent elongation (that is, when the load is released, the sample will return to its original length).
2. Yield point (yield strength)—The stress at which appreciable elongation occurs without increase in stress.
3. Ultimate strength—Maximum stress required to break a specimen.
4. Modulus of elasticity—The ratio of stress to strain within the elastic limit. It is a measure of stiffness.
5. Elongation—The ratio of the increase in gage length to the original gage length, expressed in percentage.
6. Reduction in area—The ratio of the decrease in the cross-sectional area of the test bar, after fracture, to the original area expressed in percentage.
7. Ultimate strain—The unit elongation (elongation per unit of length) at the specimen breaking-point. It is a measure of ductility.

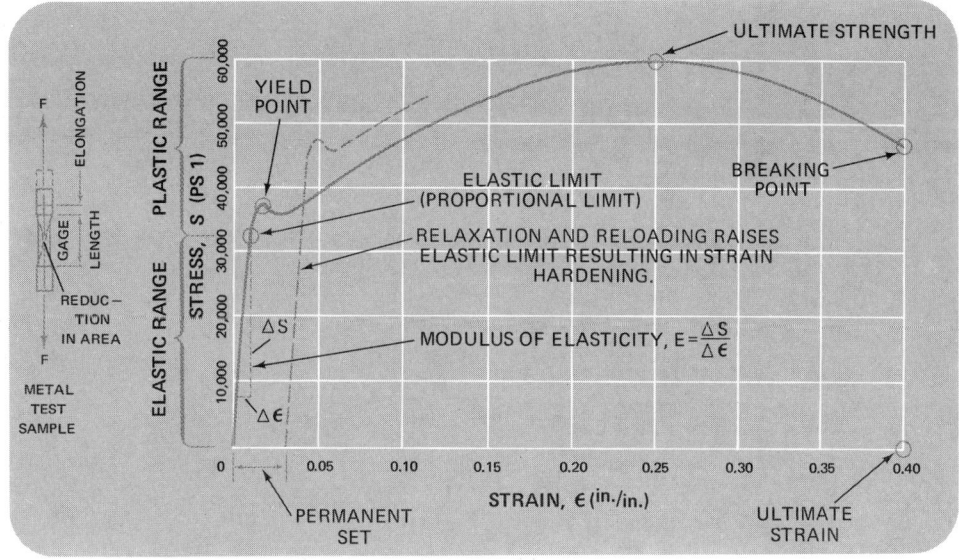

Figure 5-1 **STRESS VERSUS STRAIN CURVE FOR 1020 STEEL**

118   THE SCIENCE OF ENGINEERING DESIGN

DUCTILITY

The ability of a metal to deform plasticly without fracturing. In general, it means deformation under slow stressing instead of sudden impact. Although there are other measures, ductility is most commonly measured by means of elongation and reduction of area in the tensile test.

$$\% \text{ elongation} = \frac{\text{final gage length} - \text{original gage length} \times 100}{\text{original gage length}}$$

$$\% \text{ reduction of area} = \frac{\text{original area} - \text{final area} \times 100}{\text{original area}}$$

STIFFNESS

The ability of a metal to resist bending, stretching, shortening, and twisting. Stiffness is defined by the modulus of elasticity. The modulus applies only within the elastic range—the range within which a metal will return to its original shape if the load is removed. It is the ratio of stress to strain (E = ΔS/Δε) at any temperature, and is expressed in pounds per square inch.

INDENTATION HARDNESS

In metal-working, hardness generally implies resistance to penetration. It may, however, include resistance to scratching, abrasion, or cutting. Indentation hardness is probably the most widely-used mechanical testing procedure. It is a nondestructive test, relatively inexpensive, and can be performed by semi-skilled operators. The Brinell hardness test measures the diameter of impression of a hard ball under compressive load on the flat surface of the metal sample. The Rockwell hardness test measures the depth of penetration of a diamond point (Rockwell C) or steel ball (Rockwell B) into the sample.

FATIGUE

A description of the behavior of metals under the action of alternating (cyclic) loads as distinguished from their behavior under steady loads. Metals can fail under cyclic loads at levels below their static 0.2% yield strengths. Other factors besides alternating stress may also affect fatigue. These include the average stress, frequency of stress application, condition of the material surface, temperature, and the environment (air, water). Cyclic stressing can occur in bending, tension, compression, torsion, or in combinations of these, and may be found in such diverse applications as axles, connecting rods, springs, aircraft landing gears, and ships' hulls.

IMPACT STRENGTH

The ability of a material to withstand shock loading. With the increasing use of subzero temperatures in industrial applications, impact testing has become very important. Some metals become brittle as the temperature is reduced below zero. Others remain ductile to extremely low temperatures.

TRANSVERSE SHEAR STRENGTH

The value obtained by dividing the breaking load by the transverse shear area. It is needed in the design of rivets, bolts, pins, and similar parts.

TORSION

The twisting of a material by force which turns one end of a bar about its longitudinal axis while the other end is either clamped in a rigid fixture or twisted in the opposite direction. Torsion develops shear stresses in the material and occurs in such applications as propeller shafts, springs, and trailer leveling bars.

PROPERTIES OF PLASTICS

Plastics are synthetic resins (organic materials) that are solid in finished form but, at an early stage in their processing, are fluid enough to be shaped to a designated form through the application of heat and/or pressure.

In finished form, plastics consist of long-chain molecules called *polymers*. Components that may be combined into polymers by catalysts, heat, or pressure are known as *monomers*. The mixing of two or more polymers, a process analogous to alloying in metals, is known as *copolymerization*.

The two basic types of plastics are:

1. Thermoplastic resins—This plastic is analogous to tar which may be softened and resoftened repeatedly without undergoing a change in chemical composition.
2. Thermosetting resins—This plastic undergoes a chemical change with application of heat and pressure and cannot be resoftened.

These two types of plastics may be "alloyed" to form compounds containing both thermoplastic and thermosetting resins. Such a compound has some of the properties of each. If one were to compare plastics with metals and list their favorable and unfavorable qualities for the purpose of selection for design needs, the following list might result.

Favorable considerations of plastics versus metals are:

1. Lighter weight.
2. Better chemical and moisture resistance.
3. Better resistance to shock and vibration.
4. Transparent or translucent.
5. Tend to absorb vibration and sound.
6. Higher abrasion and wear resistance.
7. Self-lubricating.
8. Often easier to fabricate.
9. Can have integral color.
10. Often cost less per finished part.

Unfavorable considerations of plastics versus metals are:

1. Lower strength.
2. Much higher thermal expansion.
3. More susceptible to creep, cold flow, and deformation under load.
4. Lower heat resistance, both to thermal degradation and heat distortion.
5. More subject to embrittlement at low temperatures.
6. Softer.
7. Less ductile.
8. Change dimensions through absorption of moisture or solvents.
9. Flammable.
10. Some varieties are degraded by ultraviolet radiation.

The bar charts shown in Figure 5-2 compare eight important properties of plastics and metals.

GUIDE TO MATERIAL SELECTION 121

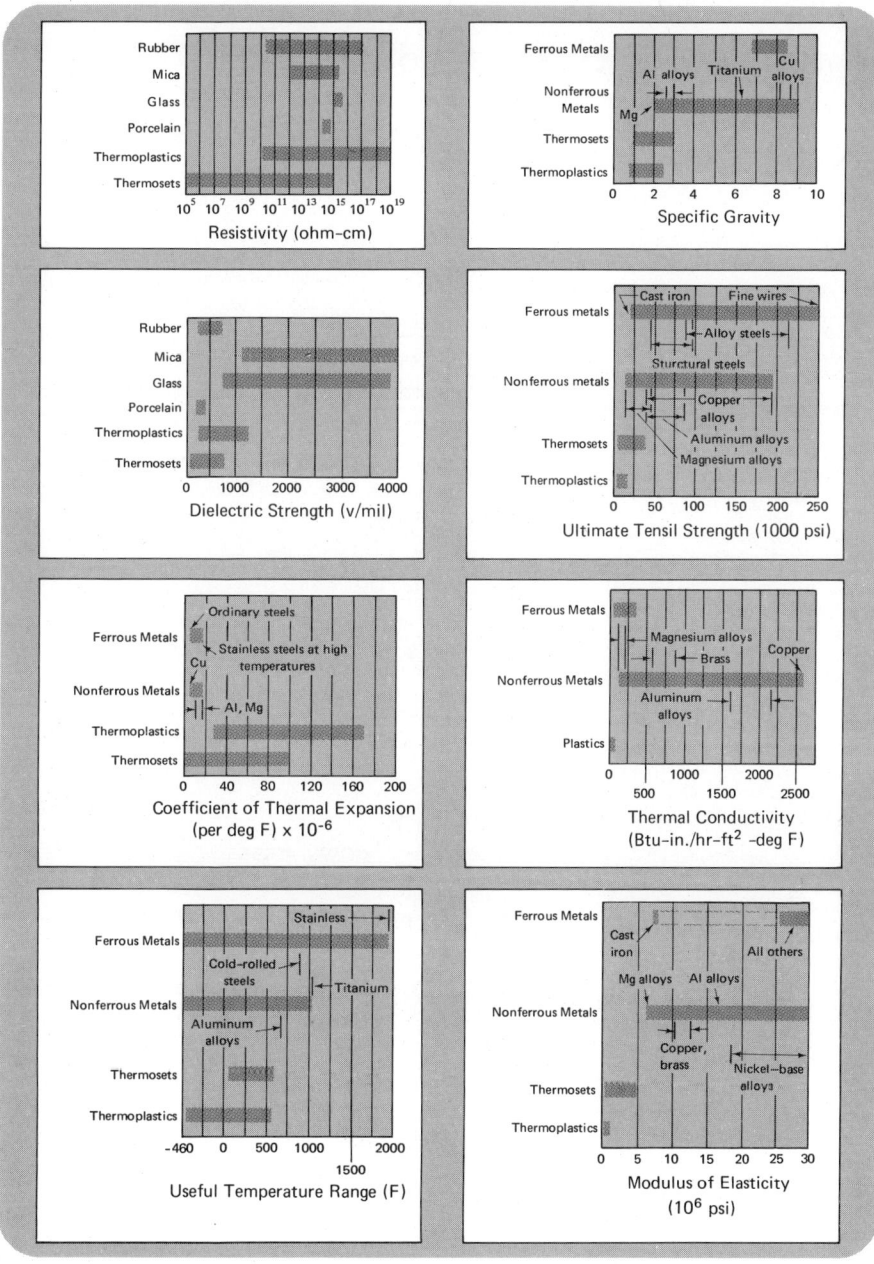

*Reprinted from MACHINE DESIGN, PLASTICS REFERENCE ISSUE, December 1968, a Penton Publication.

Figure 5-2 **PLASTIC VERSUS METALS PROPERTIES**

COMPOSITION AND DENSITY OF SELECTED MATERIALS

Figure 5-3 lists a number of widely-used engineering materials in alphabetical order with their approximate metallurgical composition in percentage and density values to facilitate weight calculations.

MATERIALS	APPROXIMATE COMPOSITION (ESSENTIAL ELEMENTS) IN PERCENT	DENSITY LB/CU. IN.
ALUMINUM ALLOYS		
1100 or 2S	Al 99 plus	.098
2011 or 11S	Al Bal, Cu 5.5, Pb .5, Bi .5	.102
3004 or 4S	Al Bal, Mg 1.0, Mn 1.2	.098
5052 or 52S	Al Bal, Mg 2.5, Cr .25	.097
6061 or 61S	Al Bal, Cu .25, Mg 1.0, Si .6, Cr .25	.098
ALUMINUM BRONZES		
Cast Grade 9A	Cu 87.5, Al 9, Fe 3.5	.267
Cast Grade 9B	Cu 98, Al 10, Fe 1	.270
Cast Grade 9D	Cu 81, Al 11, Fe 4, Ni 4	.273
Annealed Hard	Cu 92, Al 8	.281
BERYLLIUM COPPER	Cu 97.6, Be 2.05, Ni .35 or Co .25	.297
BRASS ALLOYS		
Gilding	Cu 95, Zn 5	.320
Commercial Bronze 90%	Cu 90, Zn 10	.318
Red Brass 85% (Wrought)	Cu 85, Zn 15	.316
Red Brass 85% (Cast)	Cu 85, Zn 5, Pb 5, Sn 5	.314
Low Leaded Brass	Cu 65, Zn 34.5, Pb .5	.306
Medium Leaded Brass	Cu 65, Zn 34, Pb 1.0	.306
Naval Brass	Cu 60, Zn 39, Sn .75	.304
Admiralty Brass	Cu 70, Zn 29, Sn 1	.308
COPPER	Cu 99.9 plus	.322
GOLD (PURE)	Au 99.99	.698
IRON ALLOYS		
Wrought Iron	Fe Bal. Slag 2.5	.278
Ingot Iron	Fe 99.9 plus	.284
Cast Iron	C 3.4, Si 1.8, Mn .5, Fe Bal	.26

Figure 5-3   **MATERIAL COMPOSITION AND DENSITY**

MATERIALS	APPROXIMATE COMPOSITION (ESSENTIAL ELEMENTS) IN PERCENT	DENSITY LB/CU. IN.
Malleable Iron	C 2.5, Si 1, Mn .55 max, Fe Bal	.264
Ductile Iron (containing Mg)	C 3.4, Si 2.5, Mn .7, P .1 max, Ni 1.5, Mg .06, Fe Bal	.26
Ductile Ni-Resist	C 2.8, Si 2.5, Mn 1, P .01 max, Ni 20, Cr 2, Mg .1, Fe Bal	.268
Carbon Steel SAE 1021	Fe Bal, Mn .45, Si .25, C .20	.284
Cast Carbon Steel	Fe Bal, Mn .70, Si .40, C .30	.283
Cast Alloy Steel	Fe Bal, Ni 1.75, Mn .80, Cr .75, C .30, Mo .25	.284
CORROSION RESISTANT STEEL Chromium Nickel Types		
Type 301	Fe Bal, Cr 17, Ni 7, C .11	.29
Type 304	Fe Bal, Cr 19, Ni 9.0, C .08 max	.29
Type 310	Fe Bal, Cr 25, Ni 20, C .25 max	.29
Type 316 (cast)	Fe Bal, Cr 19, Ni 9, Mo 3, C .1	.286
Type 317	Fe Bal, Cr 19, Ni 13, Mo 3.5, C .1 max	.29
Type 318	Fe Bal, Cr 18, Ni 14, Mo 2.8, Cr 10 x C min, C .08 max	.29
Ferritic Chromium Types		
405	Fe Bal, Cr 12.5, Ni .5 max, Al .2, C .08 max	.28
446	Fe Bal, Cr 25, C .35 max	.27
CHROMIUM MARTENSITIC GROUP - HARDENABLE		
403	Fe Bal, Cr 12.5, Ni .5 max, C .15 max	.28
416	Fe Bal, Cr 13, Ni .5 max, C .15 max	.28
431	Fe Bal, Cr 16, Ni 2 max	.280
LEAD		
Lead, Chemical	Pb 99.9, Cu .06, Bi .005 max	.410
Lead, Antimonial	Pb 94, Sb 6	.393
Lead, Tellurium	Pb 99.85, Te .04, Cu .06	.410
MAGNESIUM ALLOYS		
A-10	Mg Bal, Al 10.0 Mn .1	.066
Az 31 X	Mg Bal, Al 3.0, Zn 1.0	.064
Az 61 X	Mg Bal, Al 6.5, Zn 1.0	.065
Az 80 X	Mg Bal, Al 8.5, Zn .5	.065
M 1	Mg Bal, Mn 1.2	.064
NICKEL ALLOYS		
Nickel (pure)	Ni 99.99	.322

Figure 5-3 (cont.) **MATERIAL COMPOSITION AND DENSITY**

MATERIALS	APPROXIMATE COMPOSITION (ESSENTIAL ELEMENTS) IN PERCENT	DENSITY LB/CU. IN.
Nickel (wrought)	Ni 99.4	.321
Nickel (cast)	Ni 96.7, Si 1.5	.301
Low-Carbon Nickel	Ni 99.4, C .02 max	.321
Duranickel	Ni 93.5	.298
Hastelloy A	Ni Bal, Mo 21, Fe 19, Mn, Si	.318
Hastelloy C	Ni Bal, Mo 16, Cr 16, Fe 5, W 4, Mn, Si	.323
Illium G	Ni Bal, Cr 21, Fe 6, Mo 6, Cu 4, W, Mn, Si	.300
80 Ni 20 Cr Alloy (wrought)	Ni 78, Cr 19	.302
Monel (wrought)	Ni 67, Cu 30, Fe 1.4, Mn 1	.319
Monel (cast)	Ni 67, Cu 29, Fe 1.5, Si 1.25	.312
Nickel Bronze (cast)	Cu Bal, Ni 5, Sn 5, Zn 2, Pb .01 max	.318
Nickel Silver 20% (cast)	Cu Bal, Ni 20, Zn 6, Pb 5, Sn 4	.322
Nickel Silver 10% (wrought)	Cu 65, Zn 25, Ni 10	.313
Cupro-Nickel 70-30	Cu 70, Ni 30	.323
PALLADIUM		
Palladium (commercial)	Pd 99.5	.432
Palladium (hard)	Pd Bal, Ru 4, Rh 1	.432
PLATINUM ALLOYS		
Platinum (pure)	Pt 99.99	.772
Rhodium Platinum 10%	Pt Bal, Rh 10	.742
SILVER ALLOYS		
Silver (pure)	Ag 99.9 plus	.379
Easy-Flo Silver Brazing Alloy	Ag 50, Cd 18, Zn 16.5, Cu 15.5	.342
R-T Silver Brazing Alloy	Ag 60, Cu 25, Zn 15	.344
TANTALUM	Ta 99.9 plus	.60
TIN ALLOYS		
Tin	Sn 99.9 plus	.263
Soft Solder 50-50	Sn 50, Pb 50	.323
Soft Solder 60-40	Sn 60, Pb 40	.307
TITANIUM ALLOYS		
Titanium (pure)	Ti Bal, Fe .2, N .1 max, W .01 max, Mo .2 max, C .07 max	.164
Titanium-Chromium-Iron-Molybdenum Alloy	Ti Bal, Fe 2, Cr 2, Mo 2, C .07 max	.169
Age-Hardened Titanium-Chromium-Iron-Oxygen Alloy	Ti Bal, Fe 1.5, Cr 2.8, C .07 max	.167

Figure 5-3 (cont.)   **MATERIAL COMPOSITION AND DENSITY**

MATERIALS	APPROXIMATE COMPOSITION (ESSENTIAL ELEMENTS) IN PERCENT	DENSITY LB/CU. IN.
Republic Steel Alloy (RS)		
RS-55, 70 and 70-A	C .20 max, Ti Bal	.163
RS110	C .20 max, Cr 3.5, Fe 1.5, Ti Bal	.168
RS110-A	C .20 max, Mn 7.0, Ti Bal	.170
RS110-BX	C .20 max, Mn 3.0, Al 1.5, Ti Bal	.165
RS120	C .20 max, Mn 7.0, Ti Bal	.170
RS130	C .20 max, Mn 4.0, Al 4.0, Ti Bal	.165
RS140	C .20 max, Cr 2.75, Al 5.0, Fe 1.25, Ti Bal	.163
ZINC ALLOYS		
Zinc	Zn Bal, Pb .1	.258
Zilloy 15	Zn Bal, Cu 1.05, Mg .011	.259
Zilloy 40	Zn Bal, Cu 1.05	.259

Figure 5-3 (cont.)   **MATERIAL COMPOSITION AND DENSITY**

PROPERTIES OF SHEET METAL

When choosing a sheet metal to form a sculptured shape (automotive bodies or small boat hulls) or to produce a square-cornered cover for electronic equipment, it is wise to consider its mechanical as well as its fabrication properties. These properties are provided for easy reference in tabular form in Figure 5-4 and 5-5.

MATERIAL	YIELD STRENGTH, POUNDS PER SQUARE INCH	WEIGHT, POUNDS PER CUBIC INCH	MODULUS OF ELASTICITY, POUNDS PER SQUARE INCH	BRINELL HARDNESS NUMBER
1100-0 Aluminum Soft Temper	5,000	.098	10,000,000	23
1100-H14 Aluminum Half Hard	17,000	.098	10,000,000	32
2024-T3 Aluminum	46,000	.100	10,600,000	120
5052-0 Aluminum Soft Tempter	13,000	.097	10,200,000	45
5052-H32 Aluminum Quarter Hard	28,000	.097	10,200,000	62
5052-H34 Aluminum Half Hard	31,000	.097	10,200,000	67
6061-T6 Aluminum	40,000	.098	10,000,000	95
C.R. Steel, Dead Soft (1020)	25,000	.284	30,000,000	85
C.R. Steel, Half Hard (1020)	50,000	.284	30,000,000	140
Stainless Steel, Soft Temper (301)	30,000	.290	28,000,000	145
Stainless Steel, Half Hard (301)	110,000	.290	26,000,000	290
Stainless Steel, Full Hard (301)	140,000	.290	24,000,000	380

Figure 5-4   **MECHANICAL PROPERTIES OF SHEET METAL**

MATERIAL	SUITABILITY FOR MANUFACTURING							
	Shearing, Blanking, Piercing	Forming, Bending	Spinning, Drawing, Peening	WELDING (Note 7)		Brazing (Note 1)	Soldering (Note 3)	Machining
				Resistance	Fusion (Note 1)			
1100-0 Aluminum Soft Temper	Fair	Preferred	Preferred	Fair	Preferred	Note 2	Note 3	Preferred
1100-H14 Aluminum Half Hard	Preferred	Preferred	Fair	Fair	Preferred	Note 2	Note 3	Preferred
2024-T3 Aluminum	Preferred	Poor	Do not use	Preferred	Do not use	Do not use	Note 3	Preferred
5052-0 Aluminum Soft Temper	Preferred	Preferred	Preferred	Preferred	Fair	Fair	Note 3	Fair
5052-H32 Aluminum Quarter Hard	Preferred	Preferred	Fair	Preferred	Fair	Fair	Note 3	Preferred
5052-H34 Aluminum Half Hard	Preferred	Fair	Do not use	Preferred	Fair	Fair	Note 3	Preferred
6061-T6 Aluminum	Preferred	Fair	Do not use	Preferred	Preferred	Fair	Note 3	Preferred
C.R. Steel, Dead Soft (1020)	Fair	Preferred	Preferred	Preferred	Preferred	Preferred	Preferred	Fair
C.R. Steel, Half Hard (1020)	Preferred	Fair	Do not use	Preferred	Preferred	Preferred	Preferred	Fair
Stainless Steel, Soft Temper (301)	Preferred	Preferred	Fair	Preferred	Preferred	Note 5	Preferred	Fair
Stainless Steel, Half Hard (301)	Fair	Fair	Do not use	Preferred	Fair	Note 4	Preferred	Poor
Stainless Steel, Full Hard (301)	Fair	Do not use	Do not use	Preferred	Do not use	Note 4	Preferred	Poor

NOTES:
1. All Materials in the heat-treated, half hard and full hard tempers will have lower strengths after fusion welding and brazing operations.
2. Preferred, when special aluminum brazing alloys are used.
3. Fair, when soft solder is used on plated aluminum. However, heat-treated and half hard materials will have somewhat lower strength.
4. Fair, when low temperature silver solder is used and materials are light gage.
5. Preferred, when low temperature silver solder is used.

Figure 5-5 **FABRICATIONAL PROPERTIES OF STAINLESS STEEL**

PROPERTIES OF SELECTED MATERIALS

Figure 5-6 tabulates the average mechanical properties of a number of selected engineering materials. A rather broad range of materials including pure metals, alloys, ceramics, plastics, and cermets is given in the chart so that the designer may compare their properties and select the best material for a specific design need. Refer to the definition of terms listed earlier in this chapter to see how each property may affect the design application.

CHOICE OF MATERIAL FOR FORMING

In addition to the mechanical properties (strength characteristics) of a material, the designer will often base his selection on a material's ability to be formed into a desired shape. The converse is also important in selecting the best method of forming a specific shape. Figure 5-7 and 5-8 provide this information.

MATERIAL	Density lb/in³	E 10⁶ psi	μ Poisson's Ratio	σ_Y 10³ psi	σ_{ULT} 10³ psi	Melting Temp. or Range °F	Thermal Conduct. lb/sec °F	Vol. Spec. Heat psi/°F	Coef. of Thermal Expansion x 10⁻⁶/°F	Thermal Diff. in²/sec
PURE METALS										
Beryllium	0.066	44		55	90	2340 ±	19.9	294	6.9	0.0678
Cobalt (Sintered)	0.32	30		44	100	2723 ±	8.7	271	6.8	0.0312
Columbium	0.31	23		24	39	4380		172	4.0	
Copper	0.32	17	0.33	10	32	1980	51	254	9.3	0.20
Hafnium	0.41	20		22	59	3800 ±		123	3.4	
Lead	0.41	2	0.43	2	2.5	621	4.5	105	16.4	0.043
Molybdenum	0.37	42		75	100	4700 ±	1.7	193	2.7	0.0088
Nickel	0.32	30		20	70	2625 ±	7.9	356	6.6	0.022
Tantalum	0.60	27		100	110	5425 ±	7.0	185	3.6	0.038
Tungsten	0.70	50			300	6092	26.2	204	2.4	0.13
Uranium	0.68	30		25	90	2015	3.4	163		0.021
Vanadium	0.22	22		55	68	3150 ±		223	4.3	
Zirconium	0.25	12		16	36	3380		247	2.4	
Aluminum (1100-0)	0.10	10	0.33	3.5	11	1200 ±	29.0	197	13.1	0.147
ALLOYS										
Aluminum 2024-T4	0.10	10.6	0.33	44	60	1075 ±	15.8	197	12.9	0.080
7075-T6	0.10	10.4	0.33	70	80	1035 ±	17.0	197	13.1	0.086
Brass, Hard Yellow	0.31	15	0.35	60	74	1710	14.9	236	10.5	0.063
Cast Iron (25T)	0.26	13	0.2	$\sigma_f = +24$,	-120	2150	5.8		6.7	
Magnesium ZK51A-75	0.07	6.5	0.35	25	40	1175	15.1	138	14.5	0.11
Phosphor Bronze, Hard	0.32	16	0.33	65	80	1920	8.6	246	9.4	0.035
Steel: 0.2% C										
Hot Rolled	0.283	30	0.27	40	70	2760	6.5	245	6.7	0.027
Cold Rolled	0.283	30	0.27	65	80	2760	6.5	245	6.7	0.027
Steel: 1% C										
Hot Rolled	0.283	30	0.27	84	143		6.5	245	7.3	0.027
H & T, 800°F	0.283	30	0.27	138	200		6.5	245	7.3	0.027
Steel: 4640										
H & T, 800°F	0.28	29	0.30	190	202		6.5	245		0.027
Stainless Steel										
Type 302 C.R.	0.286	29	0.3	100	140	2575 ±	1.9	294	8.9	0.0065
Titanium RC130	0.163			130	140	3270	5.4	198	4.0	0.027
CERAMICS										
Crystalline Glass "Pyroceram"	0.09	12.5	0.25	$\sigma_f = +20$		~2280	0.24	147	1.0	
Fused Silica Glass	0.08	10.5	0.17			~2880	0.17	117	0.3	
Alumina Ceramics	0.14	45		$\sigma_f = +40$,	-350	3686		155	4.0	
PLASTICS										
Cellulose Acetate	0.047	0.25	0.4	$\sigma_f = +5$,	-20		0.032	145	70	
Nylon	0.041	0.41	0.4	$\sigma_f = +8$,	>-13		0.03	140	55	
Epoxy	0.040	0.65		$\sigma_f = +7$,	-30		0.10	110	33	
CERMETS										
Tungsten Carbide Carboloy 999	0.54	100	0.24	$\sigma_f =$	-600		15.1		2.2	

Figure 5-6 **AVERAGE MECHANICAL PROPERTIES OF SELECTED MATERIALS**

Figure 5-7 **CHOICE OF MATERIAL FOR METAL FORMING**

Form	Irons	Steels (Carbon, Low Alloy)	Heat & Corr Res Alloys	Aluminum Alloys	Copper Alloys	Lead Alloys	Magnesium Alloys	Nickel Alloys	Precious Metals	Refractory Metals	Tin Alloys	Titanium Alloys	Zinc Alloys
Sand Castings	■	■	■	■	■	□	■	■			□		□
Shell Mold Castings	■	□	□	■	■			□					
Permanent Mold Castings	■	□		■	□	□	■	□			□		□
Die Castings				■	□	■	■				□		■
Plaster Mold Castings				■	■								
Investment Castings		■	□	■	■		■	□	□				
Centrifugal Castings	■	■	■	□	□			□					
Open Die Forgings	□	■	■	□	□		□	□		□		□	
Closed Die Forgings Blocker Type		■	■	□	□		□	□		□		□	
Conventional Type		■	■	□	□		□	□		□		□	
Precision Type				□	□		□						
Upset Forgings		■	■	□	□		□	□		□		□	
Cold Headed Parts		■	□	■	■	□		□	□				
Impact (cold) Extruded Parts		□		■	□	□	□				□	□	□
Stampings, Drawn Parts		■	□	□	■		□	□	□	□		□	□
Spinnings		■	□	■	■	□	□	■	□			□	□
Screw Machine Parts	□	■	□	■	■			□	□	□			
Powder Metallurgy Parts [b]	■	■	□	□	■			□	□	■		□	
Electroformed Parts [c]	■			□	■	□		■	■		□		□
Cut Extrusions		□		■	■	□	■	□		□	□	□	
Sectioned Tubing		■	■	■	■		■	■		□		■	
Roll Formed Parts		■		□	□		□					□	□
Continuous Castings		□		■	■[d]	□							

[a] ■ = Materials most frequently used.
□ = also materials currently used.
[b] Iron-copper and iron-copper-carbon most frequently used.
[c] Most frequently used material is iron (99.8% pure).
[d] Particularly tin bronze and tin-lead bronze.

© Reprinted from Design Engineering, Materials Selector Issue, Mid-October 1965, Reinhold Publishing Corporation.

GUIDE TO MATERIAL SELECTION

Figure 5-8
CHOICE OF MATERIALS FOR PLASTIC AND RUBBER FORMING

■ = Materials most frequently used.
□ = Materials also currently being used.

© Reprinted from Design Engineering, Materials Selector Issue, Mid-October 1965, Reinhold Publishing Corporation

130 THE SCIENCE OF ENGINEERING DESIGN

FINISHES AND COATINGS

Another factor that has an influence on the selection of a design material is its finishing and/or coating ability. This factor is particularly relevant when the material is used as a housing that must demonstrate aesthetic appeal. The following list indicates changes in the outer layer of a material that may be brought about through some form of surface treatment.

1. Increased hardness
2. Wear resistance
3. Corrosion resistance
4. Tarnish resistance
5. Acid resistance
6. Heat resistance
7. Friction resistance
8. Rust prevention
9. Decorative effect
10. Improved appearance
11. Increased reflectivity of light
12. Decreased reflectivity of light
13. Increased material life
14. Ease of cleaning
15. Increased conductivity
16. Increased bonding ability

Methods available to provide finishes and coatings are listed below. Determining which material matches best with a finishing or coating process to provide a required function is beyond the scope of this book. Such information may be found in the literature. It is important that the designer be familiar with the majority of these processes.

1. Sprayed metal coatings
2. Electrodeposited coatings
3. Hard facings (overlays applied by welding operations)
4. Ceramic, cermet, and refractory coatings
5. Hot-dip coatings
6. Immersion coatings
7. Diffusion coatings
8. Vapor-deposited coatings
9. Organic coatings
10. Porcelain enamel coatings
11. Chemical conversion coatings
12. Rust-preventive coatings
    a. Oil
    b. Solvent
    c. Emulsifiable
    d. Wax
13. Mechanical finishes for aluminum, copper, and stainless steel
    a. Buffed
    b. Burnished
    c. Wheel- or belt-polished
    d. Wire-brushed
    e. Hammered
    f. Embossed or engraved
    g. Sandblasted

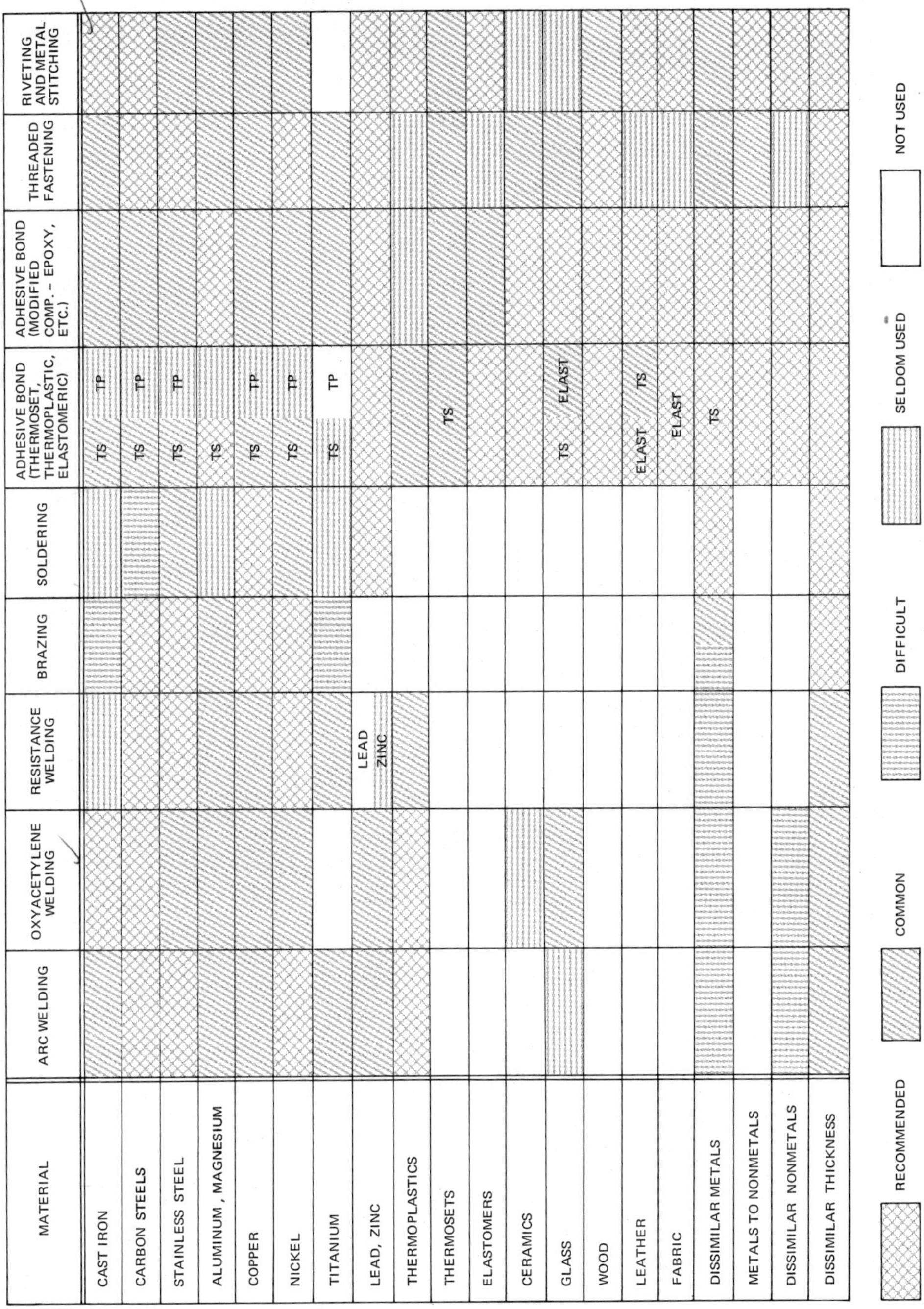

GUIDE TO MATERIAL SELECTION 131

MATERIAL	ARC WELDING	OXYACETYLENE WELDING	RESISTANCE WELDING	BRAZING	SOLDERING	ADHESIVE BOND (THERMOSET, THERMOPLASTIC, ELASTOMERIC)	ADHESIVE BOND (MODIFIED COMP. - EPOXY, ETC.)	THREADED FASTENING	RIVETING AND METAL STITCHING
CAST IRON	common	common	recommended	common	difficult	TS / TP	recommended	recommended	recommended
CARBON STEELS	recommended	recommended	recommended	recommended	common	TS / TP	recommended	recommended	recommended
STAINLESS STEEL	recommended	common	recommended	recommended	common	TS / TP	recommended	recommended	recommended
ALUMINUM, MAGNESIUM	common	common	recommended	common	difficult	TS / TP	recommended	recommended	recommended
COPPER	common	common	recommended	recommended	recommended	TS / TP	recommended	recommended	recommended
NICKEL	recommended	common	recommended	recommended	common	TS / TP	recommended	recommended	recommended
TITANIUM	common	difficult	recommended	difficult		TS / TP	recommended	recommended	recommended
LEAD, ZINC		LEAD	ZINC		recommended		recommended	recommended	
THERMOPLASTICS						TS	recommended	recommended	recommended
THERMOSETS							recommended	recommended	recommended
ELASTOMERS							recommended		
CERAMICS		difficult					recommended		
GLASS		difficult				TS / ELAST	recommended		
WOOD							recommended	recommended	recommended
LEATHER						ELAST / TS	recommended	recommended	recommended
FABRIC						ELAST	recommended	recommended	recommended
DISSIMILAR METALS	difficult	difficult	difficult	common	common	TS	recommended	recommended	recommended
METALS TO NONMETALS							recommended	recommended	recommended
DISSIMILAR NONMETALS							recommended	recommended	recommended
DISSIMILAR THICKNESS	difficult	difficult	difficult				recommended	recommended	recommended

Legend: RECOMMENDED | COMMON | DIFFICULT | SELDOM USED | NOT USED

Figure 5-9   **JOINABILITY OF MATERIALS**

JOINING AND FASTENING

Just as with finishes and coatings, the method of joining and fastening materials also influences the selection of an appropriate material. Of the methods commonly available for joining materials and parts *mechanical* fastening is the most versatile, with the threaded method classified as semi-permanent and the riveted or stapled method as permanent. *Adhesive* fastening requires a more carefully controlled environment for its application and may be classified as a permanent joining method. Soldering and welding, permanent methods of fastening or joining, are more versatile than adhesives but less versatile than mechanical methods. To assist the designer in matching a material with an effective method of fastening or joining, the table shown in Figure 5-9 will be of value.

MAKING THE SELECTION

All principal factors that have a bearing on the selection of a material to fulfill a design requirement have been discussed in this chapter with the exception of cost. Since cost is never static and often undergoes erratic change, it will not be treated here. Cost of materials may be learned from the manufacturer or distributor at a moment's notice and the designer must continually update his knowledge of changes that occur. The principal selection factors may be summarized as follows:

1. Mechanical properties
2. Composition and density
3. Formability
4. Finishing and coating
5. Joinability
6. Cost

An experienced designer should have little or no difficulty in matching or selecting an appropriate material to optimally meet the demands of the design requirement. Until this ability is acquired, it is effective to adapt the decision matrix technique from Chapter 3 to material selection and proceed in the following manner: Consider that a material is to be selected to form the outer cover or cowling for a rotary lawn mower. The designer would probably write the following specifications for the material:

1. Must be easily formed to required shape (ductile).
2. Must have a relatively high impact strength.
3. Must be light weight.

4. Must be finished to prevent rust (natural or color).
5. Cost must be appreciably lower than a functioning part.
6. Must be easily fastened to main frame to withstand vibration but yet be removable.

The designer would now proceed to construct the decision matrix chart shown in Figure 5-10.

PROPERTIES / WEIGHTING VALUE / MATERIALS	MECH. PROPERTIES	COMPOSITION & DENSITY	FORMABILITY	JOINABILITY	FINISHING OR COATING	COST	SUM
	0.3	0	0.2	0	0.2	0.3	1.0
ALUMINUM	10 / 0.30		9 / 0.18		9 / 0.18	5 / 0.15	0.81
SHEET METAL	10 / 0.30		8 / 0.16		7 / 0.14	8 / 0.24	0.84
COPPER ALLOY	8 / 0.24		9 / 0.18		8 / 0.16	3 / 0.09	0.67
THERMOSET PLASTIC	3 / 0.9		10 / 0.20		10 / 0.20	10 / 0.30	0.79

WEIGHTING VALUE — MEASURE OF RELATIVE IMPORTANCE (0 to 10, $\Sigma = 1.0$)

RATING FACTOR — RATED VALUE OF MATERIAL TO REQUIRED PROPERTIES (0 TO 10)

Figure 5-10 **DECISION MATRIX**

Properties required of the material are listed across the top of the matrix and assigned weighting values so that their sum will equal zero. These values are based on the specifications or requirements placed on the part written earlier. Notice that composition and density and also joinability were assigned zero (0) value since they were judged insignificant in making the decision. From the large number of materials available four broad classifications of materials were chosen since they seem to best meet the requirements of the lawn mower cowl. These are listed along the left column of the matrix and are assigned rating factors 0 to 10, based on their ability to measure up to the properties listed at the top of the matrix. A comparison is now made by multiplying rating factors by weighting values and arriving at a sum for each material. It is obvious that sheet steel is probably the best material to use in constructing the cowl. A decision concerning smaller details may now be made through closer investigation of the properties of sheet steel as required by the design. The final designation of material for the cowling in this case would be cold rolled sheet steel, half hard (1020).

EXERCISES

1. Name three products or technical achievements that would not have been possible without an advancement in material science.
2. List the composition of an alloy specified as SAE 3312.
3. Why is the design stress for a material chosen just below the elastic limit instead of just below the ultimate strength?
4. How does the Brinell hardness differ from the Rockwell hardness specification of a material?
5. In two opposite columns, list five uses of metals that are not appropriate for plastics and five uses of plastics not appropriate for metals.
6. What sheet metal would be most appropriate for forming the oil pan of an automobile engine?
7. What is meant by Poisson's ratio?
8. What is meant by *spinning* in the forming process of a material?
9. Write a series of specifications that will govern the selection of material to construct the handle for a refrigerator door.
10. Construct a decision matrix and conclude the optimum material to construct the refrigerator door handle of question nine.

6 HUMAN FACTORS IN DESIGN

> *"We bear in mind that the object we are working on is going to be ridden in, sat upon, looked at, talked into, activated, operated, or in some other way used by people."*
>
> *"If the point of contact between the product and the people becomes a point of friction, then the designer has failed."*
>
> *"On the other hand, if people are made safer, more efficient, more comfortable . . . or just plain happier . . . by contact with the product, then the designer has succeeded."*
>
> Source: Henry Dreyfuss

Of all the living creatures that inhabit the earth, man is probably the most fragile and ill-adapted to his environment. One can quickly name hundreds of ways nature has provided protection for animals from their environment as well as from natural enemies. Shells, fur, wings, feathers, scales, fangs, odors, camouflage, keen senses of smell and sight are all protective devices enjoyed by animals. Nature endowed man, however, with one ability unknown to other creatures: the mind which has the gift of intelligence. This gives man a sense of reason, the ability to plan, the ability to remember things past and to think into the future, and above all the ability to improve his environment. Clothes, dwellings, transportation, automation, and space travel are examples of these improvements.

The science that considers man and his reaction to familiar as well as strange environments is known by several terms including human engineering, human factor engineering, human factors, human performance, biomechanics, bioengineering, ergonomics, and engineering psychology. No matter what name the science is known by, it all means that man has not undergone physical change during the past several thousand years and will probably undergo little during the next thousand, while his environment has and will change enormously. To adapt man to fit the environment is almost impossible, so means must be devised to adapt the environment to suit man. This chapter is devoted to that idea.

DA VINCI AND GILBRETH

Probably the first known systematic work in human factors was performed by Leonardo Da Vinci. His study of the human body led him to draw in anatomical detail figures of man in order to show muscular form. He even advised making models in which copper wires represent the individual fascicles of muscle. This work can be considered the beginning of the science of biomechanics. His work dealing with proportions of the human body is known today as anthropometry. A classic example of Da Vinci's body proportions is shown in Figure 6-1. He was not interested in man's environment when studying the human body, but rather was driven by his sense as an artist to report the human body in the most accurate detail.

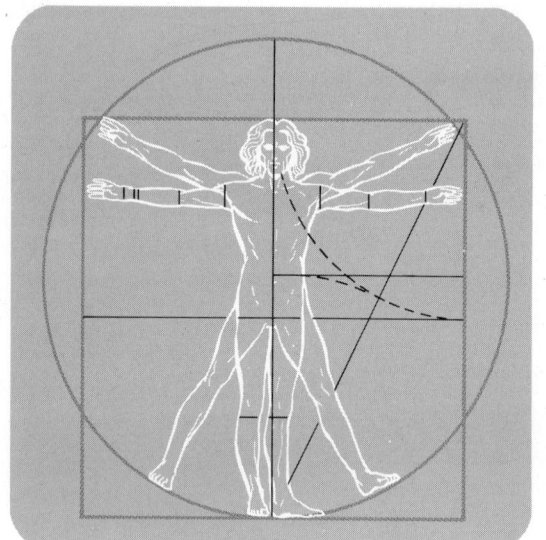

Figure 6-1 **DA VINCI'S HUMAN PROPORTIONS**

It was at the beginning of the Industrial Revolution, when the United States began depending upon high levels of production efficiency to be able to compete with foreign markets, that the talents of Frank B. Gilbreth were recognized. Around 1911 Gilbreth intuitively discovered and defined the anatomical, physiological, and behavioral variables which form the basis for ergonomics of the man-equipment-task system. He is known as the father of work simplification, and time and motion studies. With the techniques available to him at that time, Gilbreth concentrated upon three measurable quantities to establish a base upon which more productive, less tiresome, and more efficient work habits could be recommended. These quantities included performance time that was measured with a stop watch, the motion inventory demanded by a task established through therblig analysis, and shape and length of motions pathway mapped out by a chronocyclegraph (producing multiple exposure photos where lights were attached to the subject's moving parts). Figure 6-2 lists the three classes of variables considered by Gilbreth as intrinsic to each and every task. This describes and defines the field of biomechanics as early as 1911.

I VARIABLES OF WORKER	II VARIABLES OF ENVIRONMENT	III VARIABLES OF MOTION
1. Anatomy	1. Appliances	1. Acceleration
2. Brawn	2. Clothes	2. Automaticity
3. Contentment	3. Colors	3. Motion sequence
4. Creed	4. Music, reading, etc.	4. Cost
5. Experience	5. Heating, cooling, ventilating	5. Direction
6. Fatigue	6. Lighting	6. Effectiveness
7. Habits	7. Quality of material	7. Foot-pounds of work
8. Health	8. Reward and punishment	8. Inertia & momentum overcome
9. Nutrition	9. Size of unit moved	9. Length
10. Size	10. Fatigue eliminating devices	10. Necessity
11. Skill	11. Tools	11. Path
12. Temperament	12. Union rules	12. Play for position
13. Training	13. Weight of unit moved	13. Speed

Source: "Gilbreth Revisited" by E.R. Tichauer, A.S.M.E. paper No. 66-WA/BHF-7.

Figure 6-2 **GILBRETHIAN VARIABLES**

THE MAN-MACHINE SYSTEM

Every time man dials a telephone, drives a car, studies an oscilloscope, operates a computer, or forms a part on a lathe he has joined his sensing, decision-making, and muscular powers to an engineering system. If he pushes the wrong button, turns the wrong dial, panics when he must watch too many indicators, or grows uncomfortable in his chair, the system is not functioning properly. Error can sometimes result in discarded materials, costly time, or even disaster. For example, over a 22-month period during World War II, 457 U.S. Air Force accidents were caused by pilots who confused landing gear and flap controls.

Figure 6-3 shows the coupling of components necessary to consider man as an integral part of an efficient man-machine system. In order for man to operate a machine so that some work output will result, it is necessary to adapt the machine to man's capabilities. This adaptation usually takes place through a control system that magnifies, minifies, or transforms man's muscular input to that easily accepted by the machine. Man's input could be through his hands or feet or both. When man has reached the control desired, it can be understood through a feedback loop to the senses in the form of feel. (An example of this was the tendency for automobiles to oversteer and understeer when power steering was first made available. To reduce this, designers added 0.8 ft pounds of drag to the steering wheel as resistance in the form of feel-feedback to the operator). Another signal often required by the operator (man) is one of display to the senses in the form of sight or sound information. Man undergoes a physiological output in the system which is a measure of fatigue, depression, alertness, inefficiency, and so forth. The design engineer must take into account all avail-

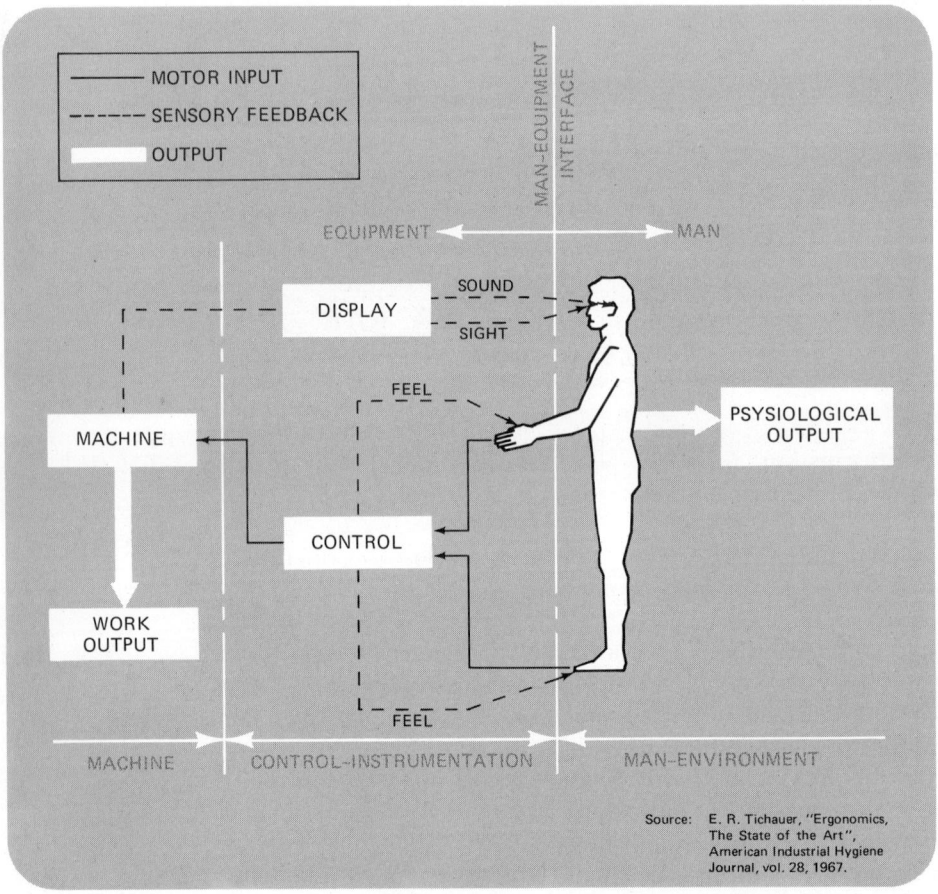

Figure 6-3   **MAN-MACHINE-ENVIRONMENT SYSTEM**

HUMAN FACTORS IN DESIGN

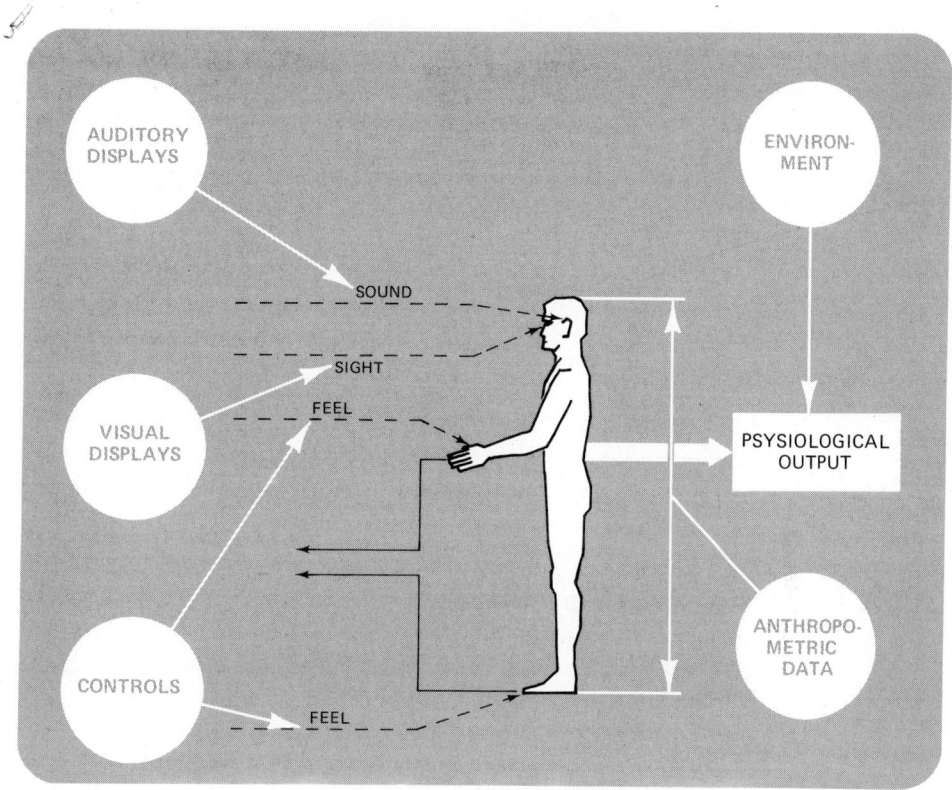

Figure 6-4 **CLASSIFICATION OF HUMAN FACTORS**

able human-factors information to be able to provide an equitable interface between man and the equipment he comes into contact with in the performance of his daily tasks.

HUMAN FACTORS

When we isolate man from the system shown in Fig. 6-3 and consider what is required for him to operate efficiently as a component, we might conclude the following: Man's input from the hands and feet are to controls and output to physiological measures concerned with *environment:* feedback to the senses in the form of feeling may be classified as a *control;* sound and sight as *auditory* and *visual* displays; and body size is known as *anthropometric data*. Each of these human factors is shown according to man's function in relation to the system of Figure 6-4.

controls

This data concerns itself with the shape and size of control knobs and levers, effort required to set the control, types of controls (for example, on-off or setting), direction of motion, manner of operation, and the like. When designing controls, the designer considers output as well as feedback information.

environment

Here the designer is concerned with physiological measures including heart rate, respiration, and blood pressure for reasonable comfort zones. The designer must make the environment as normal as possible in order for man to function efficiently over long or short periods of time. This involves consideration of such things as temperature, sound, vibration, altitude, radiation, acceleration, and pressure.

auditory display

Feedback signals in the form of sounds are also a concern of the design engineer. The most basic is that of speech. In addition, the designer is concerned with reproduction of speech through artificial means, warning and caution sounds, as well as sounds that indicate efficient and inefficient operation of machinery.

visual display

This category concerns itself with pictorial information in the form of dials, signs, numbers and letters, scale markings, counter and pointer design, and labels.

anthropometric data

Anthropometry is the science dealing with dimensions of the human body. These dimensions are divided into statistical groups known as percentiles. If 100 men are lined up from the smallest to the largest in any given respect they would be classified from the 1 percentile to the 100 percentile. The 2.5 percentile male means that designs based on this series of dimensions would allow up to 2.5% of the population to use the system. The 50 percentile means that 50% of the population (the average person) would fit into a system based on these measurements (this of course includes the 2.5 percentile previously mentioned). The remaining

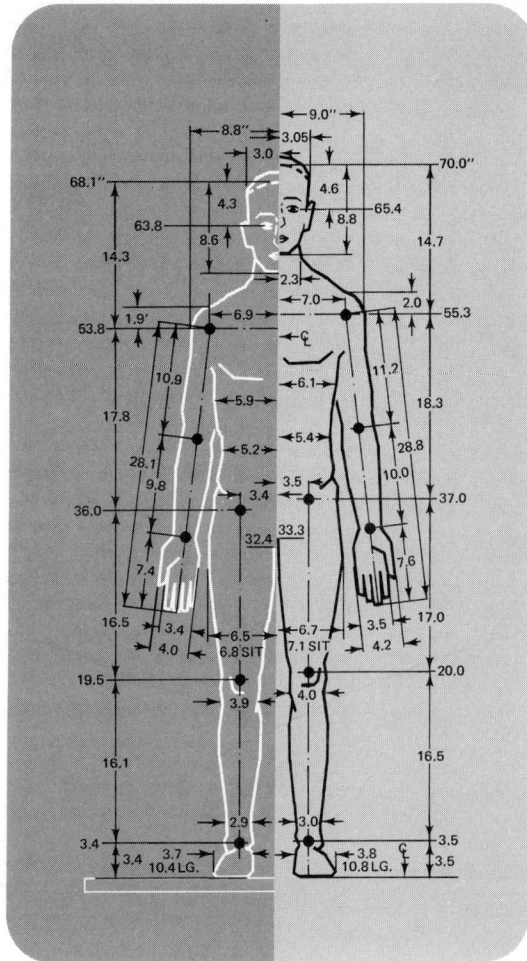

Figure 6-5 **CHANGE IN 50 PERCENTILE MAN OVER 20 YEAR SPAN**

50% would be excluded. The following examples will serve as a guide to the designer in the selection of an appropriate percentile group:

1. When designing a door return spring, design for the 2.5 percentile female. This means almost all adults will be able to open door with ease.
2. When designing a passageway, design for the 97.5 percentile male. Only $100 - 97.5\% = 2.5\%$ of population will be inconvenienced.
3. When designing commercial airliner kitchens, design for the 50 percentile female. Airlines screen out 2.5 and 97.5 percentile women as hostesses and choose those nearest the average.

142   THE SCIENCE OF ENGINEERING DESIGN

4. When designing automotive seating, design for a range of from the 2.5 to 97.5 percentiles so that only 5% will be inconvenienced and seating will fit 95% of the population.

Anthropometry is continually changing as evidenced by the chart shown in Fig. 6-5, showing the change in dimensions of the 50 percentile male over a 20-year span. If the task requires design work of an exact nature, it is important to the engineer to check the date of his data.

HUMAN FACTORS DATA

The data shown on the following pages is divided into the categories previously mentioned. It is not intended to be complete in any way but is included here to give the designer a feel for the type of data available. Please refer to the references on human factors listed in the Bibliography for complete data to be applied in design situations involving humans.

CHARACTERISTICS	TYPE OF CONTROL					
	DISCRETE ADJUSTMENT				CONTINUOUS	
	HAND PUSH BUTTON	FOOT PUSH BUTTON	TOGGLE SWITCH	ROTARY SELECTOR SWITCH	KNOB	CRANK
CONTROL TIME SETTING	VERY QUICK	QUICK	VERY QUICK	MEDIUM TO QUICK	SLOW	SLOW
RECOMMENDED NUMBER OF CONTROL SETTINGS	2	2	2-3	3-24	UNLIMITED	UNLIMITED
SPACE REQUIREMENTS	SMALL	LARGE	SMALL	MEDIUM	SMALL TO MEDIUM	MEDIUM TO LARGE
ACCIDENTAL ACTIVATION POSSIBLE	YES	YES	YES	YES	YES	YES
EFFECTIVENESS OF CODING	FAIR TO GOOD	POOR	FAIR	GOOD	GOOD	FAIR
EASE OF VISUALLY IDENTIFYING SETTING	POOR	POOR	FAIR TO GOOD	FAIR TO GOOD	FAIR TO GOOD <1 REV.	POOR
EASE OF NONVISUALLY IDENTIFYING SETTING	POOR	POOR	GOOD	FAIR TO GOOD	POOR TO GOOD	POOR
EASE OF CHECK READING IN ARRAY OF LIKE CONTROLS	POOR	POOR	GOOD	GOOD	GOOD <1 REV.	POOR
EASE OF OPERATING IN ARRAY OF LIKE CONTROLS	GOOD	POOR	GOOD	POOR	POOR	POOR
EFFECTIVENESS AS PART OF COMBINED CONTROL	GOOD	POOR	GOOD	FAIR	GOOD	POOR

Source: Cornell Aeronautical Laboratory, Inc., POCKET DATA FOR HUMAN FACTOR ENGINEERING, 1964.

Figure 6-6   **CHARACTERISTICS OF COMMON CONTROLS**

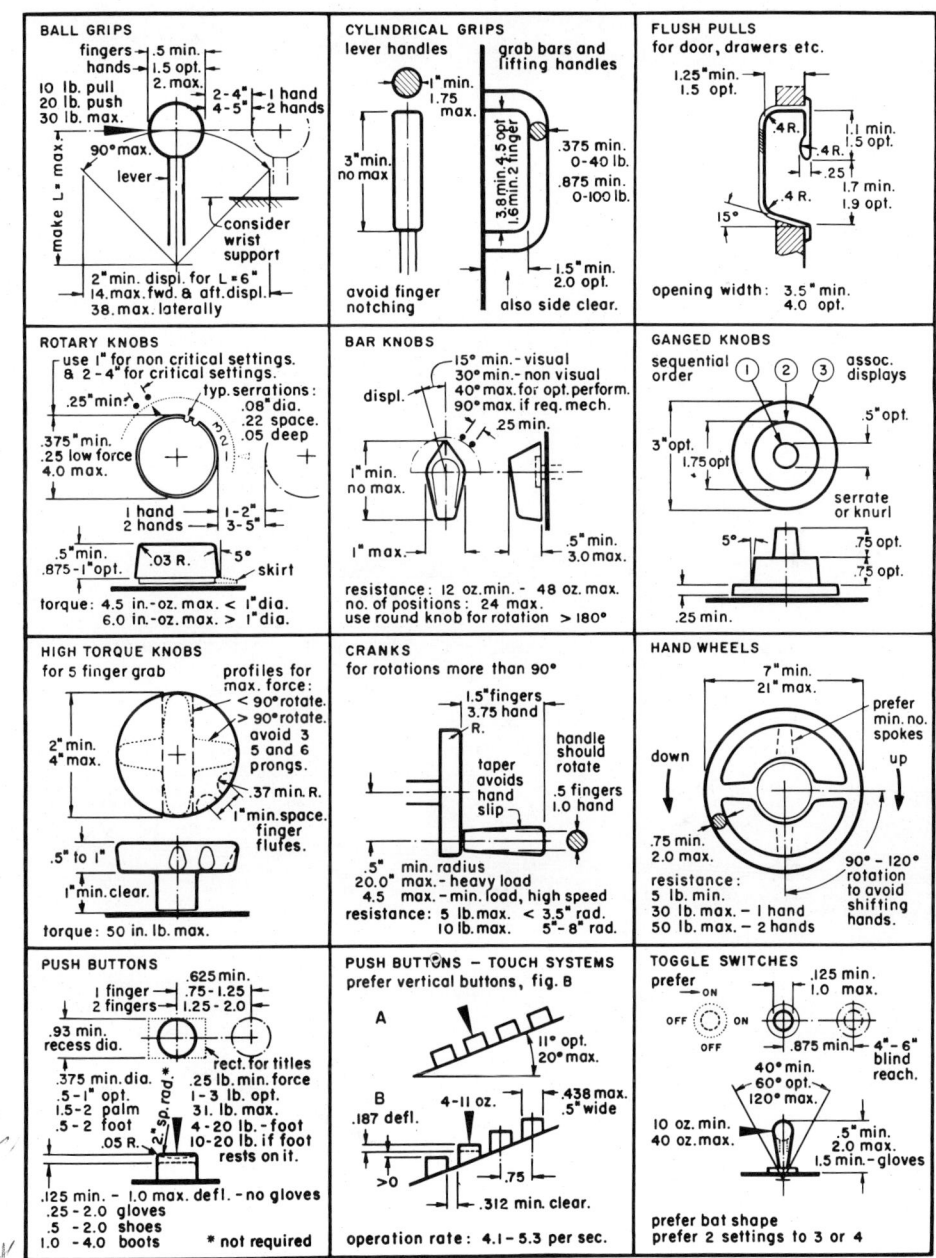

Figure 6-7 **BASIC CONTROL DATA**

144   THE SCIENCE OF ENGINEERING DESIGN

ENVIRONMENT

The first circle of environmental zones shown in Figure 6-8 is the bearable zone limit. Outside this limit great discomfort or possible damage is encountered. It is also necessary to consider infrared radiation, ultra sonic vibration, noxious gases, pollen, and heat exchange with liquids and solids.

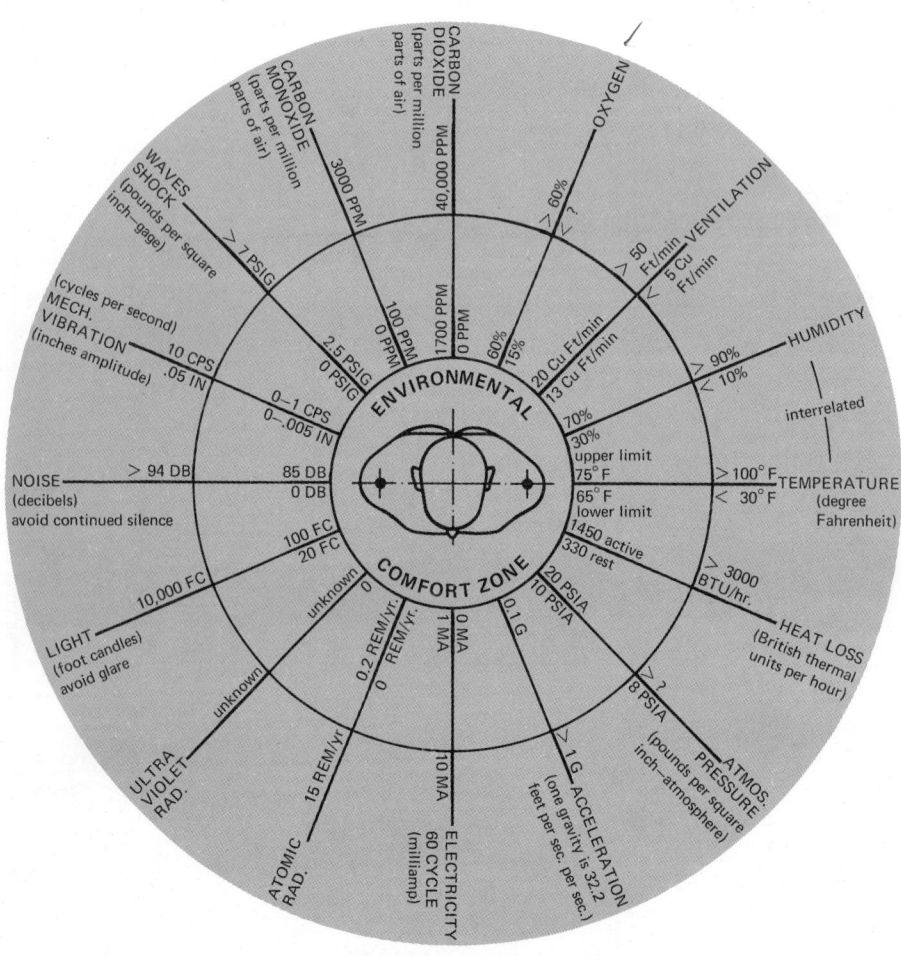

©"The Measure of Man" by Henry Dreyfuss,
Whitney Library of Design, New York, N.Y.,
1959, 1960.

Figure 6-8   **ENVIRONMENTAL ZONES**

FUNCTION	TYPE OF SIGNAL		
	TONES (PERIODIC)	COMPLEX SOUNDS (NON-PERIODIC)	SPEECH
QUANTITATIVE INDICATION	<u>Poor</u> Maximum of 5 to 6 tones absolutely recognizable.	<u>Poor</u> Interpolation between signals inaccurate.	<u>Good</u> Minimum time and error in obtaining exact value in terms compatible with response.
QUALITATIVE INDICATION	<u>Poor</u> Difficult to judge approx. value and direction of deviation from null setting.	<u>Poor</u> Difficult to judge approx. deviation from desired value.	<u>Good</u> Info. concerning disp., direction, and rate presented in form comp. with req. response.
STATUS INDICATION	<u>Good</u> Start and stop timing. Cont. info. where rate of change of input is low.	<u>Good</u> Especially for irregular occuring signals. (e.g. alarm signals).	<u>Poor</u> Inefficient, more easily masked. Problem of repeatability.
TRACKING	<u>Fair</u> Null position easily monitored; signal-response compatibility.	Required qualitative <u>indications</u>. Difficult to provide.	<u>Good</u>.
COMMENTS	1. Good for auto.comm. of limited info. 2. Meaning must be learned. 3. Easily generated.	1. Some sounds available with common meaning. (eg. fire bell). 2. Easily generated.	1. Most effective for rapid (not auto.) comm. of complex info. 2. Meaning intrinsic when std. min. of new learning req'd.

Source: Cornell Aeronautical Laboratory, Inc., POCKET DATA FOR HUMAN FACTOR ENGINEERING, 1964.

Figure 6-9 **FUNCTIONAL EVALUATION OF AUDIO SIGNALS**

146　THE SCIENCE OF ENGINEERING DESIGN

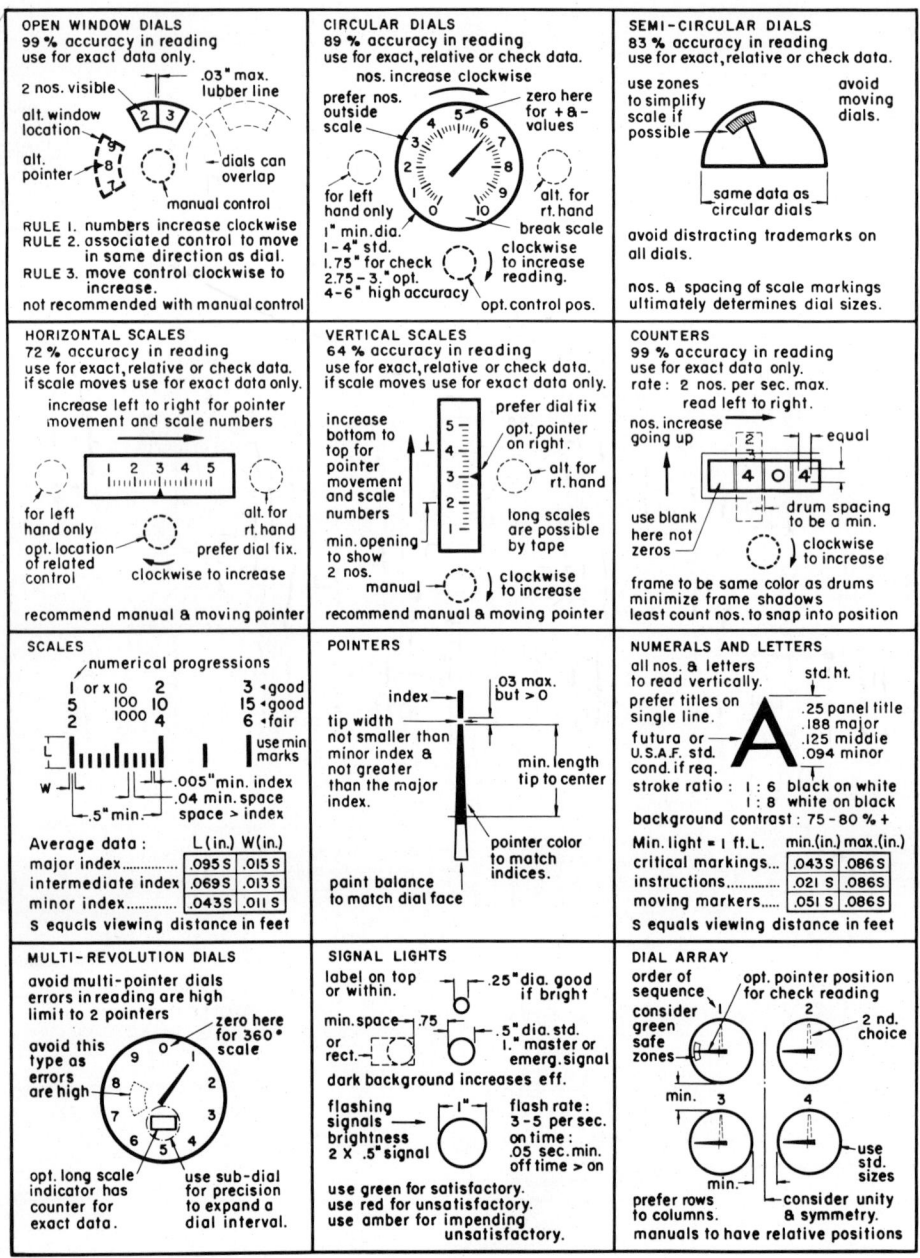

Figure 6-10　**BASIC VISUAL DISPLAY DATA**

HUMAN FACTORS IN DESIGN 147

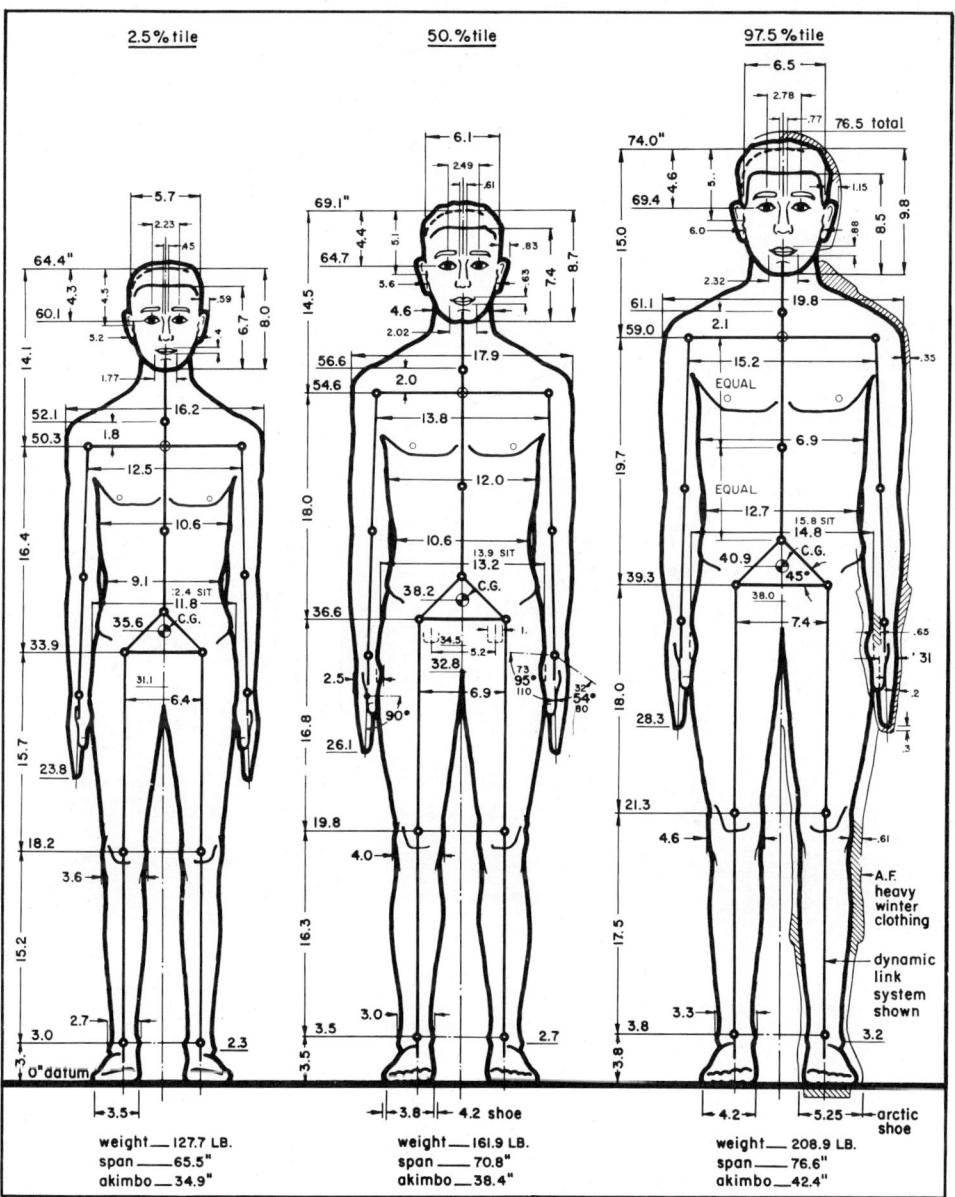

Figure 6-11 **STANDING ADULT MALE**

148 THE SCIENCE OF ENGINEERING DESIGN

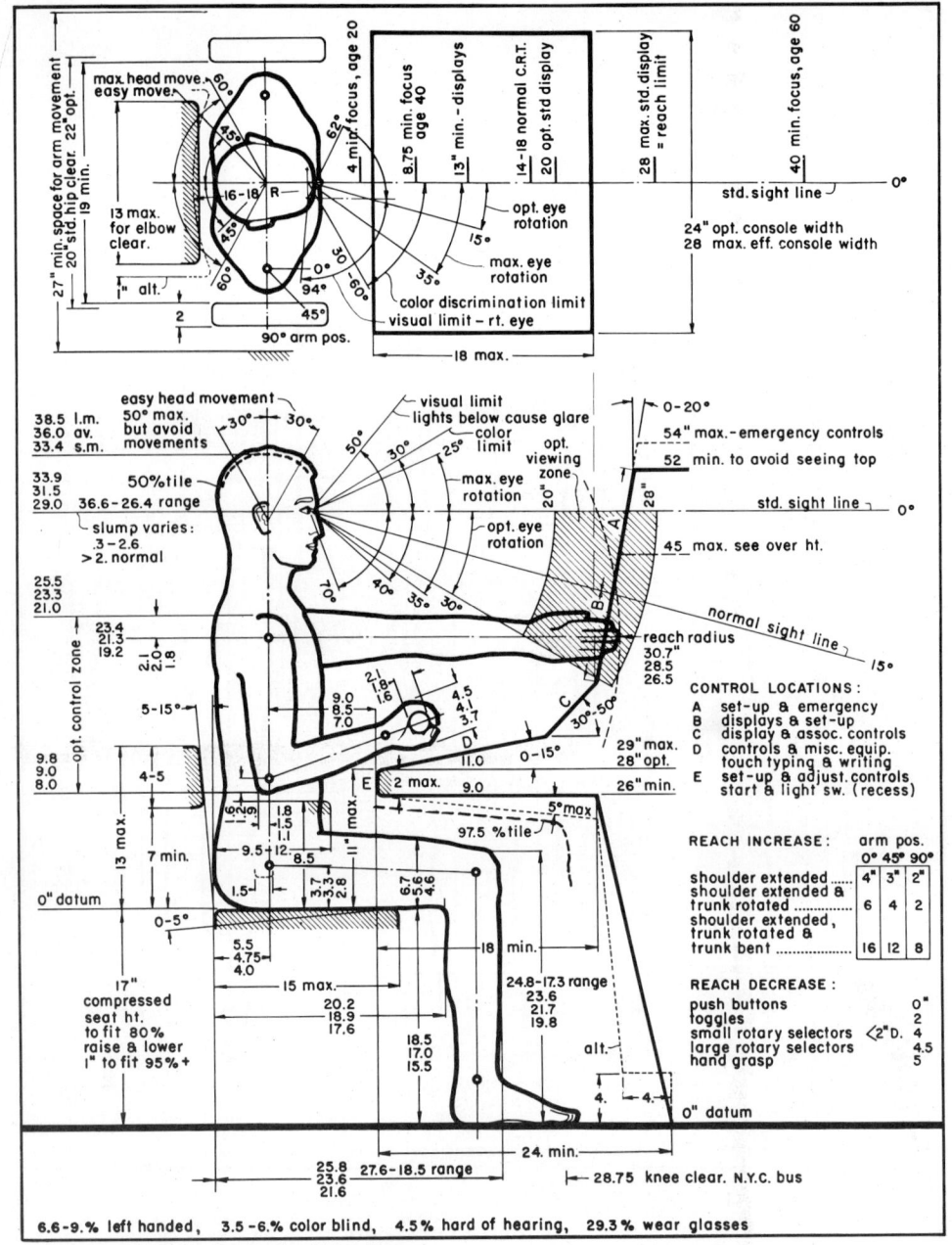

Figure 6-12 ADULT MALE SEATED AT CONSOLE

some examples of human engineering

To best demonstrate the service the human engineers perform, consider the following case study. A pilot is trained in a single engine aircraft and usually sits behind the instructor holding the throttle in his left hand and the control stick in his right. If he should undershoot the runway, he should increase throttle and pull up the nose of the aircraft (that is, forward with the left hand, back with the right). As a commercial pilot he is seated alongside a copilot and is on the left side of the aircraft. He now holds the control wheel in his left hand and the throttle levers in his right. If he now undershoots the runway, he still needs more power and lift, but this calls for pulling back with the left hand and pushing forward with the right. This is opposite to previous training. While the pilot may become conditioned to this new response through training, if an emergency arises he may panic and revert to the procedure he first learned. This could result in an accident. Here is an excellent problem for the human engineer since he must have an understanding of the human element as well as the machine man is to operate. Although this problem has not yet been solved, the following examples will serve to show effective solutions to a number of human factors problems.

fuel gage

During World War II Navy pilots flew a number of different kinds of planes. These planes differed in fuel consumption so that a long-range bomber with a supply of 500 gallons might run dangerously low while a single engine fighter with the same amount of fuel could fly an entire mission. Since the fuel gage in each plane indicated the supply in gallons, as shown in Figure 6-13, pilots often became confused. In some cases they were in danger when they thought they

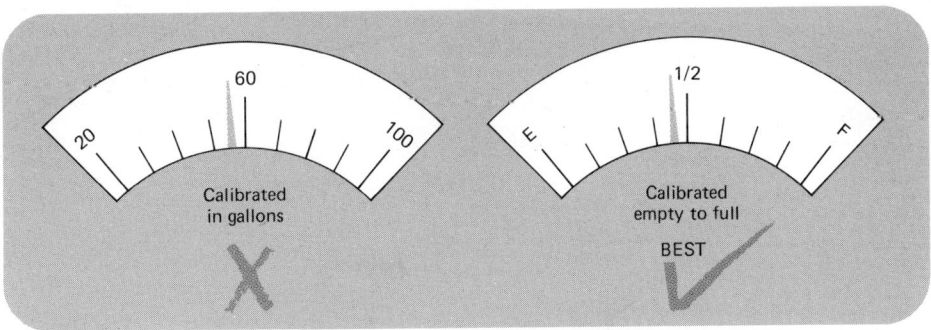

Figure 6-13   **FUEL GAGE DISPLAY**

were safe. The problem was solved through consideration of the human element by indicating, instead of number of gallons in the tank, the relative fuel level according to aircraft by simply showing empty (E), ½ full, and full (F). This method is now standard on all aircraft as well as automobiles.

altimeter

An altimeter is a dial used in aircraft to display the vertical distance (altitude) above the ground. This important dial must be read at frequent intervals when a pilot is making a landing. Three evolutions in the optimum design of this display are shown in Figure 6-14. The first generation altimeter shown at the left is of the accumulation type similar to a watch or clock. Experiments involving Air Force pilots show that this dial was read 11.7 percent of the time with an error of 1000 feet or more. An improvement in this gage is shown at the center in which the 1000 foot interval is shown on a separate concentric calibration. Tests show that this dial was read incorrectly (1000 ft or more) 4.8% of the time. The dial shown at right in which 1000-foot intervals are shown through a window was read incorrectly 0.7% of the time. This dial with modifications is now being installed on new aircraft. Unfortunately older aircraft retain the dials shown at left.

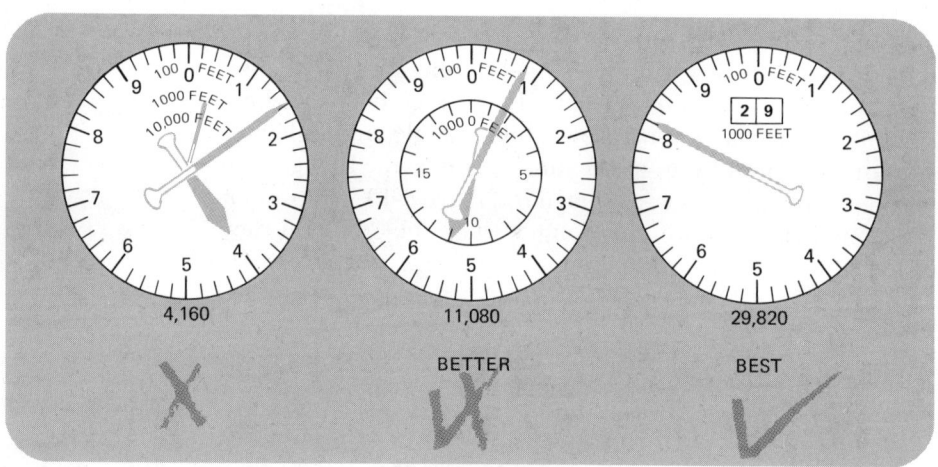

Figure 6-14   **ALTIMETER DESIGNS**

artificial horizon

The pictorial instruments shown in Figure 6-15 are designed to give information as to the attitude of an aircraft quickly. The aircraft shown is in a left bank (turn).

HUMAN FACTORS IN DESIGN 151

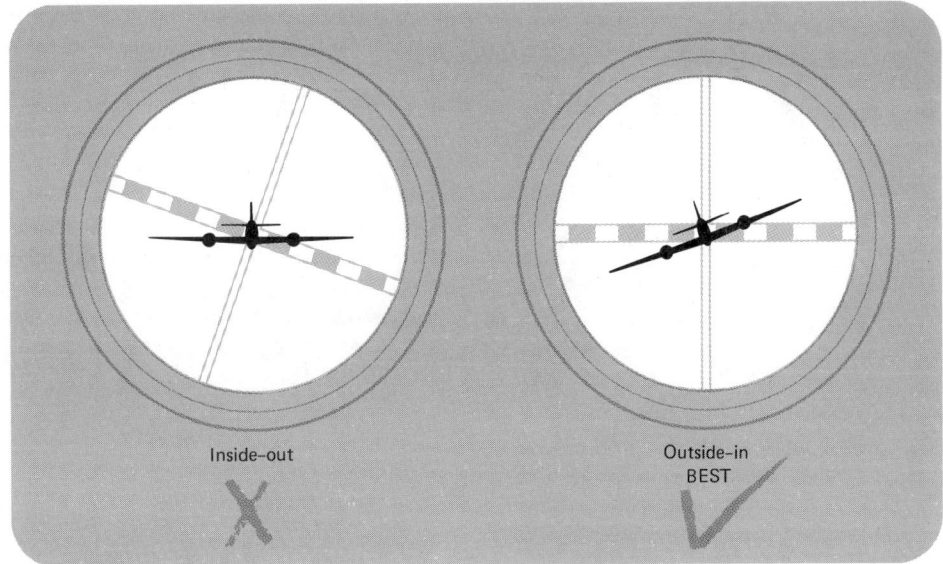

Figure 6-15 **ARTIFICIAL HORIZON INDICATOR**

Early instruments were designed on an "inside-out" basis, meaning you are inside the aircraft looking out. Naturally the wings of the aircraft seem level and the horizon moves. After many errors by pilots correcting to level flight in the wrong direction (this is especially dangerous when flying in formation), the instrument shown at right was tried. Here the "outside-in" principle was used. The pilot is outside the aircraft looking in. This means the horizon remains level and the aircraft moves. The psychological basis for this design is that man spends a larger proportion of his time in an outside-in environment where the horizon remains level and therefore more readily reacts to this type of information in a favorable way.

dial array

Dials are often used to indicate the operating condition of a remote station such as engine oil pressure and temperature. When many dials are used, it is advantageous to group them in such a pattern that when a dial moves out of the normal operating range, the information signal will be obvious. Consider Figure 6-16, showing an array of nine dials which indicate parameters in a chemical control process. Each dial is to be monitored and the individual parameter adjusted when the flow deviates from a normal range indicated by the black section.

152   THE SCIENCE OF ENGINEERING DESIGN

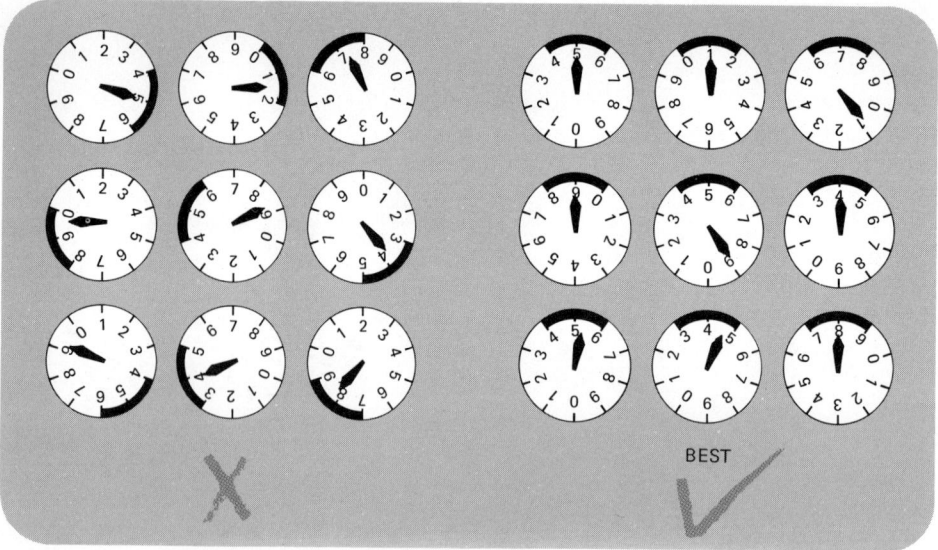

Figure 6-16   **DIAL ARRAY**

The dials shown in the top figure have been arranged at random. It is obvious that in order to monitor the chemical control process, dials must be read almost on an individual basis to determine what adjustments are required outside of the normal range. The dials shown below have been arranged so the normal range sector is always vertical (at the top of the dial). In this case the individual monitoring the system need only glance at the nine dials to detect the two indicating an unnormal reading.

This concern for the human operator in dial array has made it possible for men to fly the complex system of a jet passenger aircraft.

telephone dial

In an attempt to minimize the large percentage of wrong numbers dialed on the telephone, engineers at Bell Laboratories suggested that the dial shown at the top of Figure 6-17 was improperly designed. Their reasoning was based on the fact that one's finger covered the number or letter being dialed and one had a tendency to forget the number without the visual reminder. The scheme was changed to that shown in the second figure with numerals and letters outside the dial itself. Here the numbers could be seen at a wide angle and were not covered in dialing. Even though the amount of wrong numbers dialed was reduced, there was still a substantial number of errors. It was found that the black holes

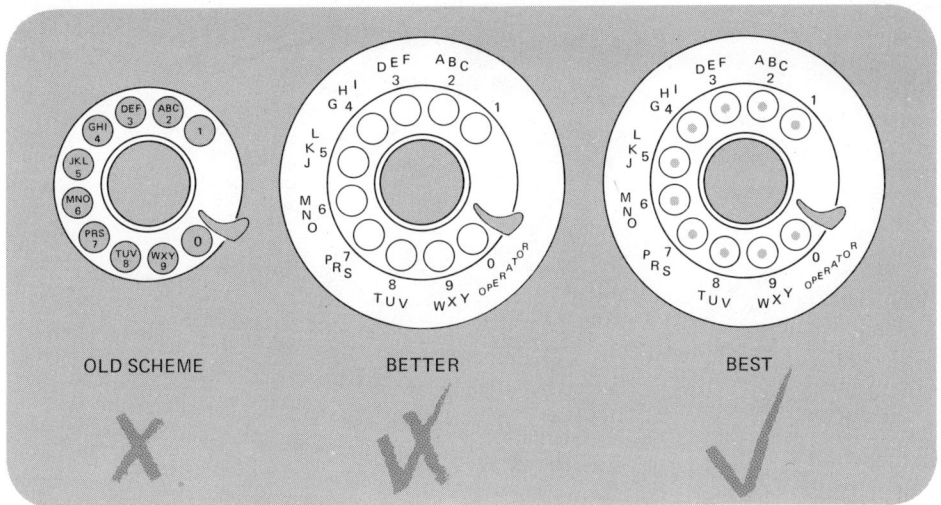

Figure 6-17 **TELEPHONE DIAL EVOLUTION**

were hard to distinguish. A target dot was then added to the holes shown in the third (bottom) figure, greatly improving dialing ease and accuracy.

It can be predicted that the next step in telephone dial evolution will be the removal of letters. One basis for this is that numbers are more quickly located visually, whereas letters take a moment's thought for proper selection. The way has already been paved for this changeover from letter codes to numerical (for example, PA 9-1585 has been changed to 729-1585). Another prediction is the change in numbers from a fixed phone base to an enlarged wheel where the number dialed would travel with the finger.

the symbolic language

Use of pictographs or the more abstract use of a symbol in place of a legend is growing in popularity in areas of industrial machine controls, automobile and aircraft controls, and highway signs. Such a scheme provides split-second information and has no language barrier. Controls identified in this way on a lathe could be constructed in Italy and operated effectively in Japan, Brazil, or the U.S.A. Visual shorthand has been used on the controls of imported sport cars.

It is unfortunate that graphics today and especially pictorial symbols are marked by a relatively high degree of amateurism, if not outright dilettantism. The choice of figure, symbol, letter size, border, shape, color, background are all

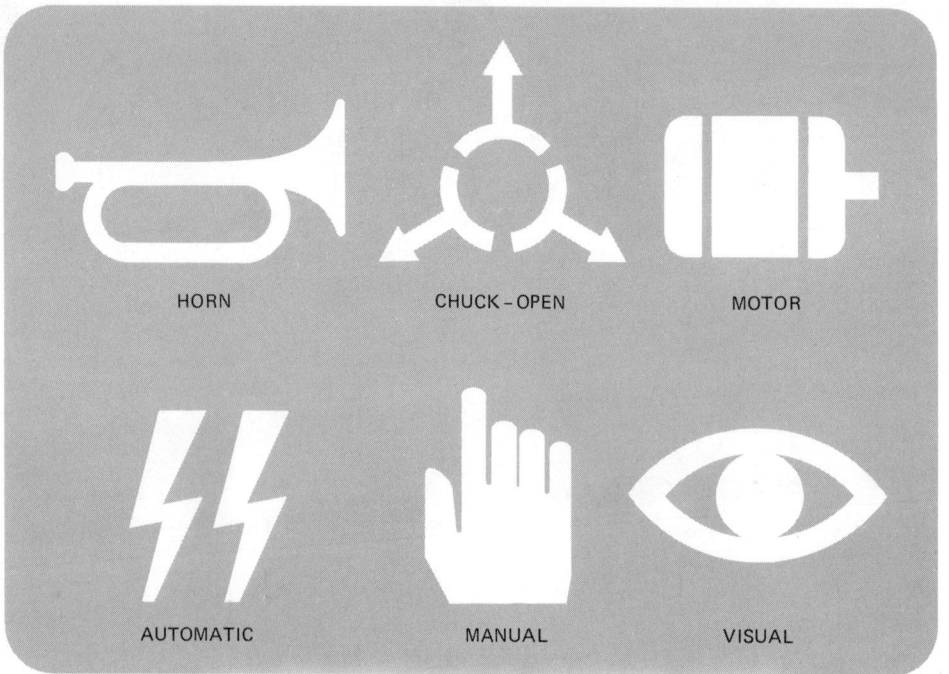

Figure 6-18 **SYMBOLS IN MACHINE CONTROL**

critical in the efficient reading of symbolic instructions. It is suggested thorough testing be done on a number of alternative symbols before a choice of the one to insure error-free interpretation is made.

Typical symbols used in machine controls are shown in Figure 6-18. These are usually moulded (depressed) into knobs and handles and painted with a contrasting color. It is obvious that once the symbol is learned it can be read at a glance, much more quickly than letters or words.

The road signs shown in Figure 6-19 were adopted as an international standard in 1949 at a United Nations convention. These symbols are now in use in most European nations. Note that "not allowed" or "no" message is conveyed by means of a diagonal slash instead of the "X" which is more traditional.

To a driver moving at 50 to 60 miles an hour the signs shown at the left in Figure 6-20 are almost useless. The print is too small, sign placement is often confusing, and arrows seem to give conflicting direction. An improved version seen throughout Europe and now beginning to appear in the U.S. is the sign shown at right which emphasizes destination rather than route number.

Figure 6-19 **SYMBOLS FOR ROAD SIGNS**

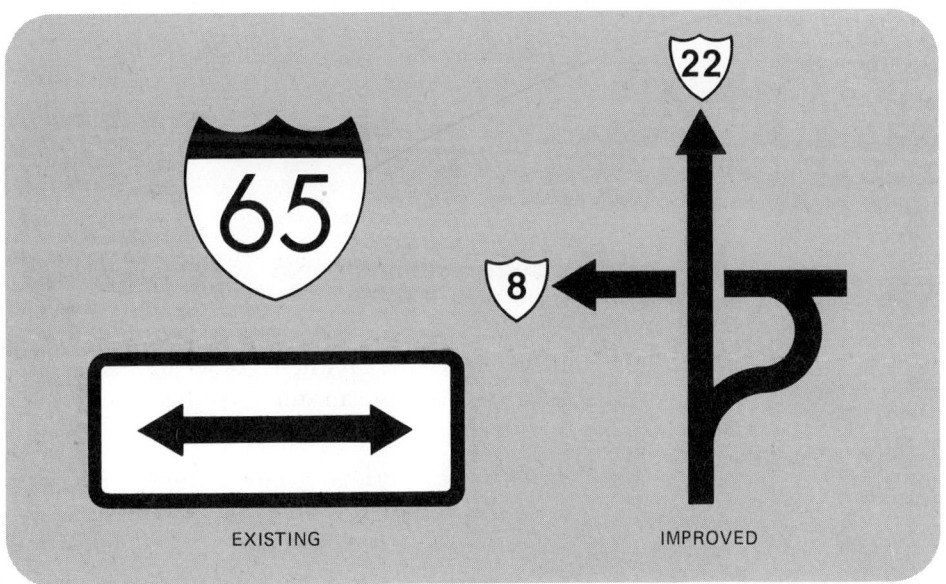

Figure 6-20 **IMPROVED ROUTE SIGNS**

color coding

Color coding is an effective means of conveying information at a glance where written instructions would take too much time or space or be impossible to read at a distance. For example, a red hexagonal sign means "stop" whether you can read the words or not. We are all familiar with some of the following color codings recommended by the National Safety Council:

Red: fire-protection equipment.
Green: safe materials, such as water and brine. Gray, white, or black are also accepted.
Blue: protective materials, such as antidotes for poison fumes.
Purple: valuable materials, caution against waste.

We are also familiar with colors used on stop lights (red, amber, green) and those used on mail boxes (red and blue). Once the colors are learned, identification as to thing, function, or instruction is made at a glance, almost without thought.

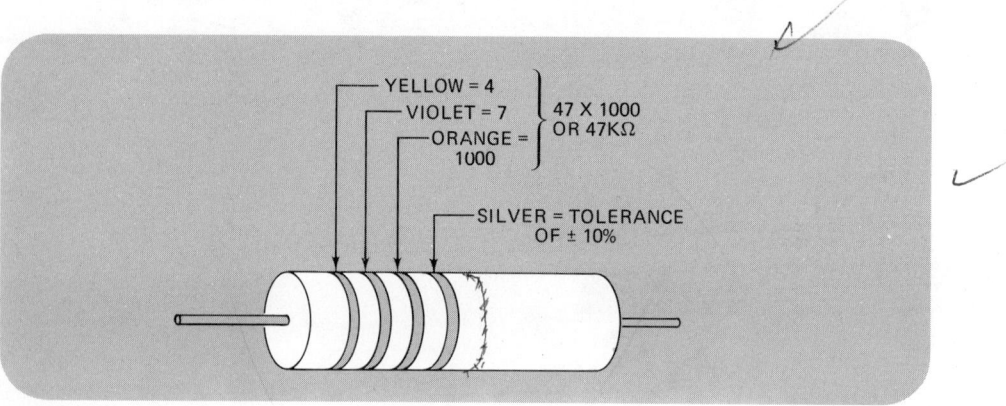

Figure 6-21   **RESISTOR COLOR CODING**

The selection of color is critical as there are only seven colors which can be identified consistently. Generally these colors should be saturated hues rather than mixtures. Often the color can still be confused unless selected with extreme care. Colors for surfaces, lights, and contrasting backgrounds are:

SURFACE COLORS—Primary color coding

    Red   Green
    Orange Brown
    Yellow Purple
    Blue

HUMAN FACTORS IN DESIGN   157

COLORED LIGHTS—Identification purposes
                  Red   Green
                  Amber Blue

BACKGROUND COLORS—Color of coded surrounding area
                  Black, White, Tan, Gray

The colored bands around the body of a color-coded electric resistor represent its value in ohms, as illustrated in Figure 6-21, and are grouped toward one

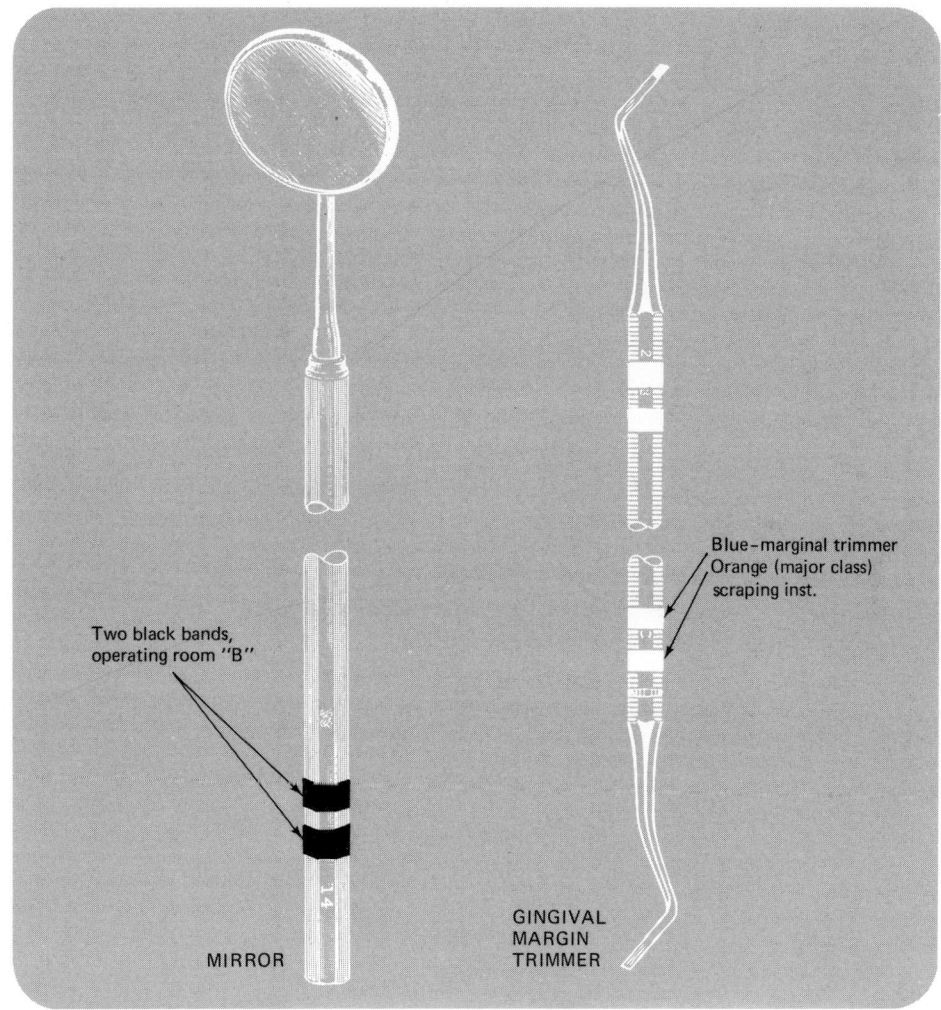

Figure 6-22   **DENTAL INSTRUMENTS**

end of the resistor body. Starting with this end, the first band represents the first digit of the resistor value; the second band represents the second digit; the third band represents the number by which the first two digits are multiplied. The fourth band of gold or silver represents a tolerance of ±5% or ±10%, respectively. The absence of a fourth band indicates a tolerance of ±20%.

Another example of color coding is that used in dental offices for the purpose of training auxiliary personnel to identify dental instruments as to function and proper care. Pressure-sensitive colored tapes specifically made for autoclaving, boiling, or cold sterilizing are affixed to the handles of instruments, as shown in Figure 6-22. The mirror is identified to its permanent location in operating room B through the use of two black tapes. One black tape would identify operating room A and three stripes C. The second instrument is first identified by the orange-colored tape as to its major classification as a scraping instrument. The second tape, blue, identifies it specifically as a gingival margin trimmer. Notice that the colors are represented at the other end, since the instrument is reversible.

ANALOG OF THE HUMAN BODY AS A DYNAMIC SYSTEM

Researchers in recent years have shifted their attention from the identification of anthropometric measurements and controls specifications to the general area of systems design, where man is considered a component of the system, and studies involving mathematical and dynamic modeling of the human body.

One technique of analogous modeling might parallel the following thinking:

Structurally the human body consists of a hard bone skeleton held together by tough fibrous ligaments embedded in a highly organized mass of connected tissue and muscle. A reasonable mechanical analog circuit closely resembling this description was invented by R. R. Coermann and is illustrated in Figure 6-23. This analog considers the human body as a linear passive dynamic system containing elastic (springs) and viscous (dash-pots) resistances so interconnected to selected masses that they describe with reasonable accuracy the important features of response of the human body to low frequency vibrations or deformations.

Another form of analog considering the abdomen-chest-mouth in the form of an electrical circuit was proposed by David E. Goldman, Captain, U.S.N. Medical Service Corps. Through studies involving vibration of the abdominal system considering exposure of a sitting or standing subject, it was concluded that at large amplitudes of vibration, speech can be modulated at the exposed frequency. This is modeled through the electrical equivalent circuit shown in Figure 6-24.

HUMAN FACTORS IN DESIGN 159

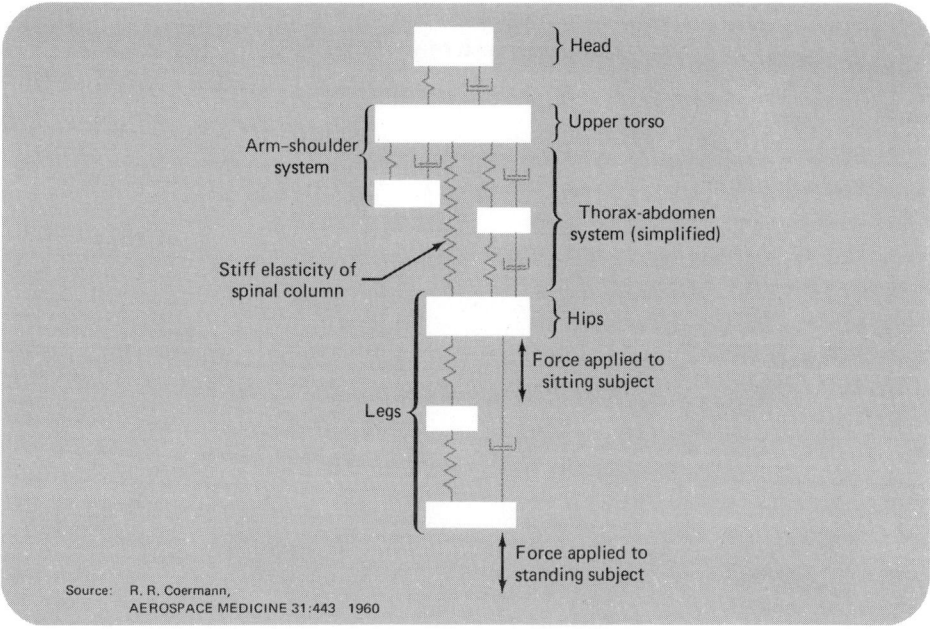

Figure 6-23   **MECHANICAL ANALOG CIRCUIT**

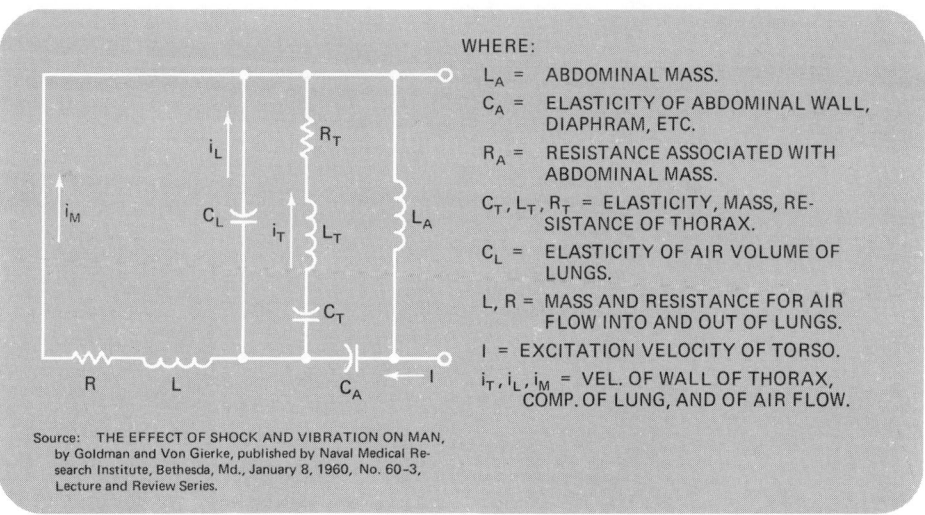

Figure 6-24   **ELECTRICAL ANALOG CIRCUIT**

Mathematical modeling of the human body related to studies of fatigue was prepared by Ezra S. Krendel in *Design Requirements* (Franklin Institute) and consists of the following differential equation written where fatigue begins:

$$M \frac{d^2y}{dt^2} = \sum F = F_{\text{exerted}} - F_{\text{viscous losses}}$$

$$M \frac{d^2y}{dt^2} = f \cdot M \cdot g - \frac{M}{a} \frac{dy}{dt}$$

This equation is based on the assumption that viscous resistance is proportional to body weight as follows:

$$\text{Viscous losses} = \frac{Mv}{a}$$

The differential equation yields the following solution:

$$y = f \cdot g \cdot a[f = a\{1 - e^{-t/a}\}]$$

and

$$v = f \cdot g \cdot a\{1 - e^{-t/a}\}$$

where: M = Mass of the human body
f = Dimensionless characteristic of build, strength, fitness $(0.5 \leq f \leq 1)$
g = Acceleration due to gravity
a = Constant
v = Horizontal forward velocity
v_o = f.g.a. = terminal velocity before the onset of fatigue
y = Distance run
t = Time measured from the moment the muscles become enervated.

Krendel's experiments and results at low frequencies have shown a remarkably close fit to the foregoing equation, so that:

$$v = v_o(1 - e^{-t/a})$$

This equation form is commonly found in dynamic systems where an energy device is in the process of discharging its energy subject to resistance losses.

DESIGN CONSIDERATIONS FOR POWERED EXOSKELETONS

One could not expect the unassisted human hand to shape the 1/4-in. steel bar shown in Figure 6-25. Nor could one expect present-day tools to shape the steel bar with the same deft touch as the hand. We realize that the hand is relatively weak and the tool, an extension of the hand, relatively clumsy. There are being developed today, however, techniques of combining the man-machine system into closer proximity than ever before. The exoskeletal hand of Figure 6-25 is an example of such a combination. The design is based on the assumption that during an average work task the thumb applies 45% of the effort, the forefinger 20%, the middle and ring fingers 10% each, and the little finger 15%. Because of the thumb's greater usage, a single series of links duplicates its action. The remaining fingers are combined according to frequency of usage into two exoskeletal links. Pins are placed at machining pivots of the human hand. Controls and power system have not been shown but would be gear, hydraulic, and/or pneumatic. This concept illustrates that any movement of the human hand would be closely paralleled by the exoskeletal hand with force and feel

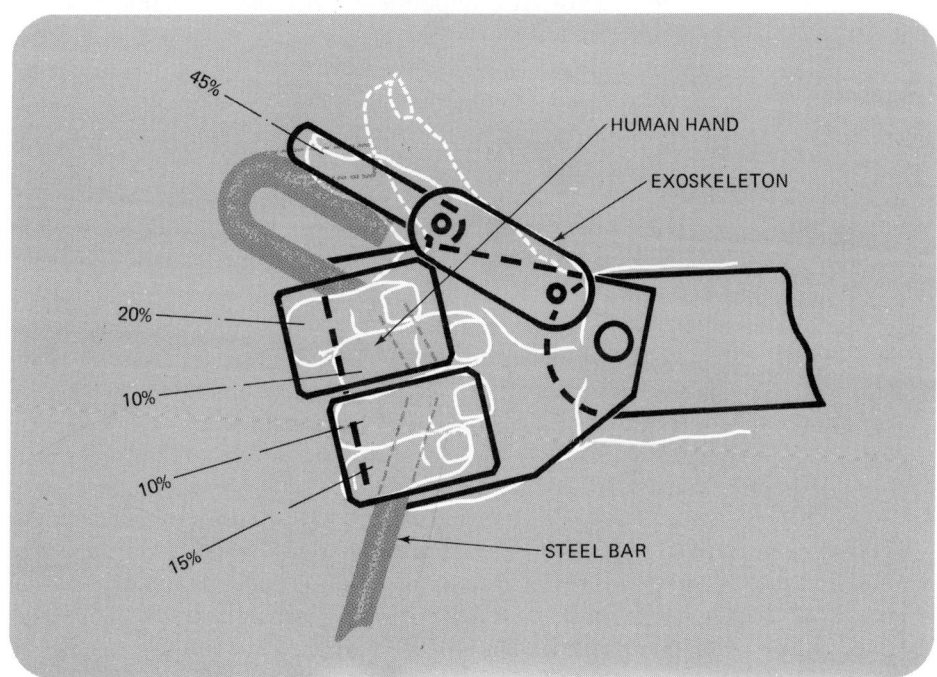

Figure 6-25   **EXOSKELETAL HAND**

feedback to the operator. These systems have been given a number of generic names, such as man-amplifiers, man-followers, exoskeletons, and cybernetic anthropomorphous machines.

One of the earliest organized approaches to exoskeleton systems was in the form of rectilinear-type remote manipulators, for example, the hydromechanical manipulator of General Electric and the Argonne electromechanical manipulator. These systems effectively integrate and mimic normal operator motion with varying degrees of simultaneity within the six degrees of freedom of human motion. Recent studies have demonstrated the engineering feasibility of powered exoskeletons enabling man to lift and carry loads of from ten to fifteen times more than his normal strength would permit. Walking machines of the type illustrated in the designer's concept in Figure 6-26 may someday provide off-road locomotion over terrain where current transportation systems are found to be inadequate.

Since these man-amplifier, man-manipulated systems are controlled by the forces and motion inputs of the human operator and must have a close spatial correspondance between operator's limb and machine member, one must look beyond anthropometric and human engineering data on which to base early design decisions. The application of kinematics, dynamics, and control system theory are heavily dependent upon biomechanics and human performance which is limited in the literature outside of the prosthetic field. There is data, however, presently being compiled by the Apollo Support Department of the General Electric Company at Daytona Beach, Florida, using photographic techniques that may prove extremely useful to designers of exoskeletons and man-manipulators in the near future. The photographs show multiple exposures of a subject performing a number of assigned tasks, taken in a plane perpendicular to the camera/subject plane, although recording of motion in more than one plane is accomplished by placing two cameras orthogonal to one another. Optical reference is accomplished by using a luminescent material fixed at each hinge joint (shoulder, elbow, wrist, hip, knee, and ankle) and a thin strip of the same material connecting these points. The subject is then photographed against a black background with a constant exposure rate (approximately 5 exposures per second in the General Electric experiments).

A graphical method of analysis was used to translate the photographic data into a form that could be used by the designer. The following examples (figures) will give the designer a feeling for the quantitative descriptions of body kinematics required by the engineers who are working on man-amplifier systems. This represents the beginning of a biomechanics reference source on a wide variety of everyday tasks that will eventually prove useful to those interested in the development and validation of mathematical models of the human body and its motions.

Figure 6-26 **BIPED WALKING MACHINE**

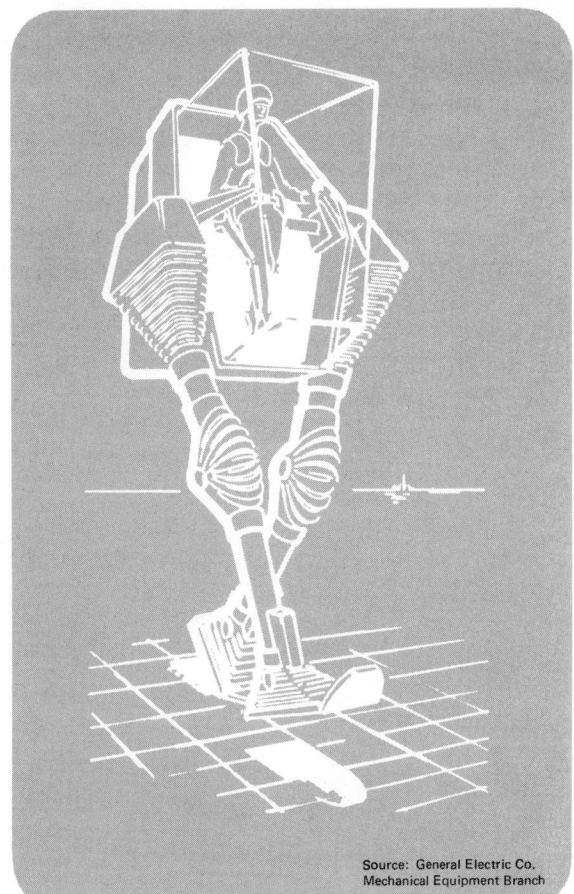

Source: General Electric Co., Mechanical Equipment Branch

Figure 6-27A and B show the plot of linear data on a 5th and 95th percentile man lifting a 30-pound load. As one would expect, the curves are quite similar, although specific values reflect the greater strength of the larger man. Both men took approximately 4.5 seconds to lift the 30-pound load to a height of 6 feet, although the 95th percentile man performed a more gradual lift while the 5th percentile man exerted more of a snap lift. As a result, while the total work done by both men was the same, as shown in Figure 6-27A, Figure 6-27B shows that because of the greater velocity of the load during the lift by the 5th-percentile man, his power output was greater.

164   THE SCIENCE OF ENGINEERING DESIGN

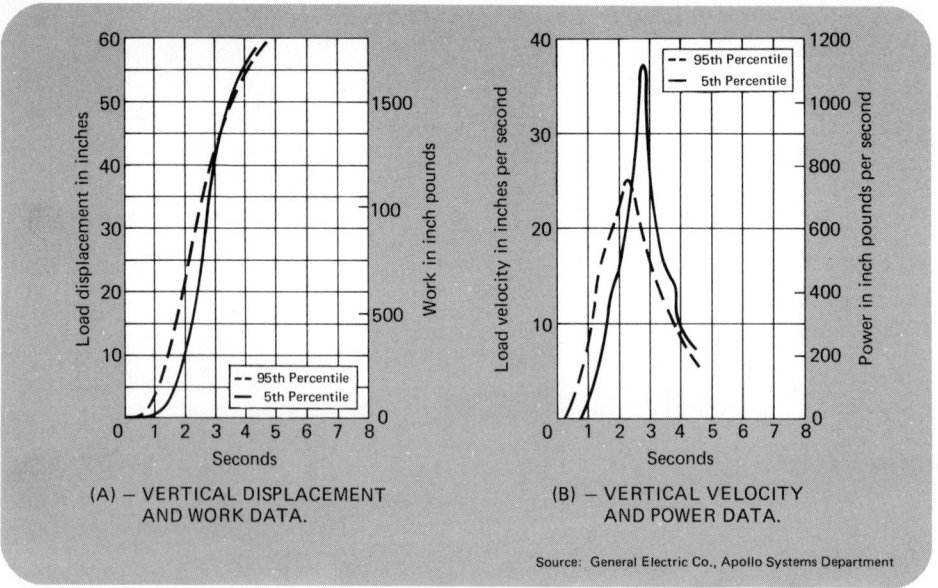

Figure 6-27   **DATA FOR A 5TH AND 95TH PERCENTILE MAN LIFTING A 30 POUND LOAD**

Figure 6-28A and B show vertical displacement, work, velocity, and power data for a 5th- and 95th-percentile man unloading a 30-pound load. The variation in curve configuration reflects the difference in height between the two populations. Load displacement remains constant during the first two seconds as the load is pulled over the simulated truck bed toward the subject. As the load slides off the edge, however, the shorter man merely slows its fall while the taller man lifts the load up off the bed to get it into a more comfortable position before lowering it to the floor.

It is interesting to note the similarity in data for the 5th- and 95th-percentile man stepping up and down with a 30-pound load, as shown in Figure 6-29A and B. The displacement curves in A are set apart by the difference in heights of the subjects while the velocity curves at B are almost identical.

HUMAN FACTORS IN DESIGN

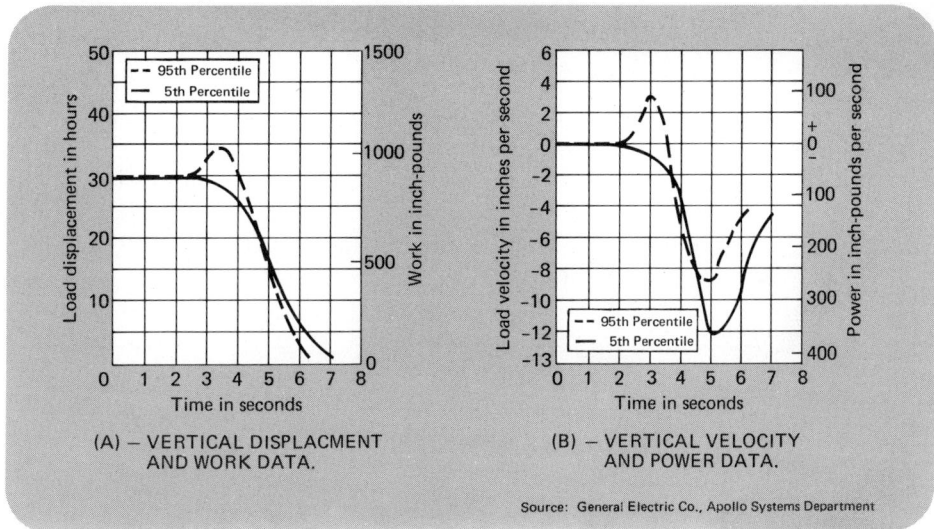

Figure 6-28   **DATA FOR A 5TH AND 95TH PERCENTILE MAN UNLOADING A 30 POUND LOAD**

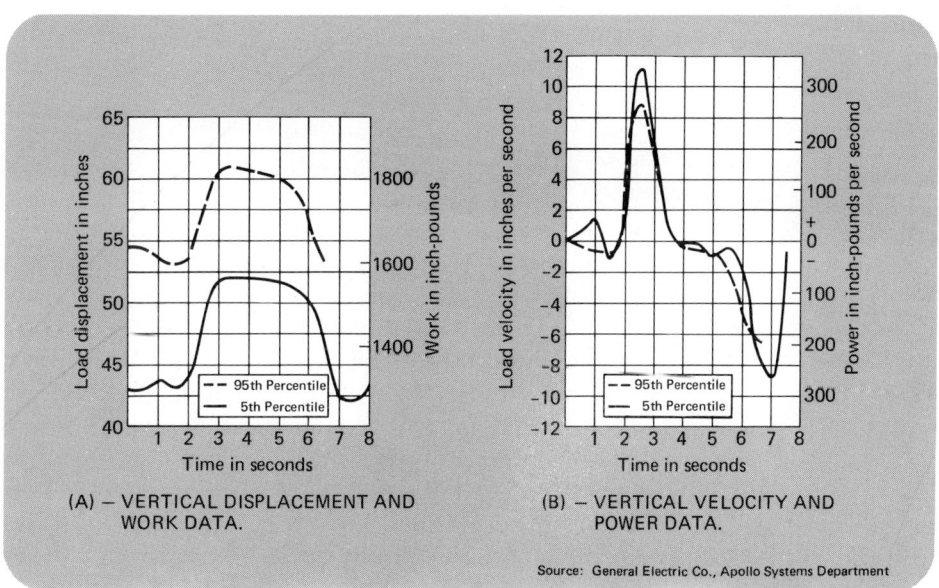

Figure 6-29   **DATA FOR A 5TH AND 95TH PERCENTILE MAN STEPPING UP AND DOWN WITH A 30 POUND LOAD**

THE INDUSTRIAL ROBOT

The word *robot* was first used in 1923 by the Czechoslovakian playwright Karel Capek when he predicted the overthrow of mankind by automatons in a play entitled *R. U. R. (Rossum's Universal Robots)*. This playwright, like Jules Verne, was able to forecast with reasonable accuracy glimpses into the machine-dominated civilization of the future. This may very well have been provoked by one of man's oldest dreams, to lift from his shoulders the burden of physical effort as a necessary condition of life.

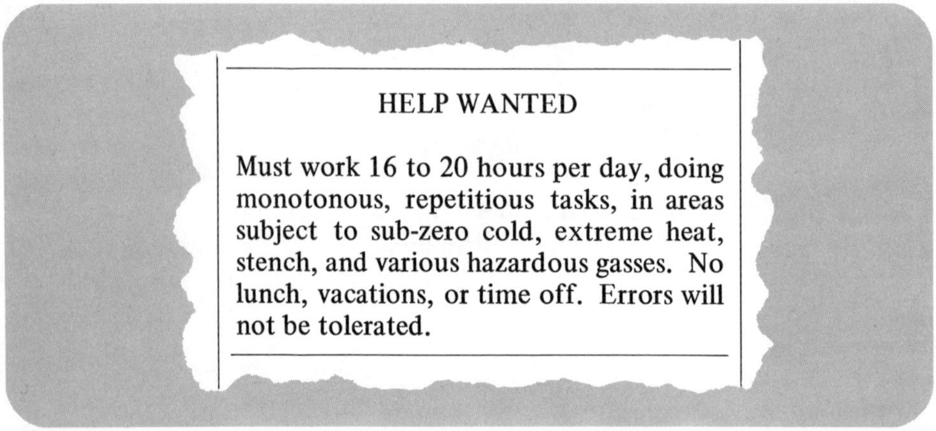

HELP WANTED

Must work 16 to 20 hours per day, doing monotonous, repetitive tasks, in areas subject to sub-zero cold, extreme heat, stench, and various hazardous gasses. No lunch, vacations, or time off. Errors will not be tolerated.

Figure 6-30   **INDUSTRIAL WANT-AD**

The want ad in Figure 6-30 will never be answered by a human and if the work described depends upon a man's effort, it will never get done. Jobs of this type are now easily performed by industrial robots of the type being manufactured by Unimation, Inc., called the Unimate. The Unimate robot is a self-contained movable machine, equipped to learn electronically and to perform automatically by means of a mechanical arm and hand various jobs otherwise handled by human labor under the most hazardous of working conditions.

Unimate's memory is of the solid-state digital type. Its mechanical arm, operated by a hydraulic and pneumatic system, is extensible to $7\frac{1}{2}$ feet, can pivot freely three-dimensionally, has a rotating wrist and hand, and can handle the lightest objects as well as weights up to 100 pounds. It can pick up an object, move it to another location, position it with accuracy, issue signals to control other equipment, or respond to external signals with a variable choice of actual patterns.

The Unimate is taught a new job by manually leading its hand once through the work program. The sequential steps are recorded in its memory by

HUMAN FACTORS IN DESIGN

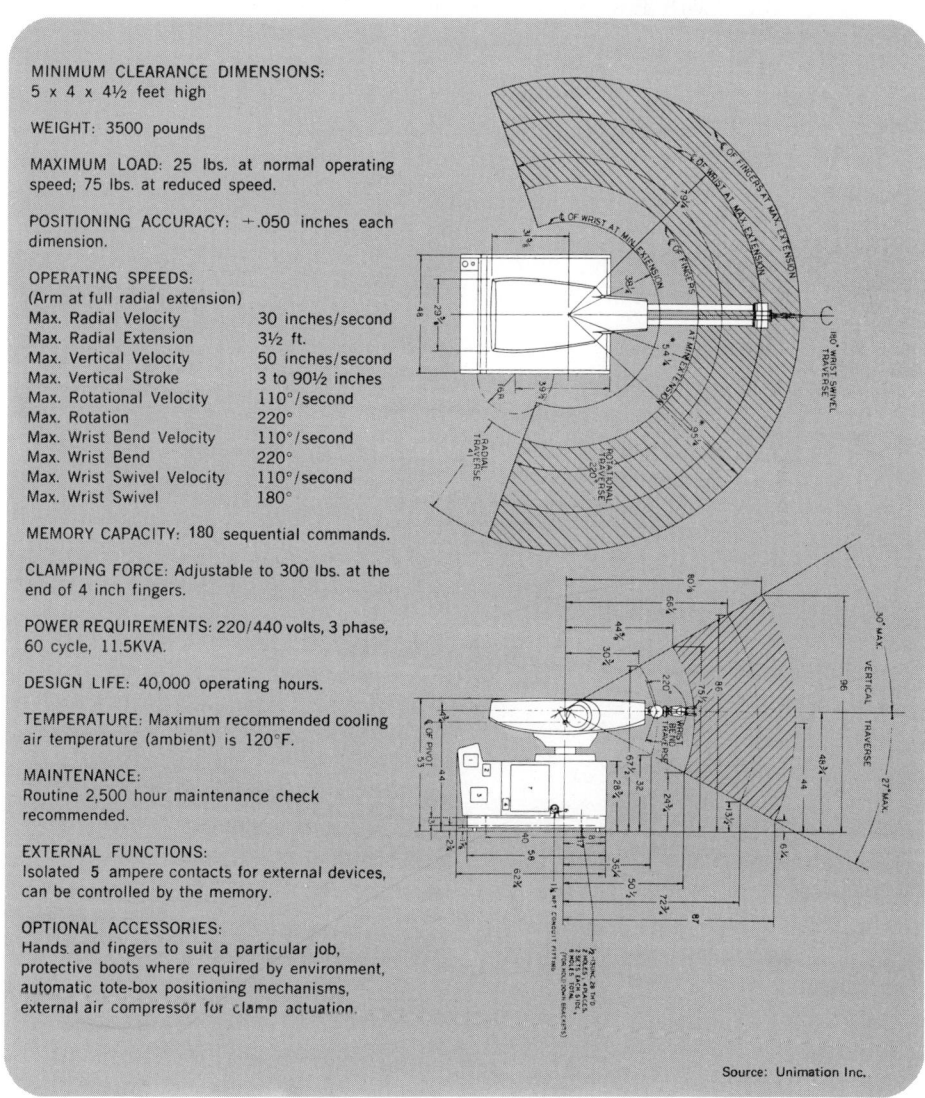

MINIMUM CLEARANCE DIMENSIONS:
5 x 4 x 4½ feet high

WEIGHT: 3500 pounds

MAXIMUM LOAD: 25 lbs. at normal operating speed; 75 lbs. at reduced speed.

POSITIONING ACCURACY: +.050 inches each dimension.

OPERATING SPEEDS:
(Arm at full radial extension)
- Max. Radial Velocity   30 inches/second
- Max. Radial Extension   3½ ft.
- Max. Vertical Velocity   50 inches/second
- Max. Vertical Stroke   3 to 90½ inches
- Max. Rotational Velocity   110°/second
- Max. Rotation   220°
- Max. Wrist Bend Velocity   110°/second
- Max. Wrist Bend   220°
- Max. Wrist Swivel Velocity 110°/second
- Max. Wrist Swivel   180°

MEMORY CAPACITY: 180 sequential commands.

CLAMPING FORCE: Adjustable to 300 lbs. at the end of 4 inch fingers.

POWER REQUIREMENTS: 220/440 volts, 3 phase, 60 cycle, 11.5KVA.

DESIGN LIFE: 40,000 operating hours.

TEMPERATURE: Maximum recommended cooling air temperature (ambient) is 120°F.

MAINTENANCE:
Routine 2,500 hour maintenance check recommended.

EXTERNAL FUNCTIONS:
Isolated 5 ampere contacts for external devices, can be controlled by the memory.

OPTIONAL ACCESSORIES:
Hands and fingers to suit a particular job, protective boots where required by environment, automatic tote-box positioning mechanisms, external air compressor for clamp actuation.

Source: Unimation Inc.

Figure 6-31 **UNIMATE MARK II SPECIFICATIONS**

pressing a button. This procedure erases prior instructions and sets the robot to repeat the new program indefinitely.

This memory which can be programmed to do a variety of tasks is one of the features distinguishing the robot from other automated equipment. A second feature is its ability to monitor and, if necessary, correct its own motion.

168 THE SCIENCE OF ENGINEERING DESIGN

Thus the Unimate can be programmed to do jobs other than its original task with little or no modification—a boon to manufacturers of expanding and changing product lines.

Specifications describing in detail the Unimate Mark II manufactured by Unimation, Inc. are shown in Figure 6-31.

Despite the hint in the foregoing discussion that the industrial robot is about to take over man's work, the device is grossly inferior to humans. The robot has no judgment, agility, or visual and tactile acuity. As mentioned earlier, the robot workers are involved in jobs that in human terms are dangerous, overburdening, unhealthy, unpleasant, or utterly boring.

The robots cost from $18,000 to $24,000 and handle a variety of industrial jobs:

> A well-known manufacturer for farm equipment placed a robot in front of a white hot furnace mouth to heat-treat tractor pins in a pick-up/heat/quench/load routine. The robot works without pause through three shifts. Not far away a pair of robots work at multiton punch presses stamping large automobile body parts.

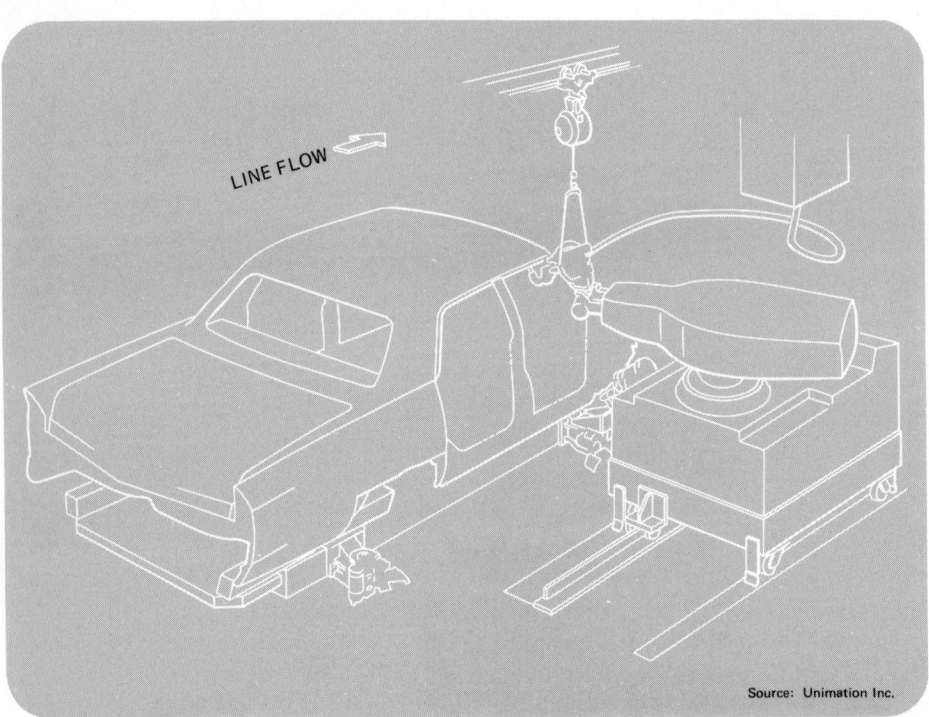

Figure 6-32 **ROBOT USED IN AUTO ASSEMBLER OPERATIONS**

HUMAN FACTORS IN DESIGN

A mid-western automobile maker employs robots on assembly lines of the type illustrated in Figure 6-32. The robot spot-welds the roof panel to side members of an automobile body. A powered truck keeps the robot moving alongside the body on the conveyor, and then returns for the next body.

Automobile companies are testing robots for paint spraying, where the big advantage is the paint saved since each job is duplicated exactly and there is a high degree of uniformity in film thickness.

General Electric Co. has ordered a robot with suction cups to pick up and carry fluorescent tubes.

Dohler Jarvis has six robots that not only unload die-casting machines, but also spray lubricant on the die cavity between operations.

Corning Glass has three robots handling Corningware products.

Chevrolet Motor Division of General Motors has a robot feeding white-hot, 2200°F steel billets to a giant forging press.

One gear manufacturer is designing his new plant especially to take full advantage of the robots.

EXERCISES

1. Design an optimum display of controls, indicators, and utility items listed below in the proposed dash layout, paying particular attention to human engineering principles. Where present dials and indicators are ineffective, redesign the display of information for efficient use by the human operator. Ask yourself the following questions about the information to be conveyed: Which is the most important? Least important? Which is the most often used? Least often used? What information must the indicator convey to the operator? Does it provide this information? Does it over-inform? Under-inform? What medium does it use? Most indicators are used visually and the eyes are often overburdened in their functional usage. This would be critical in a high-speed car. Could the same information be obtained by other means, such as light, sound, touch?

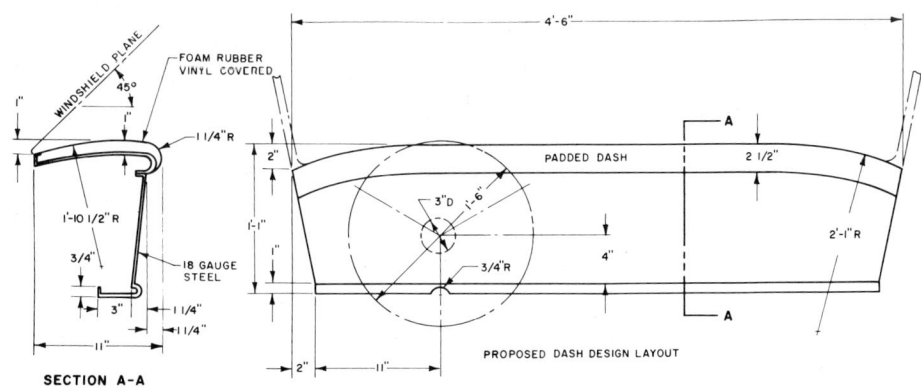

170   THE SCIENCE OF ENGINEERING DESIGN

CONTROLS	INDICATORS	UTILITY ITEMS
Lights	Speedometer	Glove compartment
Wipers	Odometer	Radio
Defroster	Fuel	Cigarette lighter
Heat	Engine temperature	Ventilation
Directional signals	Battery (C-D)	Rear-view mirror
Horn	Oil pressure	
On-off switch (starter)	Directional signals	
	High-beam lights	

2. Develop the design of one control, indicator, or utility item proposed in question one in some detail. Describe the solution through a drawing to a larger scale and in more detail than shown in question one. Label key features.

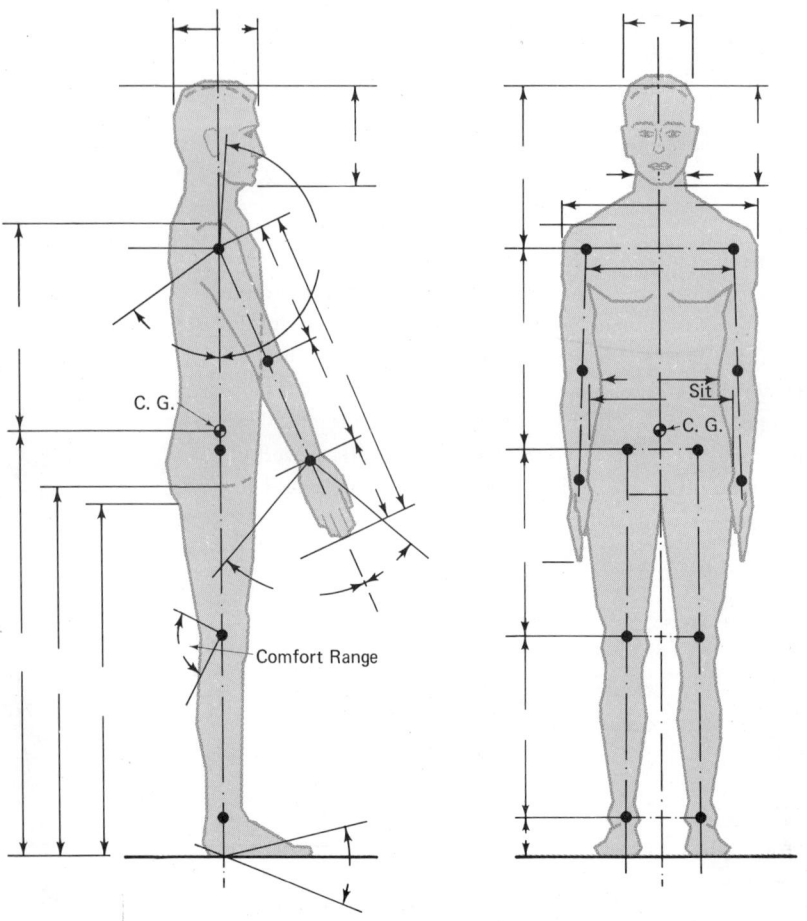

3. Consider that you have been assigned the task of designing the environment for an industrial worker who must be seated at a control console for four continuous hours each day to monitor and control plant and laboratory temperature and humidity. Describe (list) the type of environment that would be considered ideal for this operator.

4. Calling upon the assistance of a colleague, measure and record your anthropometric dimensions on the chart shown. In what percentile would you classify yourself?

5. Design a sign or signs using symbolic language to designate the men's room and ladies' room at an international airport.

6. Give an example of how color coding could be used effectively in the hardware store, the drug store, the supermarket.

7. Show a conceptual sketch of the design of an exoskeletal leg that will perform the same function as the human skeletal leg shown.

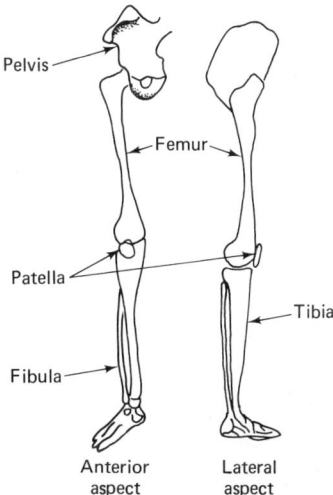

8. Make a comparison (tabular or graphical) of the ranges of comfortable operating movements between the 50-percentile man and the industrial robot, Unimate Mark II.

9. There are three methods of performing plant assembly operations; manually, automatically with conventional assembly machines, and by using robots. Make a comparison of these methods considering speed of assembly, weight-carrying capacity, cost, maintenance, flexibility, and quality of assembly. Based on this comparison, how would you rate each method?

10. List three applications of industrial robots of the Unimate type not discussed in the text. Give a reason why each application is better than present procedures.

7 CPM/PERT

"It's got the best of us. We're working for PERT instead of its working for us. I've thrown it out, after we used it for four years as a matter of stated policy."

Source: George A. Morin, Vice President
Rucker Company

"Development of the Model 900 Chromatograph spanned 18 months. We completed it within three weeks of estimated time, even though at stages along the way, we were as much as 8 weeks behind schedule. I would not now consider managing a project of more than four to six months' duration without CPM."

Source: Richard D. Condon
Perkin-Elmer

PROGRAM CONTROL TECHNIQUES

To understand the full scope of a program (job or project), to keep track of program progress, and to see that work is completed on time are among the prime responsibilities of management as well as engineering personnel. This requires a detailed knowledge of and methods of accounting for a series of sequentially-related events necessary to complete the project, as well as establishing an overview of the entire program. Here the manager really earns his way when he can detect a problem area that may be preventing project progress and, through the application of additional resources or change in personnel or equipment, affect this area, causing it to move along according to schedule. Many projects have failed through lack of attention to such detail or because the complexities of the program became overbearing. In response to the complex nature of present-day programs, from the production of the Techmatic razor by Gillette to the prototype of the Supersonic Jet Transport by Boeing, techniques have been developed to assist management in an orderly process of program coordination. A description of these techniques follows.

stored-in-the-head method

Many technical accomplishments performed by man in the early time of recorded history amaze us today as we think of how they were done. History tells us nothing of planning techniques or a systems approach to the building of the pyramids, the Colossus of Rhodes, or the Roman viaducts. One can only conclude that scheduling problems were solved by the "whip-in-hand" philosophy. There are a few men today of great mental capacity who are able to keep track of fairly complex programs entirely in their heads. There are others who prefer to write down events and milestones on slips of paper and are able to sort them in some way resembling a planned order of program progress. One such man was a plant engineer who carried slips of paper neatly arranged in the lining of his derby hat. Whenever the job seemed to be slowing down he removed his hat and, by consulting the slips arranged in the lining, identified the problem area that affected progress. When he retired, it required five men and one year of indecision to accomplish the same end. This points up one of the greatest difficulties of the "do-it-yourself" method. The complete program is never documented or it is not thought over in detail from the beginning. The program cannot be reviewed by others, and in case of illness it is almost impossible to bring in a substitute. This is to say nothing of the difficult task of finding the rare (and they are very rare) individuals who are able to cope with today's complex programs that *must* be completed on time.

Gantt (bar) chart

This chart of work versus time was introduced by Henry L. Gantt and Frederick W. Taylor in the early 1900s to aid in the administration of a program or job. In recent years this type of programming has taken the form of bar graphs or charts. The Gantt chart has little to no statistical value but is used instead for dynamic action in getting a job efficiently completed. An example of this chart's organization in designing and presenting a new product for consideration by top level management is shown in Figure 7-1. Part one shows a list of events or milestones in the program to be accomplished before the job is considered complete. This list would have been prepared in a group meeting of engineers, and drafting, manufacturing, and marketing personnel. Once the events were decided on they would have been listed in some logical sequential order, in this case by the Gantt chart in which the events are listed in one column and allocated time in another, as shown in part two. A decision must be made here as to the allocation of time for each event (beginning and end), to be represented by a bar.

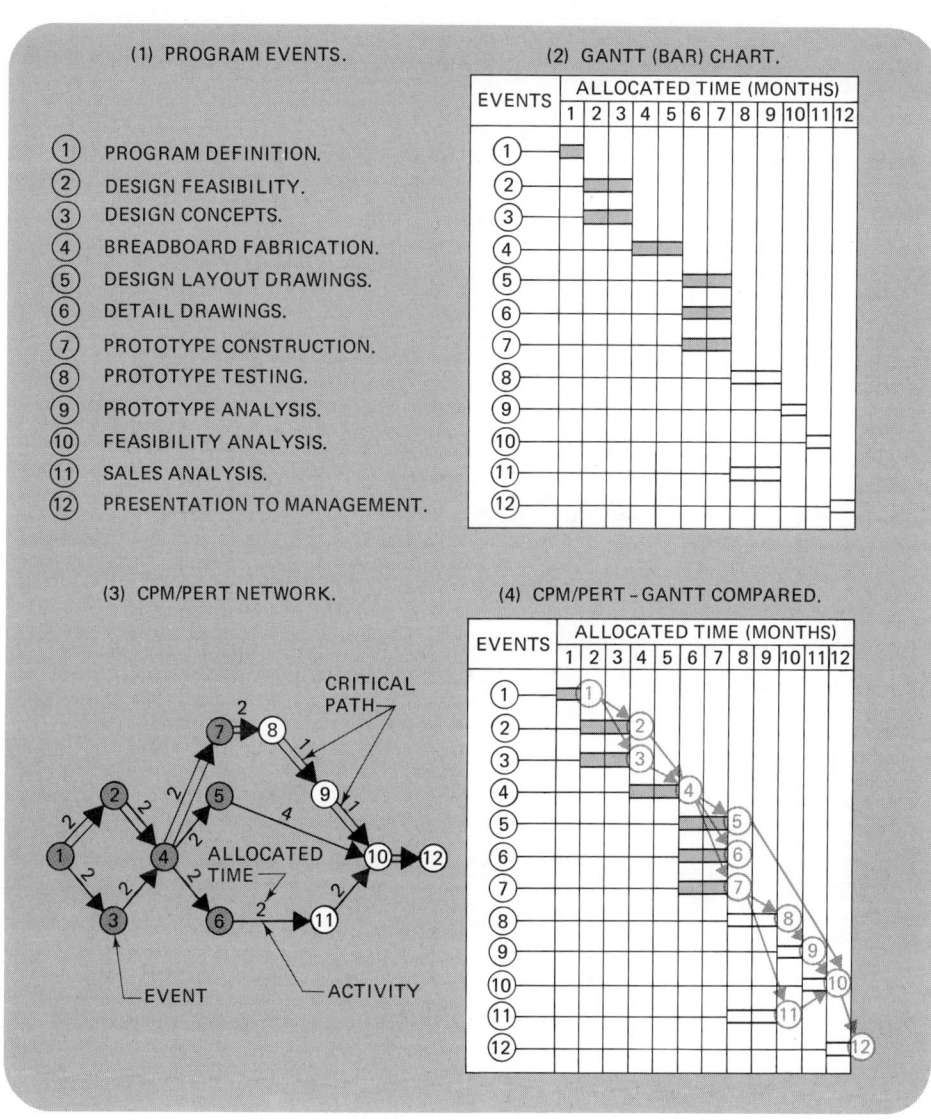

Figure 7-1 **PROGRAM CONTROL TECHNIQUES**

An open bar illustrates time allocated while a shaded or cross-hatched bar indicates actual time spent to complete the event. Such a chart forces a definition of the goals of the project and promotes detailed planning at the beginning. It provides a ready-made framework for documentation and presentation of the overall project plan. In addition to segmented planning in manageable size events, it provides the manager with a technique of determining at a visual glance what has been done in contrast to what must be done. This is an excellent means of auditing progress; one sees very clearly what is not being accomplished on time (as planned) and how it will affect the deadlines set.

cpm/pert network

The PERT (Program Evaluation Review Technique) network shown in Fig. 7-1 (3) covers the same information as the Gantt chart but in addition provides a measure of dependency and a definition of activities required to complete an event. The chart also shows the critical path. Work along this track must be completed on time for the overall job to perform according to schedule. Events are designated by circled numbers and activities by connecting arrows between events. An event is a terminal conclusion or milestone in a program while the interconnecting arrow indicates the activity necessary to accomplish the event. Time allocated for an activity is shown by numbers placed on the activity lines. Each of the terms mentioned here will be discussed in greater detail in the pages to follow. Figure 7-1(4) compares the Gantt chart with a CPM/PERT network by superimposing one upon the other. One can readily see here the great advantages of the network over the chart technique in planning and managing a program. Details explained in the following pages should point out additional advantages of the network technique.

HISTORICAL BACKGROUND

The Critical Path Method (CPM) dates back to 1956 when the E. I. duPont de Nemours Company established a group to study and recommend an effective management technique to the company's engineering function. The first area considered was the planning and scheduling of construction projects. Through the use of a Univac I the group was able to use computers in scheduling construction work. Mathematicians programmed a general approach in which the sequence of work and length of activity to accomplish the work were inputs and the computer output was a schedule of work. After a revision of the original concept in 1957, the resulting routine became the basic CPM. No significant changes have since been made in this method. CPM initially proved itself in an equipment turnaround situation involving the production of a self-deteriorating neoprene intermediate which meant that the equipment could not remain operational since the unit was on a production run. The equipment had to be shut down, purged, and maintained whenever production permitted this time interval. Previous experience showed that the average time for shutdown on a turnover was 125 hours. When scheduled according to CPM procedures, the overall turnaround was accomplished in 93 hours. Further experience with CPM reduced turnaround time to 74 hours.

Taking advantage of the CPM technique, the Special Projects Office of the Naval Bureau of Ordinance originated PERT (Program Evaluation Review Technique/Task) in January 1958. This technique was devised to manage the Polaris missile program which was already underway. The Special Projects Office (SPO) was facing the problem of monitoring and controlling more than 3000 contractors and agencies working on the program. It was obvious through the use of a Gantt chart that if all contractors delivered on schedule the program was sure to be completed on time. To keep track of and coordinate the contractors' efforts, SPO instituted weekly meetings of key personnel. There was, however, the overriding thought that a small contractor might be late in supplying a minor but vital piece of hardware that would hold up a prime contractor, thus affecting the overall program. This thought drove SPO to organize a more thorough method of keeping in touch with all divisions. A computer-orientated approach was suggested as a result of the program's complexity.

In July 1958, a team consisting of the Navy's SPO group, Lockheed Aircraft, and the management firm of Booz, Allen, and Hamilton issued a Phase

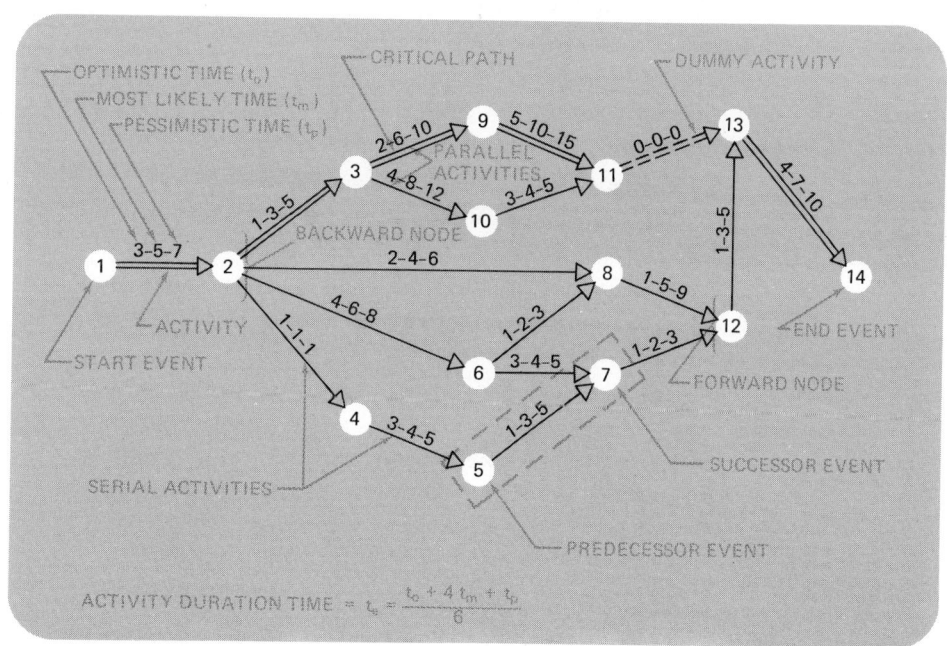

Figure 7-2 **SAMPLE PERT NETWORK**

178   THE SCIENCE OF ENGINEERING DESIGN

1 report termed PERT. This report outlined the theoretical basis for the technique and proposed a method of implementation. In September, after a revised (Phase 2) report, PERT was imposed upon the first Polaris contractors.

One of the first steps in the application of PERT to the Polaris program was the identification of major events which had to be met to keep the program on schedule and the activities necessary to cause those events to happen. The network was monitored bi-weekly to evaluate the program status.

PERT is now recognized by the Navy as the single most important feature in completing the Polaris program ahead of schedule. This is significant when one considers that the average weapons system program exceeds the predicted schedule by 36 percent.

GLOSSARY OF TERMS

The PERT network shown in Figure 7-2 is intended to give the reader a familiarity with this technique and the terms used in constructing the diagram. Definitions of the terms on the network diagram may be found in the glossary that follows.

ACTIVITY

A line on the network illustrating the dependency between two events and representing the work effort required to progress (indicated by arrows) from the predecessor to the successor event. The activity consumes a measurable amount of time and resources. There are two types of activities, real and dummy.

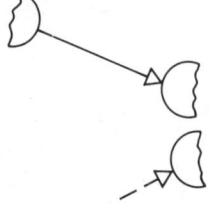

REAL ACTIVITY

An activity consuming time and resources which illustrates a dependency between two events. It is represented in the network as a solid line.

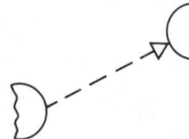

DUMMY ACTIVITY

An activity which consumes neither time nor resources but merely illustrates a dependency between two events. It is represented in the network as a dashed line.

ACTIVITY DURATION TIME (t_e)

$$t_e = \frac{t_o + 4t_m + t_p}{6}$$

A quadratic average of optimistic, most likely, and pessimistic time estimates.

OPTIMISTIC TIME

The minimum time estimate applied to an activity; the associated probability is 0.01 (one chance in 100) of being realized. t_o

MOST LIKELY TIME

A time estimate applied to an activity which, in the best judgment of the estimator, is most likely to be the time required to complete the activity successfully. The estimate does not have a probability of 0.5 but rather is a model estimate. t_m

PESSIMISTIC TIME

The maximum time estimate applied to an activity based on occurrence of unusually bad luck and only a slight chance of being realized. This time should reflect the possibility of initial failure and the requirement of a fresh start. t_p

ACTIVITY SPECIFICATIONS

A definitive statement of the work to be accomplished within an identified activity. The specification also includes the three estimates, the resources to be applied, and the person or authority responsible for the activity. (See Activity Specification Sheet later in the chapter.)

BACKWARD NODE

An event with multiple successor activities.

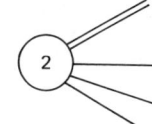

COMPLETE

A term implying successful accomplishment of an activity or the occurrence of an event.

CRITICAL PATH

That path of activities beginning at a start event and terminating at the end event which is the most time-consuming path of activities in the network; or that path of activities which has the least positive slack or most negative slack.

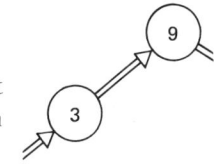

180   THE SCIENCE OF ENGINEERING DESIGN

DEPENDENCY

The PERT principle describing the constraint placed upon each specific event and activity by the requirement that all predecessor events and activities be completed prior to commencement of the next activity or the completion of the next event. Successor events are dependent upon their predecessor activities; every activity is dependent upon its predecessor event.

EVENT

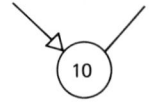

An instant in time at which the programmer can measure the plan against reality, usually the beginning point or conclusion of a specific work effort essential to the plan. The event does not consume time or resources. It is represented on the network as a square, oval, box, bubble, or other similar enclosure.

START EVENT

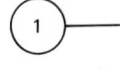

The beginning point of the plan which a network represents. Usually there is only one start event, but there may be several. Commonly the start event is entitled *now*.

END EVENT

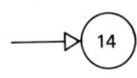

The last event in the network, the objective event, usually the project goal. There may be multiple end events, though the term usually implies a single project goal. End events must be identified as such to avoid the error of hanging events.

FORWARD NODE

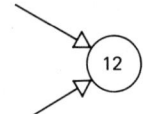

An event with multiple predecessor activities.

HANGING

A term applied to a networking error where an event or activity is left unconnected to the network on either the predecessor or successor side. Start and end events are hanging but are identified as such and not considered errors.

INTERFACE

An event in one network that connects directly to an event in another network; a point of dependency between two networks. The interface may be between networks or subnets of any indenture level.

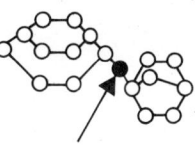

LATEST START

That point in time when an activity must start in order to be completed before the occurrence of its successor events' latest allowable time.

LEAD-TIME ACTIVITY

An activity representing waiting or delay time.

LEVEL OF INDENTURE

A term describing the level of detail or interest to which a network is drawn. The level of indenture for a subnet is lower than the indenture for a summary network.

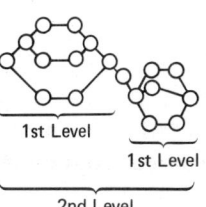

LOCALIZATION OF ESTIMATE

The principle of limiting consideration during time estimating for a given activity to that activity's specification and ignoring all others, especially those in parallel.

MILESTONE

In Gantt charting a point of accomplishment, a measuring point; also used in PERT in referring to events, it may imply a major program accomplishment as "milestone event."

NETWORK

A diagramatic representation of a complex plan, composed of events and activities.

NODE EVENT

An event which has multiple successor or multiple predecessor activities, or both; see backward node and forward node.

OVERALL NETWORK

A summary or time-phased network or a network drawn to the highest level of indenture.

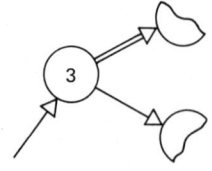

PARALLEL

Two or more activity paths which are in process at the same time.

PERT

Program Evaluation Review Technique.

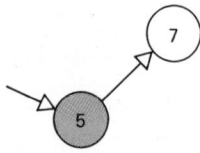

PREDECESSOR

A term referring to any event which is the origin point of a given activity. A predecessor event for one activity is the successor event for another activity.

RESPONSIBLE

The person, department, function, authority or group which assumes either the responsibility for an activity or serves as the reporting contact for the activity.

SCHEDULE DATE

T_S

A deadline date imposed on an event for any reason.

SEMI-CRITICAL PATH

S

A path of activities which is so nearly critical as to be essentially critical from a programmer's point of view. A slack path which could easily become the critical path.

SKELETON NET

A preliminary network; a network in its early stages of construction. It is also often applied to summary networks.

SLACK

The difference between the latest allowable time and the earliest expected time of an event, $S = T_L - T_E$. Slack may be positive, indicating an ahead-of-schedule situation, or negative, indicating a currently late situation.

POSITIVE SLACK

The slack resulting when the latest allowable time occurs after the earliest expected time, indicative of being ahead of schedule.

NEGATIVE SLACK

The slack resultant when the latest allowable time occurs before the earliest expected time, indicative of lateness.

SECONDARY SLACK

Slack computed from a schedule or earliest expected date at some specific event rather than from the end event.

SLACK PATH

A path of connected activities beginning at a node event and ending at a node event with a common value of slack.

SUBNET

A network of greater detail based upon or exploded from a portion of a network of lesser detail. Several subnets provide the detail support of a summary network.

SUCCESSOR

Any event which is the terminal point of a given activity. A successor event for one activity is the predecessor event for another activity.

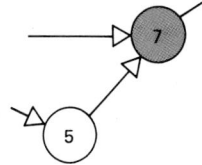

SUMMARY NETWORK

A network composed of events and activities summarized or distilled from subnetworks. Events and activities appearing in a summary network may be identical to those appearing in one or many levels of related subnets, or may be summary activities representing whole paths or even portions of a subnet.

SURPLUS PATH

A path of activities with significant slack, unlikely ever to become critical or even semi-critical.

TIME

The quantity (minutes, hours, months, or years) that measures activity duration, and whose accumulation results in an event.

EARLIEST ACTIVITY COMPLETION

T_A The earliest time an activity can be completed, derived by accumulating activity duration times along the longest path leading to the activity; when the activity terminates at a non-nodal event $T_A = T_E$; where a nodal successor occurs T_A may or may not equal T_E.

EARLIEST EXPECTED TIME

T_E The earliest time an event can be completed, derived by accumulating activity duration times along the longest time path leading to the event in question.

LATEST ALLOWABLE TIME

T_L The latest time an event may be completed without forcing a delay in the entire project.

EARLIEST START

S_E The earliest time an activity can start, equal to T_E at the activity's predecessor event.

SUCCESSOR EVENT	PREDECESSOR EVENT	t_e (DAYS)	T_E (DAYS)	T_L (DAYS)	SLACK
(END) 14	13	7	24 + 7 = 31	31	0
13	11	0	24	24	0
13	12	3	18 + 3 = 21	24	3
12	7	2	15 + 2 = 17	21	4
12	8	5	13 + 5 = 18	21	3
11	9	10	24	24	0
11	10	4	20	24	4
10	3	8	16	20	4
9	3	6	14	14	0
8	2	4	9	16	7
8	6	2	13	16	3
7	5	3	13	19	6
7	6	4	15	19	4
6	2	6	11	14	3
5	4	4	10	16	6
4	2	1	6	12	6
3	2	3	8	8	0
2	1 (START)	5	5	5	0

PATHS OF COMMON SLACK

0	3	4	4	6	7
1	2	6	3	2	2
2	6	7	10	4	8
3	8	12	11	5	
9	12		7		
11	13	SEMI-CRITICAL PATH			
13					
14	CRITICAL PATH				

Figure 7-3   **CRITICAL PATH ANALYSIS**

CALCULATING THE CRITICAL PATH

Figure 7-3 shows the tabulation required to calculate the critical path (activity path of least slack) for the PERT network of Figure 7-2. Successor events with their corresponding predecessors are listed in columns at the left of the table. The table lists events vertically and activities horizontally. Activity duration times 9(t_e) are now calculated by the following equation:

$$t_e = \frac{t_o + 4t_m + t_p}{6} \quad \left(\text{for event } \textcircled{14}-\textcircled{13} \; t_e = \frac{10 + 4(7) + 4}{6} = 7 \text{ days}\right).$$

The earliest expected time (T_E) of event completion may now be calculated by totaling activity duration times along the longest path leading to this event. Assuming time is zero at start event $\textcircled{1}$, the earliest completion time for event

② is 5 days. Earliest time for ③ is $t_e = 3$ from ② and $T_E = 5$ to ② from start or $3 + 5 = 8$. For event ⑧, $t_e = 2$ from ⑥ and $T_E = 11$ to ⑥ from start or $2 + 11 = 13$. For event ⑩, $t_e = 8$ from ③ and $T_E = 8$ to ③ from start or $8 + 8 = 16$. For event ⑫, $t_e = 5$ from ⑧ and $T_E = 13$ (longest path) to ⑧ from start or $5 + 13 = 18$. For event ⑭, $t_e = 7$ from ⑬ and $T_E = 24$ (longest path) to ⑬ from start or $7 + 24 = 31$.

If a schedule data (T_S) of 31 days is imposed on the project, we can say that the latest allowable time (T_L) for event ⑭ is 31 (see Figure 7-3). Since event ⑬ occurs $t_e = 7$ days earlier, then $T_E = 31 - 7 = 24$ days for the latest time that this event may happen.

Working backward along the network, as read from the chart of Figure 7-3, the latest allowable time for event ⑫ may be calculated by reasoning ⑬ takes 24 while ⑫ to ⑬ takes $t_e = 3$ less or $T_L = 24 - 3 = 21$. For event ⑨, ⑪ takes 24 while ⑪ to ⑨ takes $t_e = 10$ less, then $T_L = 24 - 10 = 14$. For event ⑦, ⑫ takes 21 while ⑫ to ⑦ takes $t_e = 2$ less, then $T_L = 21 - 2 = 19$. For event ⑥ there are two possibilities, ⑦—⑥ and ⑧—⑥: ⑦ takes 19 while ⑦ to ⑥ takes $t_e = 4$ less, then $T_L = 19 - 4 = 15$ and ⑧ takes 16, while ⑧ to ⑥ takes $t_e = 2$ less, then $T_L = 16 - 2 = 14$; therefore event ⑥ will take 14 days based on the shortest time of the two, indicating the shortest path to achieve the longest allowable time (most critical case). For event ③, ⑨ takes 14 while ⑨ to ③ takes $t_e = 6$ less, then $T_L = 14 - 6 = 8$.

The time difference between the latest allowable event time (T_L) and the earliest expected event time (T_E) is known as slack $(S = T_L - T_E)$. Slack is a quantity indicating how close to schedule an activity must be held to accomplish an event on time or ahead of time. Negative slack indicates a late situation and requires overtime for the program to be kept on schedule. Zero slack indicates an almost perfect situation in which no difficulties will be encountered and the program will progress according to schedule. Positive slack gives the program some breathing room in that the loss of a day or two (in the present example) will not affect the program's completion as originally planned. If events are now tabulated according to slack times, as shown in the paths of common slack of Figure 7-3, one can readily see that the path of zero slack is 1-2-3-9-11-13-14. This path of activities must command close attention by management and, if there is even the slightest indication of an activity not progressing according to schedule, additional resources must be brought to bear in order to keep that activity on schedule. This is the critical path. On the other hand, suppose there is an amount of time gained during this period in excess of four days. Then another path (path of 3 slack) will become critical and require attention to be diverted to those activities, 2-6-8-12-13.

MANAGING THE CRITICAL PATH

The management significance of the critical path is more than obvious. First choice of resources and primary attention should be directed toward the critical path since delay in any activity on the path will cause an equivalent delay in the project. Moreover, if the project is to be completed early, schedule-lightening must first occur on the critical path.

The following Description and Specification sheets are suggested as techniques for managing the critical path work effort. The examples included here are for a computer installation.

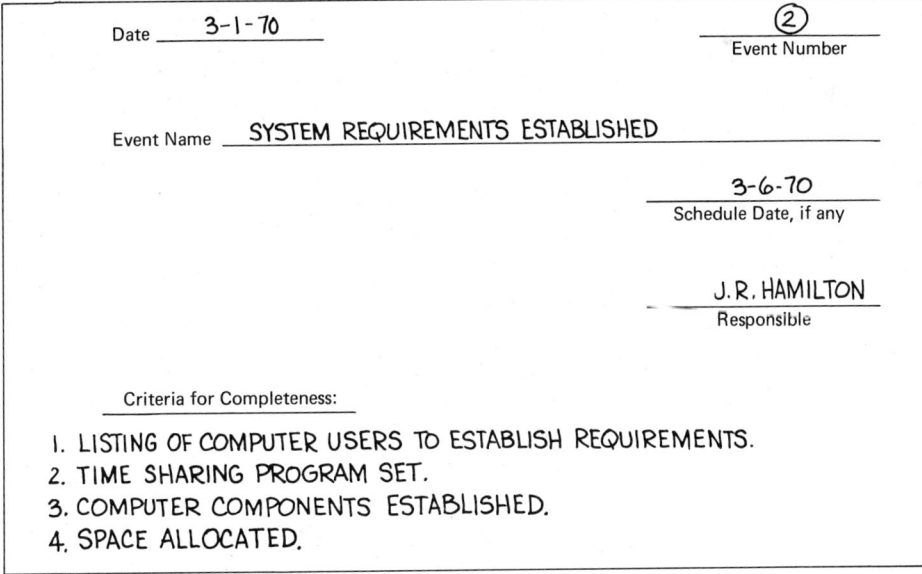

```
Date    3-1-70                                    ③
                                              Event Number

Event Name    COMPUTER ON SIGHT
                                              3-9-70
                                          Schedule Date, if any

                                              H. P. JONES
                                              Responsible

    Criteria for Completeness:

1. DATE OF ORDER SET.
2. TRANSPORT OF COMPUTER SET.
3. TIME OF ARRIVAL SET.
4. PERSON ON HAND TO RECEIVE DELIVERY SET.
```

```
Date    3-1-70                                    1-2
                                              Activity Number

                                              J. R. HAMILTON
                                              Responsible

    Activity Name    ESTABLISH SYSTEM REQUIREMENTS

    $t_o$   3 DAYS                                3-6-70
    $t_m$   5 DAYS                        Schedule Date of Successor
    $t_p$   7 DAYS                              Event, if any


    Specification of Work:                      PRIVATE CAR
                                          Resources to be Applied
1. INTERVIEW COMPUTER USERS TO ESTABLISH
   REQUIREMENTS — WORK FROM FEASIBILITY QUESTIONNAIRE.
2. VERIFY COMPUTER COMPONENTS.
3. CHECK WITH ADMINISTRATION ON SPACE ALLOCATION.
```

```
Date  3-1-70                          2-3
                                   Activity Number
                                  H. P. JONES
                                   Responsible

Activity Name  SCHEDULING OF COMPUTER DELIVERY

$t_o$  1 DAY                        3-9-70
$t_m$  3 DAYS                Schedule Date of Successor
$t_p$  5 DAYS                       Event, if any


Specification of Work:              TELEPHONE
                              Resources to be Applied
1. VERIFY DATE OF ORDER WITH I.B.M.
2. CHECK ON TRANSPORTATION OF COMPONENTS.
3. ESTABLISH AND CLEAR TIME OF ARRIVAL.
4. LOCATE RESPONSIBLE PERSON TO RECEIVE DELIVERY.
```

PREPARING A PERT NETWORK

the task

You are project director for a small company of design consultants dealing in innovative solutions to technical problems. Problem solutions provided by your engineering personnel usually take the form of a technical report and a feasibility prototype. Your company has recently been engaged by a large commercial airline to suggest methods of fast luggage-handling to and from the terminal. The system must handle passenger luggage from the check-in area to the airplane and from the airplane to the passenger at his destination. The system should be automatic if possible, harmless to luggage, and require a minimum of installation modifications. Management has agreed to accept the problem and has suggested that work begin on March 10, with a formal presentation to airline officials on July 16. Six design engineers have been assigned to the project and you as project director must prepare a PERT network so that the group will function efficiently, and so that each member understands the full scope of the program. You are also expected to have the program managed so that it will be completed on time.

outline of events

The first step in the preparation of a PERT network is to compile a list of events and arrange them in some logical order. This list is best prepared through an informal discussion with engineering and management personnel who are assigned to the project. These people should suggest events to the project director who lists them on a blackboard. The list is then revised until an outline of events results.

PRELIMINARY LIST

PROBLEM DEFINITION	IDEATION (INDIVIDUAL, BRAINSTORM, SINECTICS)	CONCEPT "E" COMPLETE	PREPARE FINAL REPORT
RESEARCH (AIRLINE)		SELECTION OF BEST CONCEPT	PRESENTATION TO CLIENT
RESEARCH (LUGGAGE)	P.E.R.T. ACCEPTED	CONCEPT ANALYSIS	~~FEEDBACK FROM CLIENT~~
~~RESEARCH (PEOPLE)~~	CONCEPT "A" COMPLETE	COST ANALYSIS	~~REDESIGN~~
TASK SPECIFICATIONS	CONCEPT "B" COMPLETE	MARKET ANALYSIS	
~~BUDGET PROPOSED~~	CONCEPT "C" COMPLETE	CONST. OF PROTOTYPE	
BUDGET ACCEPTED	CONCEPT "D" COMPLETE	TESTING OF PROTOTYPE	

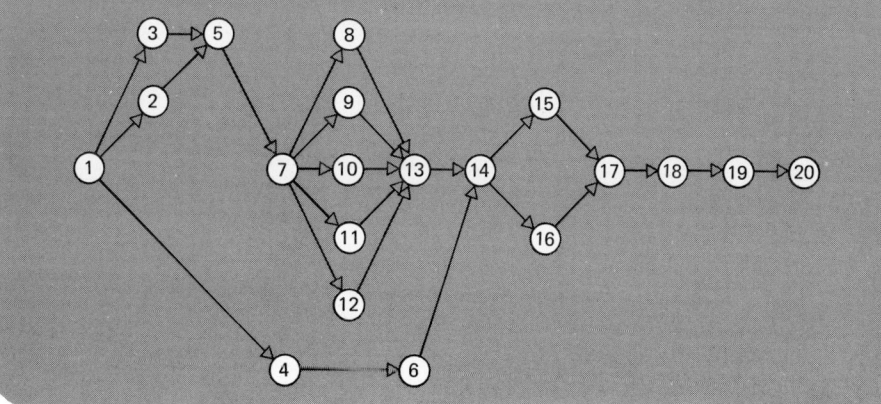

Figure 7-4 **SKELETON NETWORK**

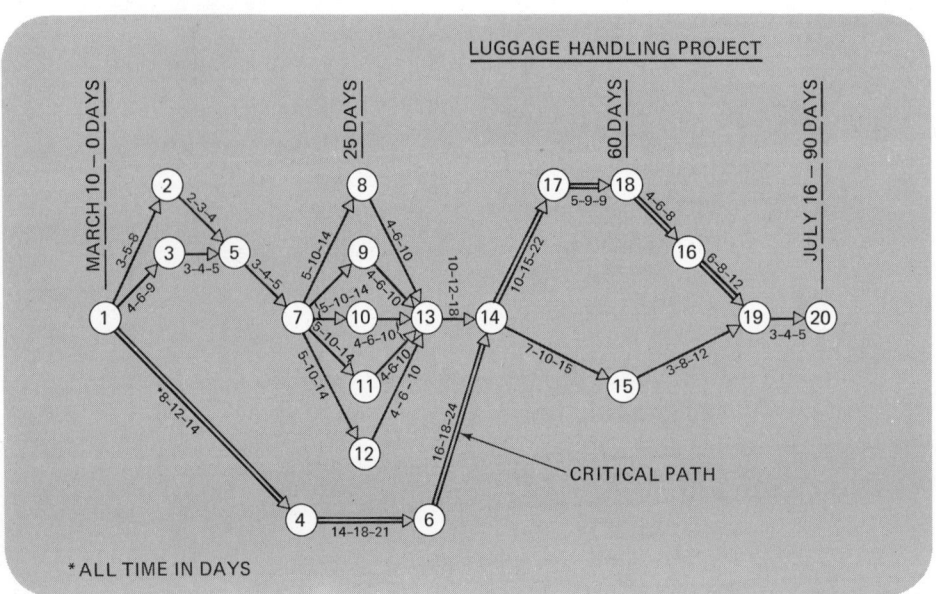

Figure 7-5 **FINAL NETWORK**

MARCH 10 TO JULY 16 = 90 WORKING DAYS

EVENT		t_o (DAYS)	t_m (DAYS)	t_p (DAYS)	t_e $\frac{t_o+4t_m+t_p}{6}$	T_E	T_L	SLACK (T_L-T_E)
SUCCESSOR	PREDECESSOR							
20	19	3	4	5	4	90	90	0
19	15	3	8	12	8	67	86	19
19	16	6	8	12	8	86	86	0
16	18	4	6	8	6	78	78	0
18	17	5	9	9	8	72	72	0
17	14	10	15	22	15	64	64	0
15	14	7	10	15	10	59	78	19
14	6	16	18	24	19	49	49	0
14	13	10	12	18	13	43	49	6
13	8	4	6	10	6	30	36	6
13	9	4	6	10	6	30	36	6
13	10	4	6	10	6	30	36	6
13	11	4	6	10	6	30	36	6
13	12	4	6	10	6	30	36	6
8	7	5	10	14	10	24	30	6
9	7	5	10	14	10	24	30	6
10	7	5	10	14	10	24	30	6
11	7	5	10	14	10	24	30	6
12	7	5	10	14	10	24	30	6
6	4	14	18	21	18	30	30	0
4	1	8	12	14	12	12	12	0
7	5	3	4	5	4	14	20	6
5	2	2	3	4	3	8	16	8
5	3	3	4	5	4	10	16	6
3	1	4	6	9	6	6	12	6
2	1	3	5	8	5	5	13	8

CRITICAL PATH (0 DAYS SLACK)
1-4-6-14-17-18-16-19-20

Figure 7-6 **CALCULATION OF CRITICAL PATH**

CPM/PERT FOR THE IBM 1130

An excellent way to understand more fully the scope of the example PERT network just explained is to process it on the 1130 computer. This is easily accomplished through a program entitled CPM/PERT FOR THE IBM 1130 FORTRAN CODED, CRITICAL PATH SCHEDULING WITH PROBABILITY ANALYSIS, as developed by John W. Burgeson, Senior Programming Systems Marketing Representative for the IBM Corporation. The program is Fortran coded for an 8K disk 1130 system with card reader and printer. Modification instructions are included to facilitate conversion to other hardware.

"Features of the program include random node numbering, both activity-oriented and event-oriented PERT reporting, simplified coding for ease of modification, maximum of 999 events, 1400 jobs, multiple start and ending nodes permitted, bar chart report, optional pre-set project completion date, and a network loop catching error routine." *

Input data and instructions to the computer are simply prepared and consist of the following cards (in part) with reference to the network described earlier: Five reports are printed as output from this program. The following are brief descriptions of each:

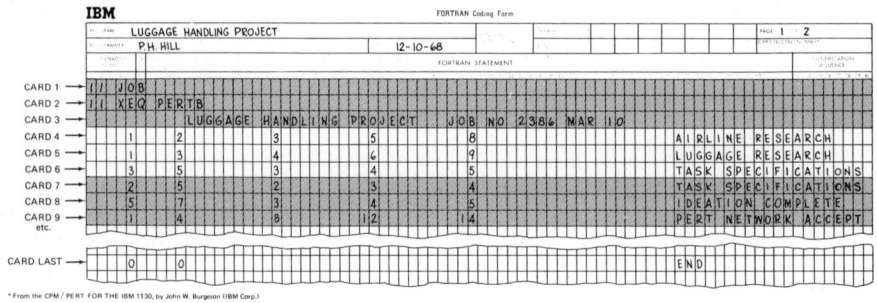

*From the CPM/PERT FOR THE IBM 1130, by John W. Burgeson (IBM Corp.)

"1. IBM 1130/1800 CPM/PERT ANALYSIS PROGRAM.
   This report is a listing of the data given to the program. The variable has been computed and attached, inconsistencies in the A, M, and B data resolved. The number of jobs, the total duration of jobs, and all beginning and ending nodes are also recorded here.
2. IBM 1130/1800 CPM/PERT OUTPUT REPORT.
   This report is the schedule of jobs within the project. The input data is repeated, the individual job variances, the scheduled times, and the floats are

---

* From the CPM/PERT FOR THE IBM 1130, by John W. Burgeson (IBM Corp.)

recorded. Critical jobs are so indicated by the letters 'CP'. Also recorded here are the total project cost and calculated project duration. If a project completion time was pre-set, this is printed along with the project slack (difference between the calculated duration and the pre-set duration).

3. PERT REPORT NO. 1. SIMPLE JOB PROBABILITIES.

This report is the job-oriented (activity-oriented) PERT analysis. Each job for which a time was given, and each job ending on the last node, are reported upon a five day spread.

4. PERT REPORT NO. 2. EVENT PROBABILITIES.

This report is the event-oriented (node-oriented) counterpart of the report above. The expected completion time of a node is the largest early finish time of all the jobs ending at it; more simply it is the earliest start time of all the jobs beginning at it. Again a five day spread is given for each number inputted.

5. BAR CHART OF THIS PROJECT.

This is a very simple report. Dummy jobs are not listed. The entire project is scaled to fit a range of 8.8 inches. For each job a series of asterisks and/or minus signs is printed. A series of asterisks indicates critically—or near critical. A series of minus signs indicates float time available." *

---

* John W. Burgeson, from the CPM/PERT FOR THE IBM 1130, (IBM Corp.).

```
PERT REPORT NO. 1.  SIMPLE JOB PROBABILITIES.
PROJECT TITLE -       LUGGAGE HANDLING PROJECT   JOB NO 2386 MAR 10

JOB    7    12 (CONCEPT E COMPLETE  ) HAS AN EF DATE OF   24.  ITS PROBABILITY OF COMPLETION BY DAY   23 IS   28.6 PCT.
                                                                                             BY DAY   24 IS   50.0 PCT.
                                                                                             BY DAY   25 IS   71.3 PCT.
                                                                                             BY DAY   26 IS   86.9 PCT.
                                                                                             BY DAY   27 IS   95.3 PCT.

JOB    7    11 (CONCEPT D COMPLETE  ) HAS AN EF DATE OF   24.  ITS PROBABILITY OF COMPLETION BY DAY   23 IS   28.6 PCT.
                                                                                             BY DAY   24 IS   50.0 PCT.
                                                                                             BY DAY   25 IS   71.3 PCT.
                                                                                             BY DAY   26 IS   86.9 PCT.
                                                                                             BY DAY   27 IS   95.3 PCT.

JOB    7    10 (CONCEPT C COMPLETE  ) HAS AN EF DATE OF   24.  ITS PROBABILITY OF COMPLETION BY DAY   23 IS   28.6 PCT.
                                                                                             BY DAY   24 IS   50.0 PCT.
                                                                                             BY DAY   25 IS   71.3 PCT.
                                                                                             BY DAY   26 IS   86.9 PCT.
                                                                                             BY DAY   27 IS   95.3 PCT.

JOB    7     9 (CONCEPT B COMPLETE  ) HAS AN EF DATE OF   24.  ITS PROBABILITY OF COMPLETION BY DAY   23 IS   28.6 PCT.
                                                                                             BY DAY   24 IS   50.0 PCT.
                                                                                             BY DAY   25 IS   71.3 PCT.
                                                                                             BY DAY   26 IS   86.9 PCT.
                                                                                             BY DAY   27 IS   95.3 PCT.

JOB    7     8 (CONCEPT A COMPLETE  ) HAS AN EF DATE OF   24.  ITS PROBABILITY OF COMPLETION BY DAY   23 IS   28.6 PCT.
                                                                                             BY DAY   24 IS   50.0 PCT.
                                                                                             BY DAY   25 IS   71.3 PCT.
                                                                                             BY DAY   26 IS   86.9 PCT.
                                                                                             BY DAY   27 IS   95.3 PCT.

JOB   17    18 (PROTOTYPE TESTED    ) HAS AN EF DATE OF   72.  ITS PROBABILITY OF COMPLETION BY DAY   58 IS    0.0 PCT.
                                                                                             BY DAY   59 IS    0.0 PCT.
                                                                                             BY DAY   60 IS    0.0 PCT.
                                                                                             BY DAY   61 IS    0.0 PCT.
                                                                                             BY DAY   62 IS    0.0 PCT.

JOB   19    20 (PRESENTATION FORMAT ) HAS AN EF DATE OF   90.  ITS PROBABILITY OF COMPLETION BY DAY   88 IS   28.1 PCT.
                                                                                             BY DAY   89 IS   38.6 PCT.
                                                                                             BY DAY   90 IS   50.0 PCT.
                                                                                             BY DAY   91 IS   61.3 PCT.
                                                                                             BY DAY   92 IS   71.8 PCT.


PERT REPORT NO. 2.  EVENT PROBABILITIES.
PROJECT TITLE -       LUGGAGE HANDLING PROJECT   JOB NO 2386 MAR 10.

EVENT  12 HAS AN EXPECTED COMPLETION TIME (TE) OF   24.  THE PROBABILITY OF COMPLETION BY DAY   23 IS   28.6 PERCENT.
                                                                                       BY DAY   24 IS   50.0 PERCENT.
                                                                                       BY DAY   25 IS   71.3 PERCENT.
                                                                                       BY DAY   26 IS   86.9 PERCENT.
                                                                                       BY DAY   27 IS   95.3 PERCENT.

EVENT  11 HAS AN EXPECTED COMPLETION TIME (TE) OF   24.  THE PROBABILITY OF COMPLETION BY DAY   23 IS   28.6 PERCENT.
                                                                                       BY DAY   24 IS   50.0 PERCENT.
                                                                                       BY DAY   25 IS   71.3 PERCENT.
                                                                                       BY DAY   26 IS   86.9 PERCENT.
                                                                                       BY DAY   27 IS   95.3 PERCENT.

EVENT  10 HAS AN EXPECTED COMPLETION TIME (TE) OF   24.  THE PROBABILITY OF COMPLETION BY DAY   23 IS   28.6 PERCENT.
                                                                                       BY DAY   24 IS   50.0 PERCENT.
                                                                                       BY DAY   25 IS   71.3 PERCENT.
                                                                                       BY DAY   26 IS   86.9 PERCENT.
                                                                                       BY DAY   27 IS   95.3 PERCENT.

EVENT   9 HAS AN EXPECTED COMPLETION TIME (TE) OF   24.  THE PROBABILITY OF COMPLETION BY DAY   23 IS   28.6 PERCENT.
                                                                                       BY DAY   24 IS   50.0 PERCENT.
                                                                                       BY DAY   25 IS   71.3 PERCENT.
                                                                                       BY DAY   26 IS   86.9 PERCENT.
                                                                                       BY DAY   27 IS   95.3 PERCENT.

EVENT   8 HAS AN EXPECTED COMPLETION TIME (TE) OF   24.  THE PROBABILITY OF COMPLETION BY DAY   23 IS   28.6 PERCENT.
                                                                                       BY DAY   24 IS   50.0 PERCENT.
                                                                                       BY DAY   25 IS   71.3 PERCENT.
                                                                                       BY DAY   26 IS   86.9 PERCENT.
                                                                                       BY DAY   27 IS   95.3 PERCENT.

EVENT  18 HAS AN EXPECTED COMPLETION TIME (TE) OF   72.  THE PROBABILITY OF COMPLETION BY DAY   58 IS    0.0 PERCENT.
                                                                                       BY DAY   59 IS    0.0 PERCENT.
                                                                                       BY DAY   60 IS    0.0 PERCENT.
                                                                                       BY DAY   61 IS    0.0 PERCENT.
                                                                                       BY DAY   62 IS    0.0 PERCENT.

EVENT  20 HAS AN EXPECTED COMPLETION TIME (TE) OF   90.  THE PROBABILITY OF COMPLETION BY DAY   88 IS   28.1 PERCENT.
                                                                                       BY DAY   89 IS   38.6 PERCENT.
                                                                                       BY DAY   90 IS   50.0 PERCENT.
                                                                                       BY DAY   91 IS   61.3 PERCENT.
                                                                                       BY DAY   92 IS   71.8 PERCENT.


BAR CHART OF THIS PROJECT
       LUGGAGE HANDLING PROJECT   JOB NO 2386 MAR 10

SCALE FACTOR - 1 INCH =   20 DAYS
START  END
NODE  NODE                         JOB  0    0    1    1    2    2    3    3    4    4    5    5    6    6    7    7    8    8
                DESCRIPTION             1    5    0    5    0    5    0    5    0    5    0    5    0    5    0    5    0    5    8
   1    2  AIRLINE RESEARCH             ***----
   1    3  LUGGAGE RESEARCH             ***----
   3    5  TASK SPECIFICATIONS          **----
   2    5  TASK SPECIFICATIONS          **----
   5    7  IDEATION COMPLETE            **---
   1    4  PERT NETWORK ACCEPT          ******
   4    6  BUDGET ACCEPTED                    *********
   7   12  CONCEPT E COMPLETE                 ******---
   7   11  CONCEPT D COMPLETE                 ******---
   7   10  CONCEPT C COMPLETE                 ******---
   7    9  CONCEPT B COMPLETE                 ******---
   7    8  CONCEPT A COMPLETE                 ******---
  12   13  BEST CONCEPT SELECT                       ***---            { * = CRITICAL OR NEAR CRITICAL TIME.
  11   13  BEST CONCEPT SELECT                       ***---            { - = FLOAT OR BLACK TIME AVAILABLE.
  10   13  BEST CONCEPT SELECT                       ***---
   9   13  BEST CONCEPT SELECT                       ***---
   8   13  BEST CONCEPT SELECT                       ***---
  13   14  CONCEPT ANALYSIS                             *******---
   6   14  CONCEPT ANALYSIS                             *********
  14   15  MARKET ANALYSIS                                     *****---------
  14   17  PROTOTYPE CONSTRUCT                                 *******
  17   18  PROTOTYPE TESTED                                           ****
  18   16  COST ANALYSIS COMP                                             ***
  16   19  FINAL REPORT PREPARE                                           ****
  15   19  FINAL REPORT PREPARE                                           *****---------
  19   20  PRESENTATION FORMAT                                                            **
```

PROJECT MANAGEMENT WITH PERT/CPM

It almost goes without saying that the PERT/CPM technique is an excellent scheme for managing a design program, product production, or process formulation. One of the most important aspects of any program in industry is the efficient and effective utilization of personnel and materials to get a job done on time. Of all management techniques known, PERT/CPM is probably the best-known aid to this end. In summary, PERT/CPM is intended to accomplish the following objectives:

1. PROVIDE ORGANIZATION FROM THE ONSET
    a. Force definition of goals.
    b. Promote detailed planning at the beginning.
    c. Provide a ready-made framework for documentation and presentation.

2. BE A PLANNING TOOL
    a. Segment planning problems to an easily-handled size.
    b. Force detailed planning.
    c. Require definition of all necessary work.
    d. Display relationship of each work effort to the entire plan.
    e. Promote consideration of alternate methods.
    f. Make resource deficiencies apparent.
    g. Aid in delegating and fixing responsibility.

3. BE A COMMUNICATION TOOL
    a. Standardize vocabulary, precise terms.
    b. Use pictures whenever possible.
    c. Explain status and changes quickly.

EXERCISES

1. Explain the difference between a Gantt chart and a PERT network.
2. How did CPM and PERT combine to form a unified technique for planning and managing complex programs?

3. Construct a PERT network to facilitate the planning required to bid and schedule work to contract to paint a two-story house. The network is to be constructed of the following events which are listed in random order:

    a. Preparation of surface complete
    b. Trim coat inspected
    c. Work scheduled in event of bad weather
    d. Removal of staging from job
    e. Workman's insurance set
    f. Contract signed
    g. Base coat complete
    h. Base coat inspected
    i. Final inspection
    j. Purchase of paint and brushes
    k. Final clean-up
    l. Ladders and staging at site
    m. Trim coat completed
    n. Overall job inspected

4. Estimate optimistic, pessimistic, and most likely times for the activities established in question three and calculate activity duration times (t_e).

5. Calculate the critical path for the network constructed in question three.

6. Construct an alternate PERT network for the example case involving the design of a luggage-handling system.

7. Write an Event Description Sheet for event (17) of the luggage-handling system project.

8. Write an Activity Description Sheet for activity (4)—(6) of the luggage-handling system project.

9. Considering yourself as as Project Director of the luggage-handling system project, what is your estimate of the probability of this program being completed on time? If the program cannot be completed on the exact date established, what is your best estimate (number of days) of the job being completed early or late?

10. List three advantages of CPM/PERT in addition to those listed just previous to the exercises.

8 VALUE ENGINEERING

"The value of a thing is based on the price it will bring."
Source: Raymond Corey, Graduate School of Business Administration, Harvard University.

Value engineering is a systematic, creative program using proven methods to get the same or better performance from a product or service at substantially lower cost. It is function-orientated as opposed to the well-known part-orientated, cost-reduction programs. VE is an organized effort to attain optimum value in a product, system, or service, by providing the necessary function at the lowest cost. VE does not lower the quality of the product.

VE is one of the best techniques available for increasing profits and meeting the challenge of competition. Reported returns as high as ten dollars and more for every dollar invested in a VE program are far above the average investment return of 10 to 15 percent in the business world. These returns are not easily achieved, for the methodology of VE requires training and a unique point of view. It requires enthusiasm and a belief in the philosophy of value engineering. A successful VE program must be based on broad understanding and acceptance of its techniques within the organization. It must be given its own corporate identity and backed by top level management. It must not, however, become the panacea which replaces conventional cost-reduction programs. VE can provide management with the additional tool of measuring cost effectiveness of all people concerned including engineering personnel who are generating a significant percentage of their total cost. VE can also provide reasonable insight into the design process.

No one has the time or the inclination to apply all the techniques of value engineering. In fact a number of these are beyond the scope of this book. However, the VE process will be examined closely and the engineering designer can gain some helpful short-cut methods of applying it.

THE VALUE ENGINEERING PROCESS

Although engineers are expected to apply the principles of value engineering both at the beginning and during the design process, it is the second look at the product to which this chapter is devoted. The engineer can only expect to approach optimum results in his design, since he is predominantly occupied with a workable solution to the design task. The VE specialist, being function-cost orientated, can take an outside look at the design product or service, thereby assisting the engineer in achieving an optimum end-product.

Figure 8-1 shows the process by which this can be accomplished. The process is shown by means of a curve plotted about the time axis that reaches the optimum VE solution at point ④. Even though the curve crosses this axis at other points, the optimum solution was not known at the time. Action that takes place when moving from one phase to another constitutes a search for ideas and an evaluation of those ideas. This assists the engineer or VE specialist in understanding the problem enough to reach the optimum solution. The VE process

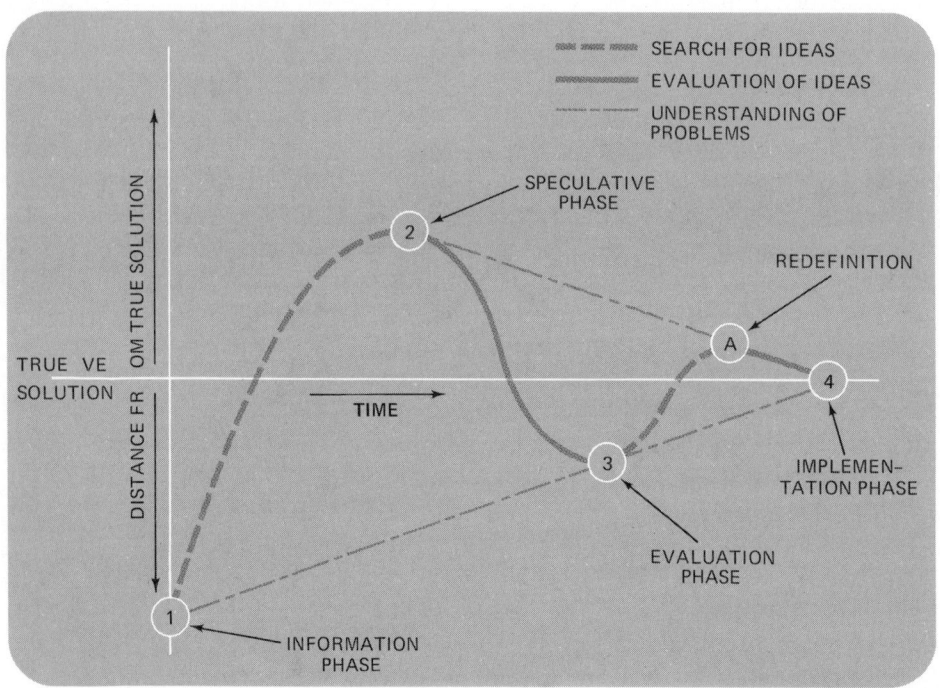

Figure 8-1 **THE VALUE ENGINEERING PROCESS**

curve is often interrupted by additional peaks such as (A), requiring a redefinition of any or all of the previous phases if additional or contradictory information is discovered.

information phase

This is the first phase of the VE job plan. The problem is defined in rather precise terms and all facts that relate to the problem are collected. These facts come from middle management to top-level thinking. Unless the VE specialist obtains more facts than anyone else has previously uncovered, he will risk untold opportunities. By dividing product or assembly into functional areas, the engineer can relate cost for each area or part to the function it performs. Here a price can be put on function by asking: What is the part? What does it do? What does it cost? What else will do the job? What will the alternative cost? Single out the individual items and use a systematic approach for searching and evaluating each to find the appropriate cost. VE, therefore, is an intensive study dealing with specifics, item by item.

speculative phase

This phase involves creative thinking where ideas are wanted in abundance. Once the function of the product has been clearly established and expressed in the simplest of terms, ideas take the form of alternate products, materials, or methods and are listed without judgment of their true worth. No barrier should be introduced to stifle creative thought.

evaluation phase

The third phase of the VE job plan is the evaluation and refinement of all ideas generated in the speculative phase. This refinement of ideas is done on an economical basis to obtain a functional, operating part within a true value perspective. Evaluate the cost of the main idea, place a value on the key tolerance and spend the company's dollars as you would spend your own. Use standard parts throughout if possible.

implementation phase

The fourth and final phase documents the VE recommendation and follows it through to completion. Costs are recorded, and savings and responsible

202   THE SCIENCE OF ENGINEERING DESIGN

authority are documented to insure that the recommendations will be carried out within reasonable time.

VALUE ENGINEERING CASE HISTORY

One of the best ways to gain a fuller understanding of the four phases of value engineering is through seeing their application to a real problem. The following case history is printed by permission of Microwave Associates and involves the value engineering of a switch body and cover. The assembly is used to provide a connection between conductors, thereby guiding microwaves and serving as a container to support or hold components. It is presently being manufactured in

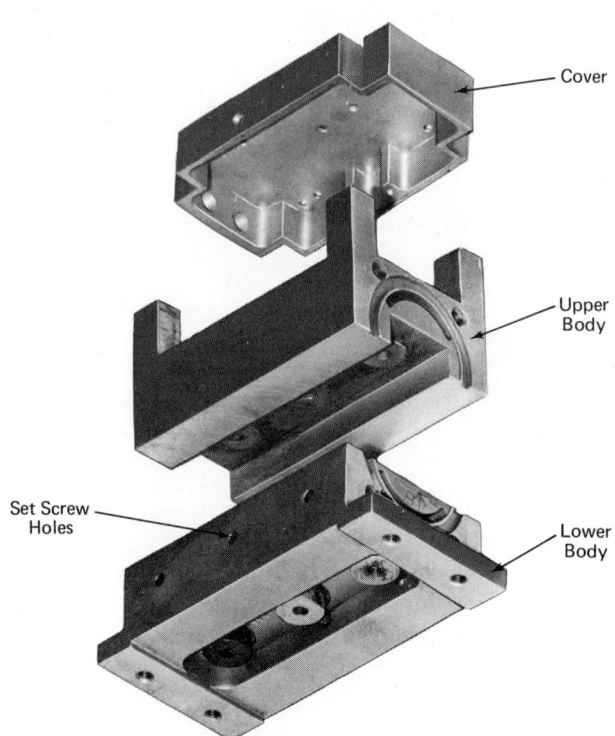

Figure 8-2   **INITIAL DESIGN OF SWITCH BODY AND COVER (BEFORE VE)**

three parts as shown in Figure 8-2 in order that the rectangular slot may be machined. The expense of fabricating the assembly is beginning to price the switch out of the market. The body is machined in two halves as follows:

1. Vendor machines two halves, partially.
2. Vendor packages and ships halves to Microwave Associates.
3. Microwave receives and inspects.
4. Microwave aluminum-dip-brazes halves.
5. Microwave inspects.
6. Microwave ships assembled body to vendor.
7. Vendor finishes machining and ships to Microwave.
8. Microwave receives and inspects.

The following pages are printed reproductions of forms used by VE specialists as they progress through the job plan in a systematic way, applying the value engineering process in an attempt to place the Microwave switch body and cover in a better market position.

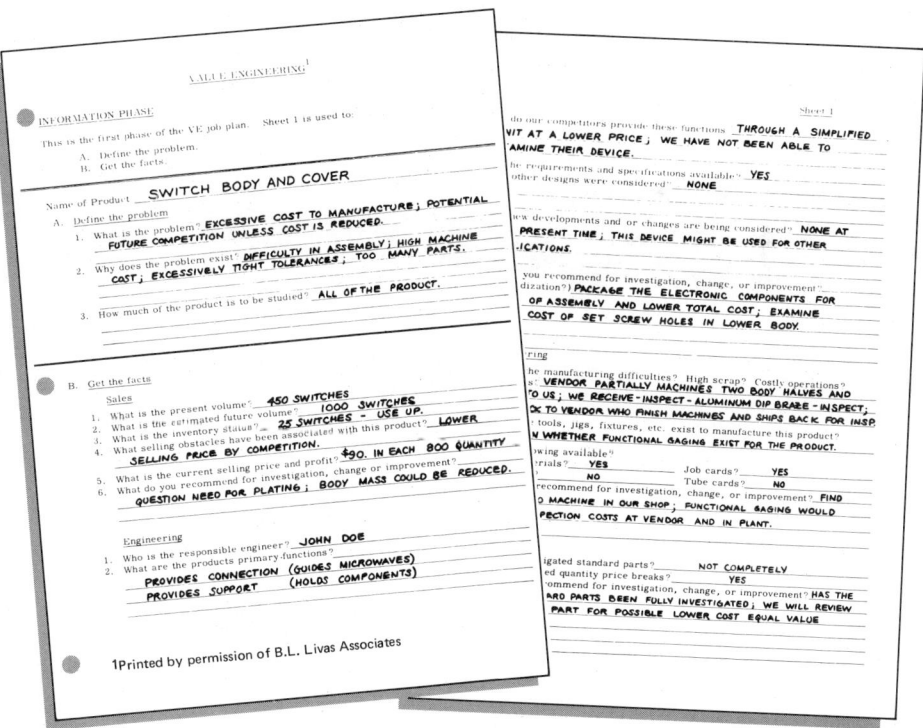

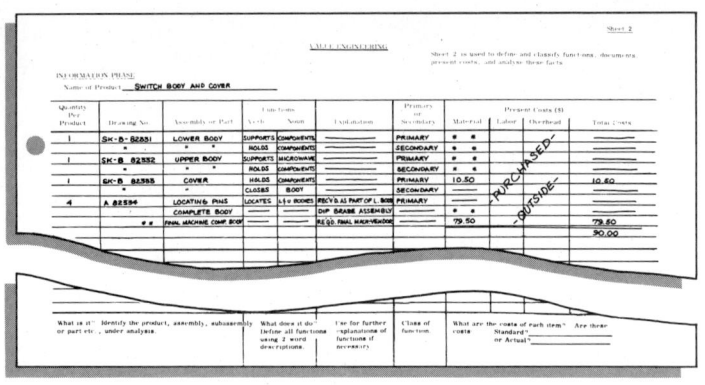

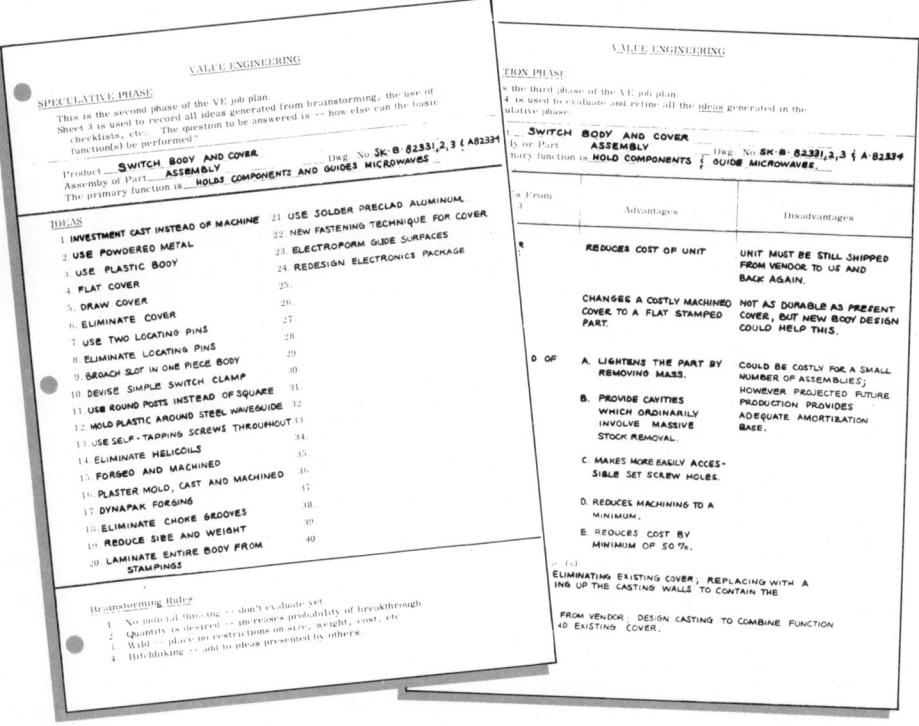

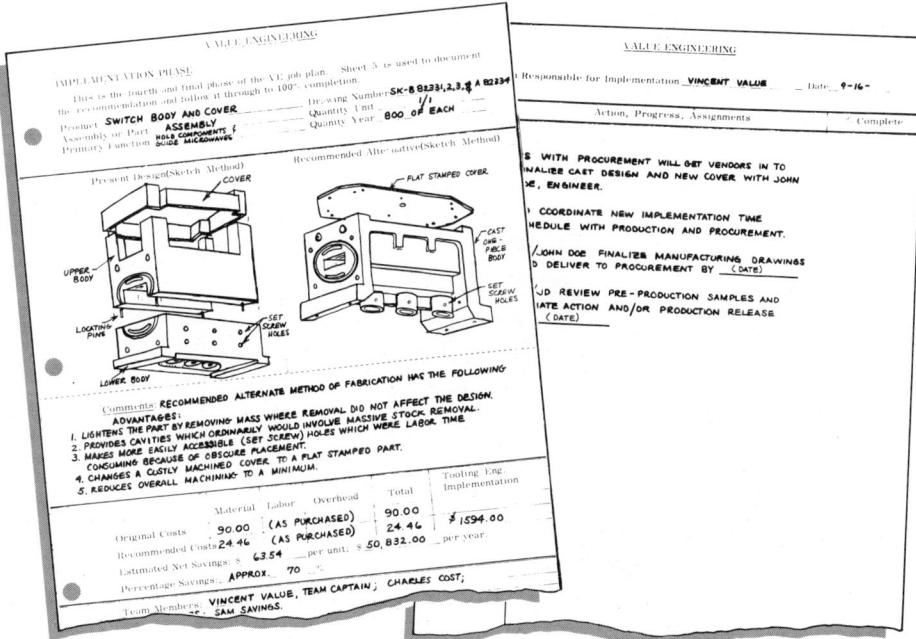

Figure 8-3 shows the switch body and cover constructed according to VE recommendations. The former design of the lower and upper bodies has been combined into a single investment casting with considerable mass removed. This design not only reduces appreciably the cost of the body but produces a part that is much more attractive. The cover is a single flat stamping on which electronic components may be easily assembled and then dropped into the body, whereas before these components had to be installed within the confines of the hollow cover. Relocation of the set-screw holes provides ease of assembly in the new design which will reduce labor cost. A summary of cost comparison between the initial and final designs shows a 70% reduction brought about through value engineering.

	Initial design		VE Design
COVER	$10.50		$ 1.16
BODY	79.50		23.30
IMPLEMENTATION COST			2.00
TOTAL (800 units)	$90.00	70% Reduction ⟶	$26.46

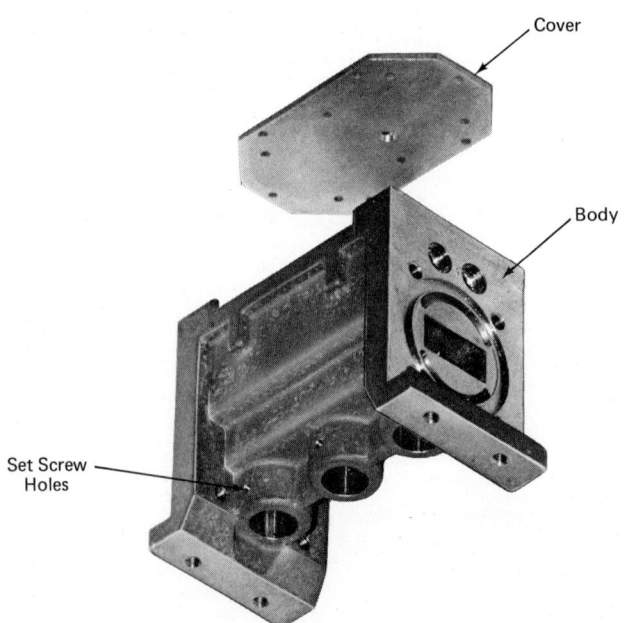

Figure 8-3 **FINAL DESIGN OF SWITCH BODY AND COVER (AFTER VE)**

PRICING AND THE BREAK-EVEN POINT

After discussing a technique of functional analysis whereby the cost of a product may be reduced while its value is retained, it seems only appropriate to consider the strategy involved in setting the price of a product. Much is written on this subject and there are often conflicting viewpoints relative to the theory, or art, of product pricing. Pricing can be defined simply as the technique of arriving at a selling price for a product or service that will interest the customer while returning the greatest profit. Pricing techniques are clouded, however, by the following questions.

    Will the price fit the competition?

    How many units do we expect to sell?

    Will materials, labor, or overhead increase during the product life cycle?

If it is a new product, what segment of the population or industry will it interest?

Who will buy the product?

One basic rule of pricing is to set the lower limit on the cost of the product (at the shipping-room door) and the upper limit on the competition. This is easier said than done, since it involves a great deal of forecasting on the part of marketing experts to translate into quantitative terms (price) the value of the product to the customer at a point in time.

Methods are available to assist management in arriving at the optimum price for a product—optimum in that the price will return the desired profit, sell an estimated number to a predicted population, and enhance the company's image. One such technique is known as a *break-even chart,* designed to show graphically profit-and-loss based on the selling price and manufacturing cost of the product.

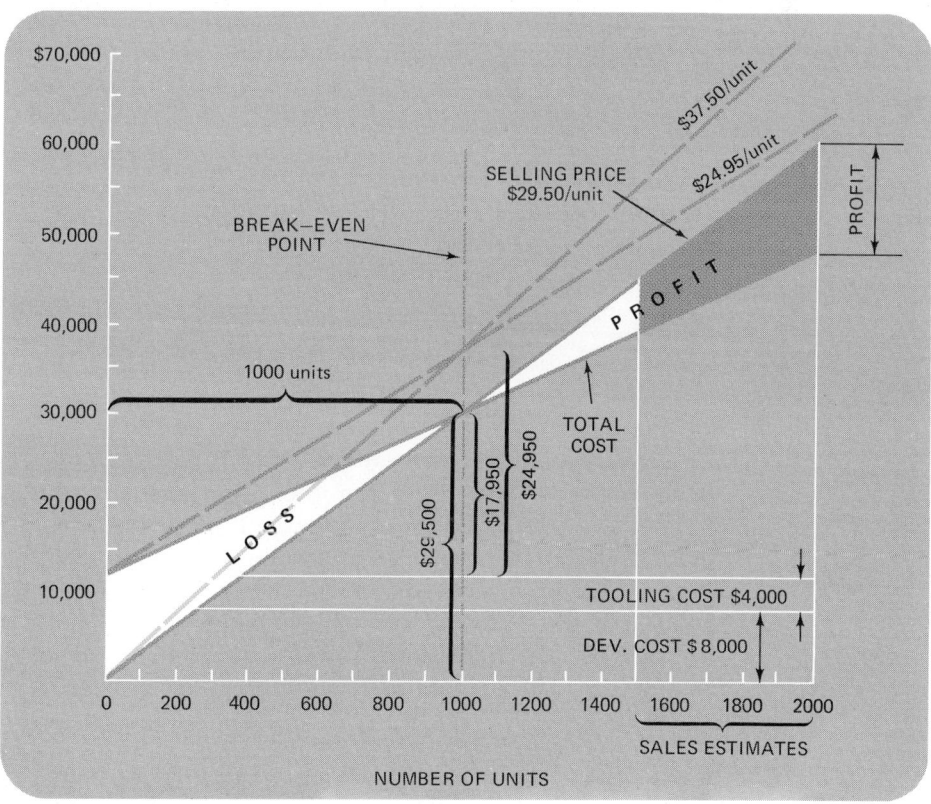

Figure 8-4 **BREAK-EVEN CHART**

Consider the break-even chart shown in Figure 8-4. The number of product units is plotted along the horizontal axis and dollars along the vertical. If the product development cost is $8000, it can be shown on the chart as a constant amount or fixed cost. This includes engineering, drafting, building models and prototypes, preliminary testing and other pre-production costs, exclusive of tools, plant overhead, and administrative services. Like development cost, a tooling cost of $4000 (representing new tools and expansion to produce the product) is plotted on the chart as a fixed amount. If the best sales estimate is between 1600 and 2000 product units and manufacturing cost is $17.95 per unit, the total cost line can be plotted on the chart. Consider now that a selling price of $29.50 per unit is established; this line can also be plotted on the chart, as shown in the figure. The selling-price line and total-price line will intersect at the break-even point, clearly showing a profit area to the right of this point and a loss to the left.

If, for some reason, manufacturing cost increased to $24.95 per unit (shown by dashed line), the product must be priced at $37.50 per unit to return approximately the same profit. The chart shows clearly the decrease or increase in profit by holding a fixed manufacturing cost and moving the break-even point to the left or right, thereby establishing a choice of selling prices based on estimated profits. The inverse is also true—using the chart to analyze manufacturing cost against profit. This can be seen by holding a fixed selling price and moving the break-even point to indicate change in this cost as well as change in the number of units that must be sold to return the required profit.

PRICING POLICIES FOR NEW PRODUCTS

Pricing policy on new products must be geared to the dynamic nature of their competitive status. New products are unique in pricing strategy in that they begin very slowly following invention and patent protection when the markets are still unexplored. Rapid expansion follows once market acceptance is gained. The product now becomes a target for competitive encroachment. The cycle begins again when the gap of distinctiveness between the product and its substitutes is narrowed through competitive innovation. Finally the new product fades into a pedestrian commodity which is so little differentiated from other products that the seller has limited independence in pricing.

Changes that occur in promotion and price elasticity and in costs of production and distribution throughout the cycle call for adjustments in price policy. Competitive pricing tactics over the cycle depend on the development of three different aspects of maturity which usually move in parallel time paths:

Technical maturity—Indicated by declining rate of product development, increasing standardization, and increasing stability of manufacturing processes.

Market maturity—Indicated by consumer acceptance and widespread belief that the product will perform satisfactorily.

Competitive maturity—Indicated by increasing stability of market shares and price structures.

Pricing problems begin when a company designs a product that is a radical departure from existing ways of performing a service and is temporarily protected from the competition by patents, trade secrets, or other barriers. Examples of this type of product are the ballpoint pen, Polaroid camera and film, Xerox copying, aluminum siding for dwellings, pre-cooked frozen foods, and the air turbine in dental drills. Major steps followed by the manufacturer in setting the price on a product during this pioneer stage include the following.

step 1: estimate of demand

The problem of estimating demand for a new product involves (a) whether the product will sell at all (assuming a competitive price), (b) what price range will make the product attractive to buyers, (c) what sales volume can be expected at various price ranges, (d) what reaction the price will produce in manufacturers and sellers of displaced substitutes.

step 2: decision on market targets

To decide on market objectives requires answering several questions: What ultimate market share is wanted for the new product? How does it fit into the present product line? What production methods should be used? What are the possible distribution channels? A basic factor in answering all of these questions is the expected behavior of production and distribution costs.

step 3: design of promotional strategy

Initial promotion expenses are an investment in the product and cannot be recovered until the market has been established. One basic strategic problem is finding the right mixture of price and promotion to maximize long-term profits. This can be accomplished by choosing a relatively high price during the pioneering stages, together with extravagant advertising and dealer discounts and recovering promotion costs early; or by using low prices and lean margins from the beginning in order to discourage potential competition, especially when the barriers of patents, distribution channels, or production techniques are inadequate.

step 4: choice of distribution channels

Placing a value on the cost of moving the new product from the factory through the channels of distribution to the final customer enters into the pricing procedure. These costs must cover the distribution costs of warehousing, handling, order-taking, profit, and so forth.

After considering those factors, the final strategic decision in pricing a new product is a choice between **A,** a policy of fixing a high initial price that "skims the cream" of demand, and **B,** a policy of fixing a low price from the beginning to serve as an active agent for market penetration.

A: skimming price

This technique of setting a relatively high price coupled with heavy promotional expenditures in the early stages of market development with lower prices at later stages has proven successful for new products that represent a drastic departure from accepted items. Examples of these products include the electric blanket, disposal (electric pig), and dishwasher, where the consumer is at first ignorant of their value as compared with the value of conventional alternatives. The theory is that initial high price is based on skimming the cream of the market that is relatively insensitive to price, giving the product exposure. Subsequent price reductions tap successively more elastic sections of the market. This pricing strategy is exemplified by the systematic succession of Polaroid cameras which started near $200 in price and can now be purchased for $16.50.

B: penetrating price

The opposite extreme to the skimming price technique is to set a low price on the product to allow for early mass-market penetration. This requires fairly large sales and once the volume is reached the price is gradually increased. Here quick sales to the many buyers at the low end of the income, or preference, scale is sought. This active approach in probing possibilities for market expansion by early penetration pricing requires research, forecasting, and courage. One example of this technique is *Playboy* magazine which sold for 50¢ per copy when first introduced. Following an initial exposure and much publicity (much of it free of charge), it has penetrated the market sufficiently to now sell for $1.00 per copy and has established a multi-million dollar business for its founder.

Once a product has reached maturity and cannot be revitalized through model changes, style, or attachments, this is evidence of deterioration and the first step for the manufacturer is to reduce the real prices promptly.

This price reduction does not mean price slashing or open price war, which is disastrous to any business. It means a reduction in price in order to keep the product returning a profit and keep the product market active as a stalling tactic so that product improvement may be studied. However, a product's price at the low end of the industry's real prices is usually associated with a product mixture showing less service or reputation. People still believe "you pay for what you get."

LINEAR PROGRAMMING

Linear programming is a recent tool of operations research in which a mathematical model is derived to represent a situation in which some optimum goal is sought, such as maximizing profit, minimizing cost, or minimizing materials. Mathematically it involves finding a solution to a system of simultaneous linear equations and linear inequalities which is optimized in a linear form. The equations are linear and their solution is referred to as the program. An example of the mathematical statement of a typical linear program problem follows.

$$\begin{aligned} \text{Minimize: } & X = 10A + 8B + 12C \\ \text{Where: } & A, B, C \geqq 0 \\ \text{Subject to: } & 6A + 9B + 7C \geqq 6000 \\ & 2A + 6B + 3C \geqq 3000 \\ & A + 5B + 2C \leqq 4000 \\ & A + B + C = 1500 \end{aligned}$$

One would expect linear programming to have begun around the year 1758 when economists began to describe economic systems in mathematical terms. In fact a crude example of a linear programming model can be found in the *Tableau Économique* of Quesnay, which attempted to interrelate the roles of the landlord, the peasant, and the artisan.

In spite of its wide applicability to everyday problems, linear programming was unknown until 1947. Dr. George B. Dantzig, part of an Air Force research group later given the title SCOOP (Scientific Computation of Optimum Programs), is credited with the invention of the linear program method. As with PERT/CPM, interest in this new area of operations research began to spread rapidly after June 1951 when the First Symposium in Linear Programming was held in Washington, D.C. Following this symposium, numerous applications were made in private industry.

The definition of a linear mathematical model and its optimum solution in a linear form can best be understood through the following example. Although the example is relatively simple and could be solved by either trial-and-error or marginal analysis, the knowledge acquired here should enable one to solve problems of a general and more complex nature.

A manufacturer of desk staplers and other stationery items produces two different stapler models. One is manual in operation, retails at $12.50 and returns $2.00 to profit; the other is automatic (uses a battery operated solenoid) and sells for $31.25, returning $5.00 to profit. The manufacturer is concerned with how to utilize his production facilities in order to maximize profits. One constraint placed upon the maximizing of profit can be written in the following form

$$P = 2Q_M + 5Q_A$$

where: P = Profit function,
Q_M = Quantity of manual stapler, and
Q_A = Quantity of automatic stapler.

Assume the manufacturing operation is such that labor skills and plant facilities are completely interchangeable and so varying quantities of manual and automatic staplers may be produced on the same day. Also assume that present demand is such that the company can sell as many as it can produce. There are, however, certain production restraints. Most of the major components are prepared on separate assembly lines prior to movement to the final assembly area. There is sufficient inventory of components for both staplers to satisfy daily production with the exception of solenoids for the automatic model, which has a daily production capacity of 200. This constraint may be written as $Q_A \leq 200$ and is plotted on the chart shown in Figure 8-5.

The shaded area under this line is the quantity that may be produced within the given limitation. Therefore, 200 automatic staplers is the maximum that may be produced per day; there is no limit on the number of manual models.

Among other duties of employees, fabrication and assembly require 18 man-minutes for each manual model and 54 man-minutes for each automatic stapler. The plant employs 45 men for this task on an eight-hour shift. Since there is only one shift, the capacity is $(8 \times 45 \times 60)$ or 21,600 man-minutes per day. This may be defined mathematically by

$$18Q_M + 54Q_A \leq 21{,}600.$$

The equation is now graphed in Fig. 8-5, narrowing the area of technical feasibility to that shown in the shaded area (in the form of a convex polygon).

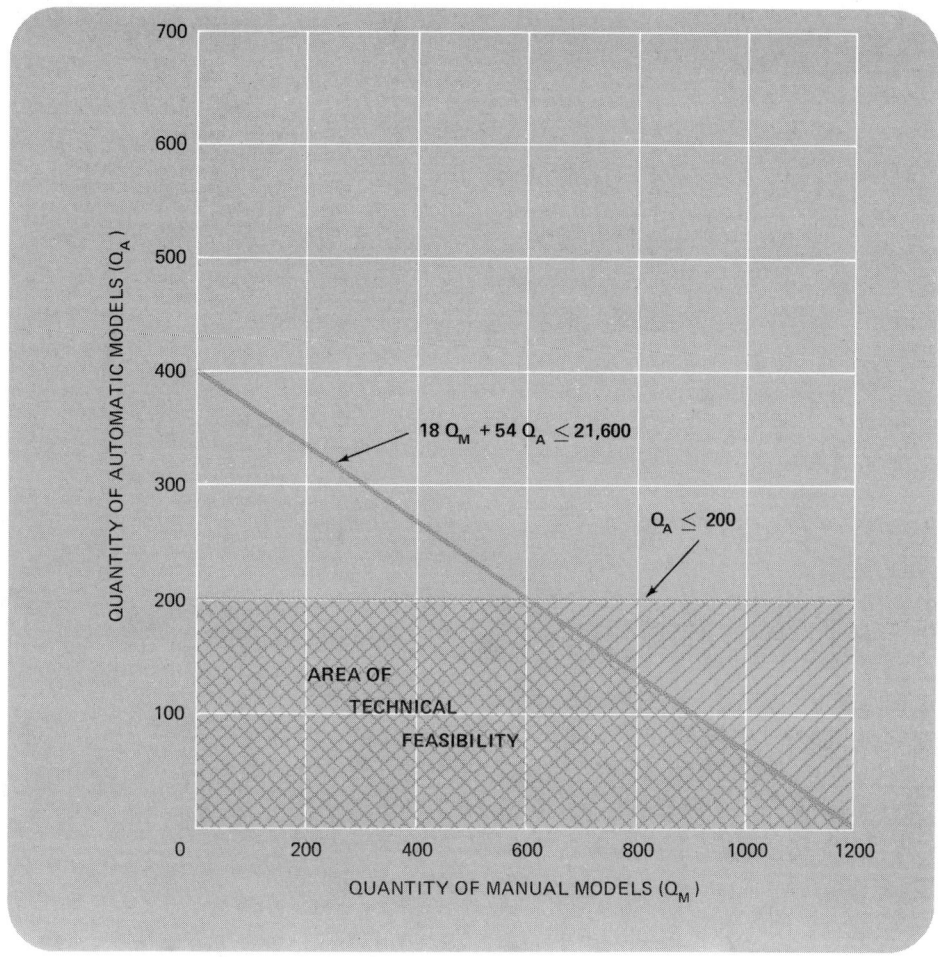

Figure 8-5 **LINEAR PROGRAM WITH SOLENOID SUPPLY AND ASSEMBLY CONSTRAINTS**

Final adjustment in the manual stapler requires 3 man-minutes on the assembly line, whereas the automatic model requires 5.4 man-minutes. Since this phase of assembly requires 6 men, the daily capacity is $(8 \times 6 \times 60)$ or 2880 man-minutes. This constraint is expressed as

$$3Q_M + 5.4Q_A \leq 2880.$$

Final inspection for workability requires 1 man-minute for manual models and 1.5 man-minutes for the automatic model. There are 4 half-time inspectors assigned to this task; therefore, the inspection capacity is $(8 \times 2 \times 60)$ or 960 man-minutes. The inspection constraint is

$$1.0Q_M + 1.5Q_A \leq 960.$$

Adjustment and final inspection constraints are now plotted on the chart as shown in Figure 8-6. It can be seen here that the adjustment constraint narrows down the area of technical feasibility while the final inspection constraint has no effect on maximizing costs and may be considered extraneous and no longer a boundary of the convex polygon.

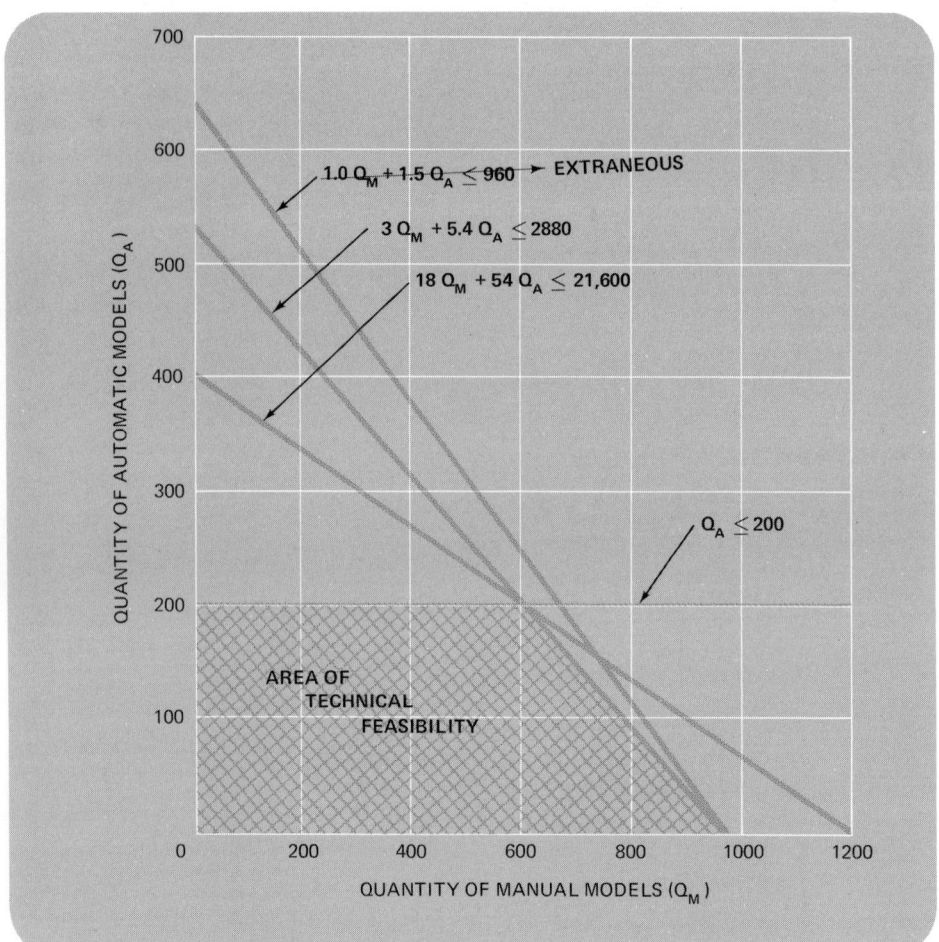

Figure 8-6 **LINEAR PROGRAM WITH SOLENOID SUPPLY, ASSEMBLY, ADJUSTMENT, AND INSPECTION CONSTRAINTS**

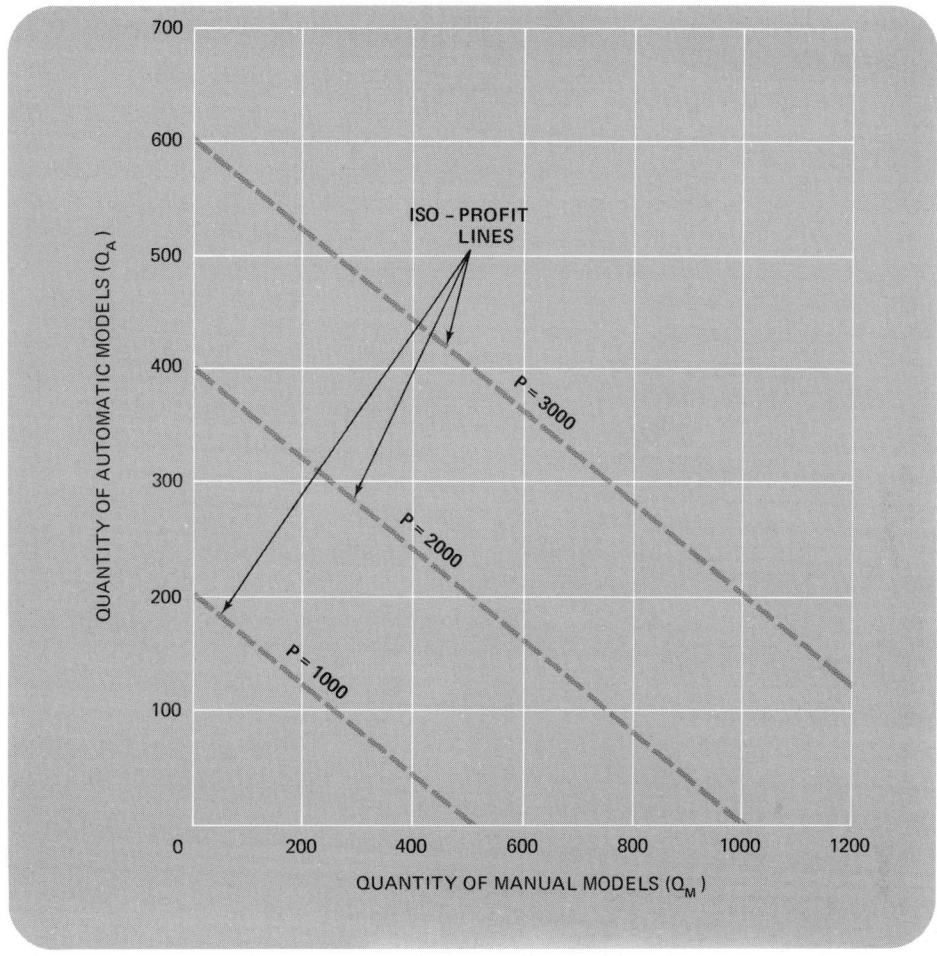

Figure 8-7 **ISO-PROFIT LINES**

We now realize that any combination of Q_A and Q_M within the envelope or area of the convex polygon of technical feasibility is a possible solution, but the optimum solution must satisfy the profit function defined earlier:

$$P = 2Q_M + 5Q_A.$$

This function is plotted as a series of parallel lines such that the value of P increases as the line moves farther from the origin, as shown in Figure 8-7. These lines are called iso-profit lines because the profit is constant at any point along a single line.

216 THE SCIENCE OF ENGINEERING DESIGN

The linear program is finally solved by determining which iso-profit line is farthest away and still contains at least one point that is technically feasible. Such a line is shown in Figure 8-8 and is derived graphically knowing the slope of the line. Thus, for maximum profit, point X is the point at which production should be carried out. Reading the chart of Figure 8-8, it is seen that the optimum solution would be to produce 600 manual staples and 200 automatic models per day at a resulting profit of $2200.

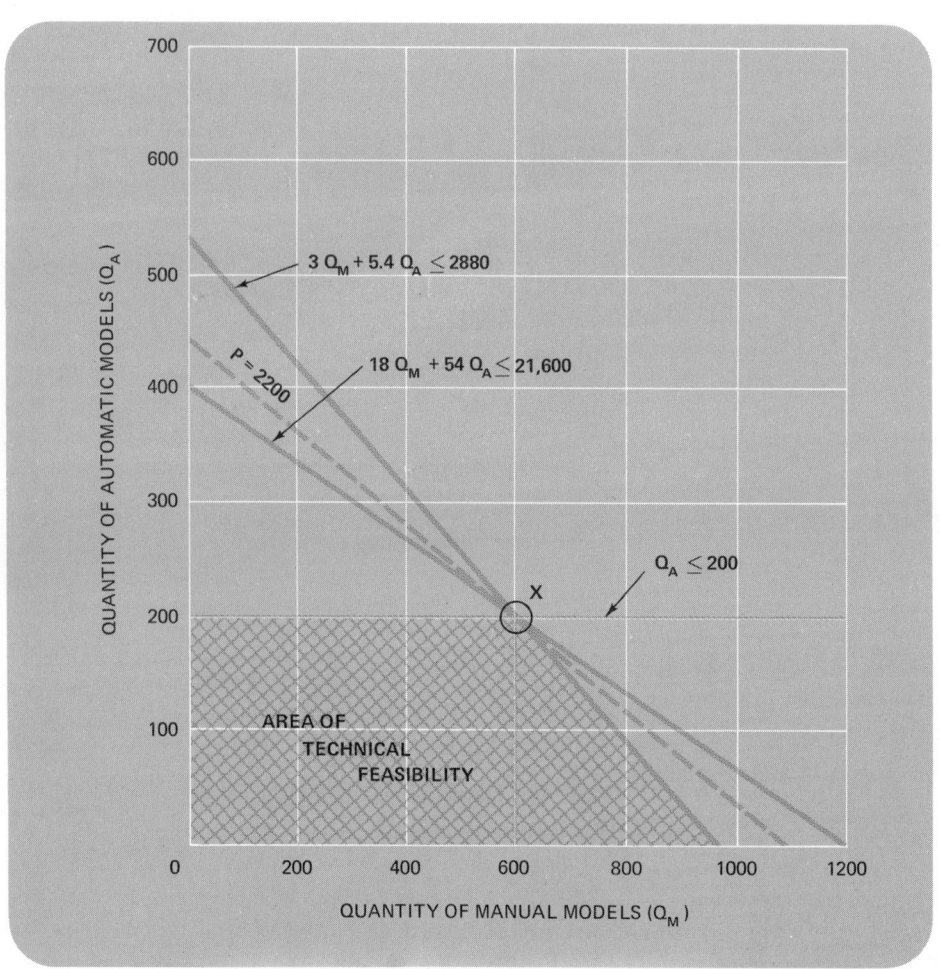

Figure 8-8   **SUMMARY LINEAR PROGRAM**

For purposes of review, the linear program to optimize profit is given below.

Maximize profit: $P = 2Q_M + 5Q_A$
Solenoid supply: $Q_A \leq 200$
Fabrication and assembly capacity: $18Q_M + 54Q_A \leq 21{,}600$
Final adjustment capacity: $3Q_M + 5.4Q_A \leq 2880$
Final inspection capacity: ~~$1.0Q_M + 1.5Q_A \leq 960$~~ → EXTRANEOUS

Although this program was used to solve a maximization problem, the technique works equally well in an attempt to minimize a solution. It can also be used by management in manipulating resources and personnel to gain additional profits, since the chart gives a clear picture of the interrelationships of variables that must be changed to increase the area of technical feasibility. This may be accomplished by managing certain constraints within their jurisdiction.

EXERCISES

1. Draw a parallel between the creative process as discussed in Chapter 2 and the value engineering process discussed in this chapter.
2. Considering the value engineering case history, what further cost reductions can you visualize involving the Microwave switch body and cover, as shown in its final form in Fig. 8-3?
3. Value engineer the simple paper matchbook consisting of cardboard cover, paper matches, a striking surface, and a wire staple. The end product should consist of a list of measures to reduce cost and maintain or even improve value.
4. List all the changes you can think of that must be levied against the price of a product before profit may be added on.
5. Outside of in-house cost, list five factors that influence the price of a product.
6. Consider a product you are involved with and estimate its development, tooling, and manufacturing costs. Assume sales estimates and a selling price. Plot a break-even chart that will show estimates of profit and loss based on the number of units sold.
7. Name two products that have reached technical, market, and competitive maturity.
8. Name one product that uses skimming price and another that uses the penetrating philosophy as a market introduction. Products listed should be other than those referred to in this chapter.
9. Consider that the constraint of final inspection of the linear program model (Figs. 8-5 to 8-8) takes the form of $2.0Q_M + 1.5Q_A \leq 960$ and all other relationships remain the same. What effect does this have on the ultimate optimum solution?

10. Write a linear program model to describe the manufacture of color and black-and-white TV sets so as to optimize profits based on the following data:

   a. Black-and-white sets retail for $198 and contribute $15 to profit and overhead.
   b. Color sets retail for $499.95 and contribute $45 to profit and overhead.
   c. The color cathode-ray tube is limited in production capacity to 50 tubes while all other components have no limits.
   d. Chassis assembly requires 6 man-hours for each black-and-white set and 18 man-hours for each color set. The plant employs 225 workers for an eight-hour shift to perform chassis assembly operations.
   e. A black-and-white set spends 1.0 man-hours on the assembly line, a color set 1.6 man-hours. There are 30 men on a single eight-hour shift assigned to assembly.
   f. Final inspection requires 0.5 man-hours for black-and-white sets and 2.0 man-hours for color. The plant employs 20 full-time inspectors and one part-time for 2 hours per day.

9 PATENTS, PRINCIPLES, AND PROTECTION

> *"Ideas are not inventions. Neither are patents. Ideas are the prelude to invention and are the tools of the inventor. They are not patentable. Patents are merely the legal documents that describe and claim inventions."*
>
> Source: D. W. Karger,
> Department of Management Engineering,
> Rensselaer Polytechnic Institute

"On April 10, 1790, President George Washington signed the bill which laid the foundation of the modern American Patent System. Three years earlier, at Philadelphia, the Constitutional Convention had given Congress the power to promote the progress of science and useful arts by securing for limited times to authors and inventors the exclusive right to their respective writings and inventions.

"For 180 years the Patent System has encouraged the genius of hundreds of thousands of inventors.

"It has protected the inventor by giving him an opportunity to profit from his labors, and it has benefited society by systematically recording every new invention and releasing it to the public after the inventor's limited rights have expired.

"The Patent Office has recorded and protected the telegraph of Morse, the reaper of McCormick, the telephone of Bell, and the incandescent lamp of Edison.

"It has fostered the genius of Goodyear and Westinghouse, of Whitney and the Wright Brothers, of Mergenthaler and Ives, of Baekeland and Hall."*

---

* "The Story of the United States Patent Office," U.S. Government Printing Office, 4th ed., Feb., 1965.

Under the Patent System, American industry has flourished. New products have been invented, new uses discovered for old ones, and employment given to millions.

Under the Patent System a small, struggling nation has grown into the greatest industrial power on earth.

The Patent System is one of the strongest bulwarks of democratic government today. It offers the same protection, opportunity, and hope of reward to every individual. For 180 years it has recognized, as it will continue to recognize, the inherent right of an inventor to his government's protection. The American Patent System plays no favorites. It is as democratic as the Constitution which created it.

COPYRIGHTS, TRADEMARKS, PATENTS

The *copyright* protects the writings of an author against copying. Literary, dramatic, musical, and artistic works are included within the protection of the copyright law, which in some instances also confers performing and recording rights.

There is no examination by the agency (the Library of Congress) for novelty or originality of the work. The copyright goes to the form of expression rather than to the subject matter of the writing. A description of a machine could be copyrighted as a writing, but this would only prevent others from copying the description—it would not prevent others from writing a description of their own or from making or using the machine.

Copyrights are registered in the Copyright Office in the Library of Congress; the Patent Office is not involved in any way with copyrights. An example of a copyright notice for books is shown in Figure 9-1. This notice usually appears on the title page and consists of three elements: the word "Copyright," the abbreviation "Corp." or the symbol ©; the name of the copyright owner; and the year of publication. Statutory copyright in published works lasts for 28 years from the date of first publication and may be renewed for a second 28-year term. Information concerning copyrights may be obtained by writing to Register of Copyrights, Library of Congress, Washington, D.C. 20504.

> Copyright ©, 1970, by HOLT, RINEHART AND WINSTON, INC., New York, N.Y. All rights reserved. No part of this book may be reproduced in any form, without permission in writing from the publisher

Figure 9-1   **COPYRIGHT NOTICE FOR BOOKS**

A *trademark* refers to the name or symbol used to indicate the source or origin of goods. Trademark rights will prevent others from using the same name or mark, but does not prevent others from making the same goods. A trademark used in interstate or foreign commerce may be registered in the Patent Office provided it has been used on the product itself or container prior to filing an application. Registered trademarks never expire although they may be lost through nonuse or misuse. Examples of trademarks attached to some successful products are shown in Figure 9-2. Such trademarks are indispensible in sales promotion and advertising by providing a visual image of either the product or the company in an easily remembered medium.

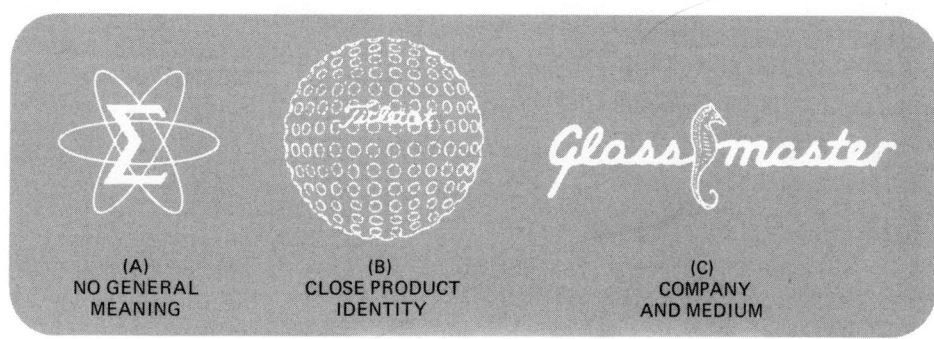

Figure 9-2 **TRADEMARK TYPES**

From a legal point of view, the best trademarks are those coined or made up as a fanciful combination of characters or symbols (see (A) in Figure 9-2) and convey no primary or genetic meaning. In this way the nature of the product can be flexible and still take advantage of the known mark under which it is sold. This ideal, however, is in direct contradiction to the type of trademark ordinarily preferred by advertising managers (see (B) in Figure 9-2).

Both the procedure related to the registration of trademarks and other information concerning trademarks are given in a pamphlet entitled "GENERAL INFORMATION CONCERNING TRADEMARKS." which may be obtained from the Superintendent of Documents, U.S. Government Printing Office, Washington, D.C. 20402.

A *patent* is a grant by the Government of the United States to an inventor of the right to exclude others for a limited time from making, using, or selling his invention in this country. It is a printed document in which the invention is fully disclosed and the rights of the inventor are defined. When an inventor secures a patent he has the opportunity to profit by manufacture, sale, or use of the invention in a protected market or by charging others for making or using it. In patents granted for invention of new processes, machines, manufactures, compositions of matter, or plants, the patent rights run for 17 years from the date the patent is granted and then become the property of the general public. A patent granted for the ornamental design for an article of manufacture can run to 14 years. For publications describing in detail information related to the U.S. Patent Office, refer to the Bibliography at the back of this book. Address inquiries to the U.S. Department of Commerce, Patent Office, Washington, D.C. 20420.

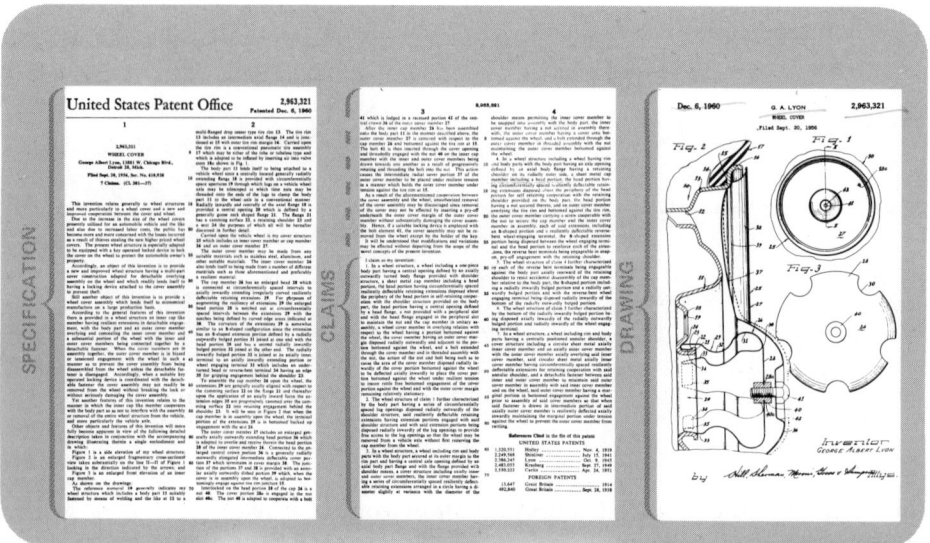

Figure 9-3 **PATENT APPLICATION**

The application shown in Figure 9-3 is an example of the document that must be filed to obtain a patent. In this case a patent on the described Wheel Cover was granted to G. A. Lyon on December 6, 1960, and recorded as patent number 2,963,321. The application consists of a *specification* of the invention which describes the object of the device, and precise statements disclosing its functioning and advantages, correlated with the drawing. All drawings must be made in accordance with specifications set forth by the Patent Office and symbols based on those specified or generally accepted standards. Both description and drawings must be such that they are readily understood by someone with reasonable skill in the field of the invention.

The third component of a patent application is known as *claims*. It is here that the patent is either allowed or disallowed. Claims are the legally enforceable part of the patent application and the ensuing patent when issued. Claims, when properly written, may be said to be the simplest and clearest overall definition of the invention consistent with the prior art that would be enforceable by the patentee or owner of the patent. Normally more than a single claim is solicited in an attempt to arrive at this perfection. Wording of the claims is of the utmost importance since they are the only basis for patent protection. They should be stated in the broadest of terms so it will be difficult or impossible for someone else to design around the idea. If even one part of each claim can be eliminated, another person may be able to produce a comparable device without infringement of the patent. For this reason claims are normally considered to be the sole responsibility of the attorney with advice from the inventor.

CONDITIONS OF PATENTABILITY

The fundamentals of the modern American Patent System are both simple and brief. "In general, any person who has invented any new and useful process, machine, manufacture or composition of matter, or any improvement thereof, may obtain a patent."

But the following conditions must be met for the invention to be patentable:

1. NEW—The invention must have a novel difference from any preceding object.
2. USEFUL—The invention must perform a function useful to mankind. It must not be frivolous, contrary to public policy, or inimical to the public welfare.
3. UNOBVIOUS—The invention cannot be something that is obvious to anyone having ordinary skill in the art at the time of the invention.
4. ORIGINAL—It must be the invention of the inventor (person) who applies for the patent.
5. NOT KNOWN OR USED IN THE UNITED STATES OR ELSEWHERE—The invention must not have been previously known or used in the United States, or described in any printed publication anywhere in the world prior to its inception.
6. NOT SOLD OR USED IN THE UNITED STATES—The invention must not have been in public use or on sale in the United States by others before the applicant made the invention or by the applicant more than one year prior to filing a patent application.
7. NOT PATENTED OR DESCRIBED—The invention must not have been patented or described in any publication anywhere more than one year before an application for patent is filed.

8. NOT ABANDONED—Ceasing to work on an invention for a time and then later finishing it would not be considered sufficient diligence in completing the invention to permit patentability.

9. PRIORITY OF INVENTION—A person is entitled to a patent unless the invention was made in this country by another who had not abandoned, suppressed, or concealed it before the applicant's invention. In determining priority of invention there will be considered not only the respective dates of conception and reduction to practice of the invention, but also the reasonable diligence of the one who was first to conceive and last to reduce to practice, from a time prior to conception by the other.

PROTECTION AND DISCLOSURE

An invention begins with its mental visualization or conception; however the conception must be complete and include the result as well as the means for bringing about the result. Because the conception is a mental process, it must be communicated to others who understand it before it can be proven satisfactory. The date of conception is the earliest date to which an inventor can be entitled for priority purposes. If the inventor can demonstrate reasonable continuous diligence in carrying out (constructing and testing) the conceived invention, for purposes of priority he may be considered as having made the invention when he began the continuous diligence. If this diligence began immediately after conception, then the date to which the inventor is entitled is the date on which the invention was conceived.

The act of transforming an inventive concept into physical reality (constructing and testing) is referred to as *reduction to practice.* The general rules of reduction to practice for the four most important classes of invention are as follows.

FOR A PROCESS—when it is successfully performed; this normally requires a test of results to demonstrate the success.

FOR A MACHINE—when it is assembled and tested or used.

FOR AN ARTICLE OF MANUFACTURE—when it is completely manufactured and tested or used.

FOR A COMPOSITION OF MATTER—when it is completely composed and tested or used.

Until the inventor discloses the invention he never provides a basis for obtaining a patent. He should, therefore, disclose his invention promptly in some form of writing, together with corroborating proof, normally by witnesses capable of understanding the technology of the invention, as to what the invention is, how it operates, when it was conceived, and when it was reduced to practice. The *disclosure* should include at least the following when appropriate:

WHY THE INVENTION WAS MADE—the problem it is supposed to solve.

WHAT HAS BEEN DONE BEFORE BY OTHERS ALONG THIS LINE—if known.

HOW THE INVENTION FUNCTIONS.

WHAT THE INVENTION CONSISTS OF—use a sketch, if practicable.

THE REAL ADVANTAGE OF THE INVENTION.

THE SIGNATURES, DATES, AND PLACE OF SIGNING IN RESPECT TO BOTH THE INVENTOR AND THE CORROBORATING WITNESSES.

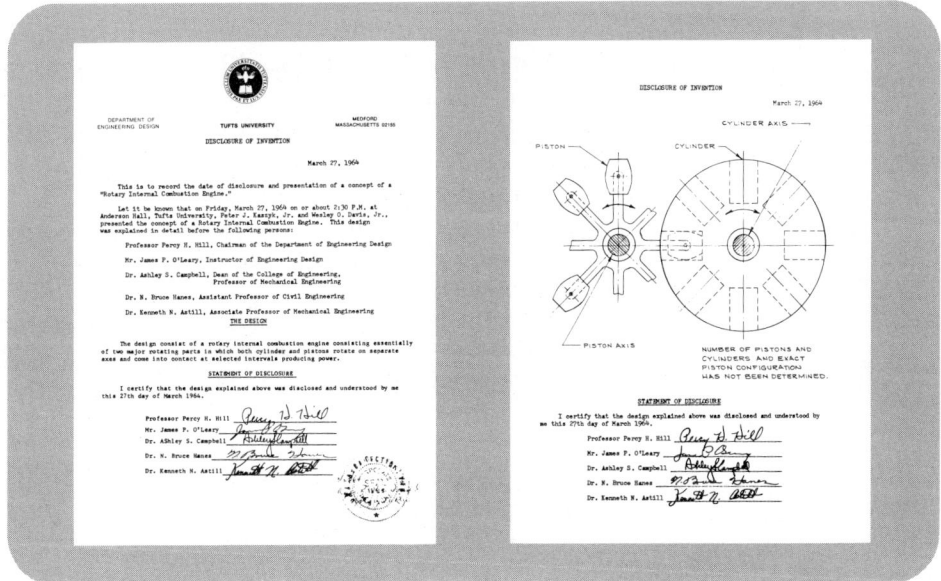

Figure 9-4 **DISCLOSURE OF INVENTION**

The disclosure of invention shown in Figure 9-4 contains most of the items listed above and is included here to give the reader a ready reference to the form used.

One of the conditions of patentability is to show that the invention was not abandoned at any time during its evolution. Reasonable continuous diligence can be demonstrated through accurate record-keeping. Such records are running accounts of the technical projects upon which the inventor is working and are often referred to as engineering notebooks. Essentially engineering notebooks should be kept on the pages of bound notebooks, with no pages being torn out, and the entries should proceed in daily chronological order, each entry for each day being signed and dated by the person recording his activities. The notebook should be a working notebook, not a "fair copy" written some days later. No blank spaces or pages should be left between entries. Periodically, if not daily, corroborating witnesses should sign and date the entries, the corroborating parties being those capable of understanding the technical aspects of the entries. They should be personally familiar with the work as it proceeds. Having reviewed the notebook, those witnesses should sign and date each page as illustrated below on a rubber-stamped line at the foot of the reading:

Witness: _____ Date: _____
Witness: _____ Date: _____

SELF-INTERROGATION BEFORE FILING

Although this chapter is dedicated to fundamental principles and advice on the obtaining of a patent, it is not always wise to file even some of the best ideas with the Patent Office. It is vital for an inventor to ask himself the following questions before taking any official or costly steps to obtain a patent.

SHOULD I TRY TO OBTAIN A PATENT?

The great majority of new ideas are marketed profitably without patent application. Often the get-in-and-get-out method will yield the larger profit. The securing of a patent is time-consuming and can be expensive.

IS MY INVENTION OR IDEA REALLY USEFUL AND NEW?

This requires some soul searching on the part of the inventor and the ability to be honest with oneself. Often ideas are thought good through the emotions of the inventor rather than by the opinions of his peers. A little preliminary market research could be useful here.

IF I DECIDE TO TRY TO OBTAIN A PATENT, WHAT STEPS CAN I TAKE TO SECURE THE BEST POSSIBLE PATENT PROTECTION?

WHAT STEPS CAN I TAKE TO IMPROVE MY
CHANCES OF DEVELOPING AND MARKETING MY
INVENTION SUCCESSFULLY?

BASIC STEPS TO OBTAIN A PATENT

Now that we know something about the Patent Office and have a fair understanding of what constitutes a patent and how it may be protected, the next step involves securing a patent. Assume that you are now ready to proceed with a patentable invention; the series of steps to follow is intended as a check list.

first step: make certain it is practical

Many persons believe they can profit from their inventions merely by patenting them. This is a mistake. No one can profit from a patent unless it covers some feature providing an improvement for which people are willing to pay. Therefore, try to make sure the invention will provide this kind of advantage before spending money in an attempt to patent it.

second step: witness, records, and diligence

1. Importance of Witnesses. A person will not be able to prove the date when he first conceived the idea of the invention to the satisfaction of the Patent Office or a United States court unless his own testimony is supported by one or more persons who have knowledge of these facts from firsthand observation.
2. Importance of Good Records. A person should prepare a record in the form of a sketch or drawing or written description promptly after he first gets the idea of an invention. He should ask one or more trusted friends to read and understand, then sign and date this record as witnesses.
3. Letter to Oneself is Not Protection. Many persons believe they can protect their inventions from later inventors merely by mailing to themselves a registered letter describing the invention. This is not true. One's priority right against anyone else making the same invention independently cannot be sustained except by testimony of a third party corroborating the inventor's testimony as to all important facts.

third step: the search

1. Why the Search is Important. If a person decides that his invention is valuable enough to patent, he should next make a careful search through patents already issued to learn if it is new as compared to those. This is important since it is less expensive to make a search than to try to obtain a patent. If it is found through the search that the invention cannot be patented, money can be saved.
2. Search in Patent Office Search Room. The search should be made in the Search Room of the Patent Office in Washington.
3. Get Help From Patent Office Roster. The Patent Office has a roster of all registered practitioners who are available to prepare and prosecute patent applications for inventors and someone from this roster may be chosen to make the search. A copy of the roster may be purchased from the Superintendent of Documents, U.S. Government Printing Office, Washington, D.C. 20402.
4. Searcher Will Furnish Estimate. The search to determine whether the invention is new is called a preliminary search, as it is preliminary to the possible preparation and filing of a patent application. The inventor asks the practitioner he has chosen to furnish an advance estimate of the cost of making such a search.
5. Explain Invention to Searcher. This explanation may be made through drawings or sketches, models, written description, oral discussion, or any combination of these.
6. Keep Correspondence for Evidence. All bills, pictures, letters, replies from inquiries regarding the invention, sound tapes, or a notebook of failures help to show diligence.

fourth step: studying patents found in the search

1. Study Search Results. The decision of whether to try to get a patent is primarily a business decision. It should be based on consideration of the practical advantages of the invention over the closest patents found in the search.
2. The person should ask himself again if his important features are new. Remember, no patent can cover old features.

fifth step: preparing the patent application

1. A patent application must include a written description.

2. Employ a registered attorney or agent.
3. Only registered persons may legally represent the inventor.
4. An application consists of a petition, a power of attorney, and an oath. Since these are very much the same in every application they are called the formal papers. Attached to these will be a description of the invention, called the specification, which ends with definitions of the invention, called claims. There will also be a drawing if the invention can be illustrated.
5. Importance of care. If the practitioner is not supplied with enough information to write a good specification and claims, the patent obtained may be so restricted that it has little value, or the right to obtain a patent may be lost.
6. Patent specification must describe the invention.
7. Do not limit a patent unnecessarily.
8. Critical importance of breadth of claims. The claims are the most important part of the patent application. They define the boundaries of the patent rights and fix the amount of protection granted by the patent.
9. Claims must distinguish the invention. The difficult job of the practitioner is to first find the features distinguishing the invention from earlier ones and then to prepare claims defining it in language broad enough to provide protection while still including one of the features which distinguishes it.

sixth step: patent office prosecution.

1. The Patent Office examiner's task. This search by the examiner is similar to the preliminary search already made, but the examiner's search will be more far-reaching in most cases.
2. Patent Office letter of rejection. The examiner will make an office action in the form of a letter in which he will reject your claims if he finds earlier patents or publications showing the features you are claiming. He will also reject them even though they include some new features if he decides the new features would be obvious to a person having skill in the field of the invention.

CLAIMS

It has been stated earlier that the claims part of a patent application is the most important element in securing and protecting a patent. The claims define the boundaries of the patent rights and fix the amount of protection granted to the inventor by the patent.

230 THE SCIENCE OF ENGINEERING DESIGN

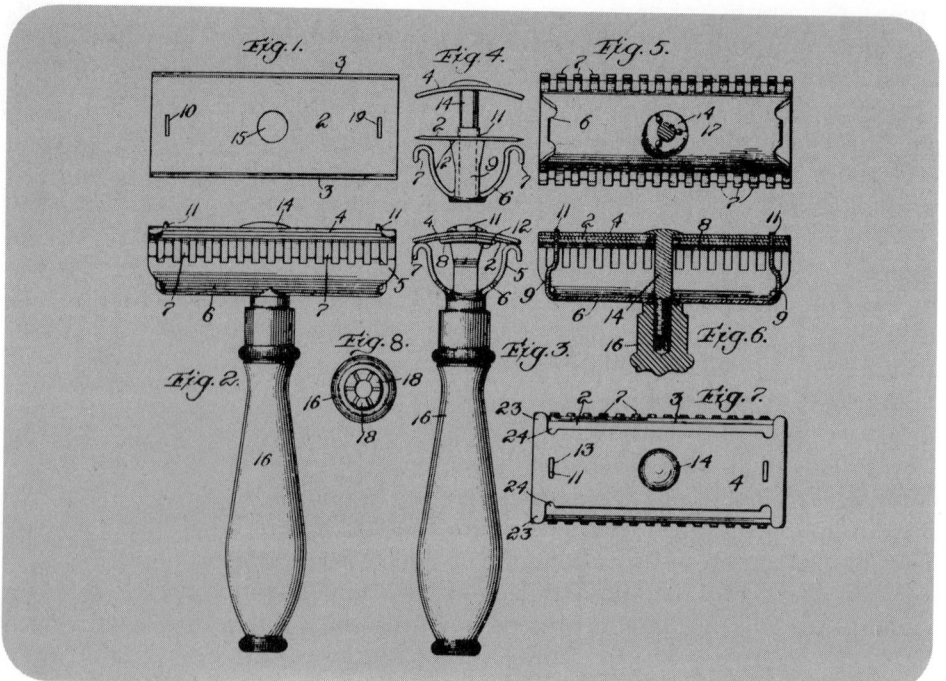

Figure 9-5 **GILLETTE'S SAFETY RAZOR PATENT NO. 775,134 (1901)**

Even though the patent attorney is responsible for the writing of claims, the inventor can be of great assistance in conveying claim facts to the attorney and in reviewing the explicitness of the written claims. One of the classic patents of all times was written by King Gillette in the year 1901 on the safety razor. He was able to write thirty claims on the invention as shown in Figure 9-5. Each claim was so carefully written that the patent protected the invention over a long enough period of time for Gillette to build a company that today ranks among America's industrial giants.

A number of claims from the Gillette razor patent are reproduced below with appropriate comments to give the reader a feeling for claim writing:

This series of claims plus 20 more protect every component of the razor from handle to backing to guard to blade. Notice the pattern of claims beginning to develop in the 10 listed: the transverse flexibility of the blade, with longitudinal edges, double, flexible sheet steel, thin, fitted to backing, backing bends blade, guard provides for blade adjustment, handle is fitted to backing.

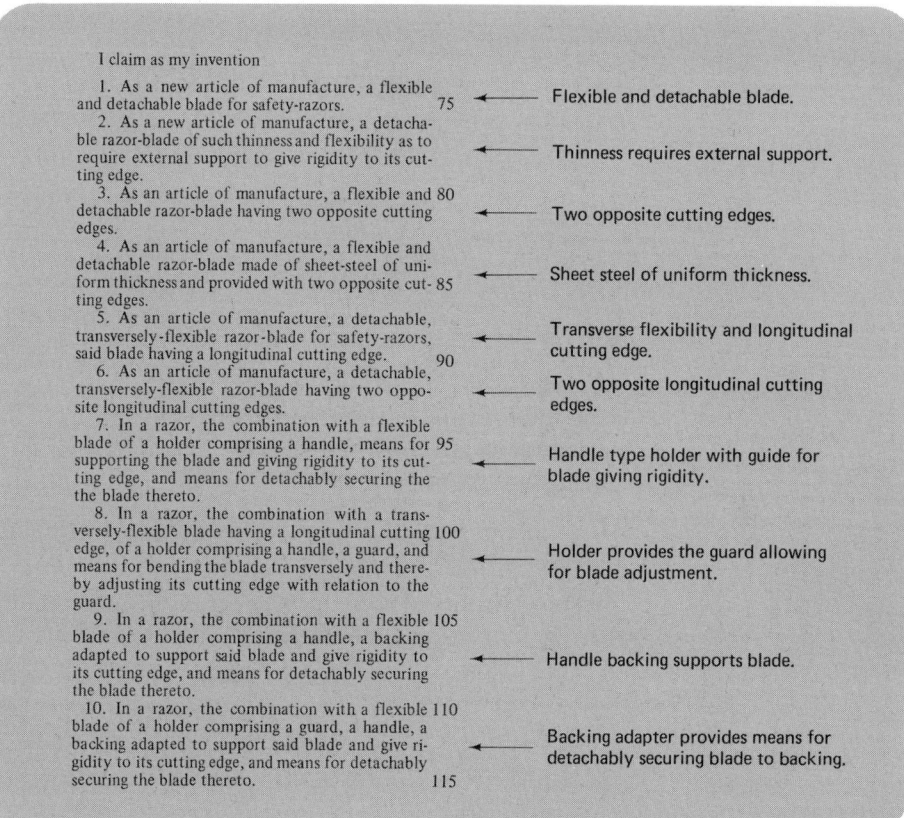

Another example of the writing of claims, although a bit crude, involves the assumption that there has been no previous invention of a three-legged stool.

Inventor A invents a three-legged stool, files an application, and a patent is issued showing the drawing in Figure 9-6.

Inventor A's patent contains but a single claim which reads: "I claim as my invention–A horizontal platform having a plurality of supporting vertical members."

It is obvious that someone wishing to manufacture, use or sell a four-legged chair would infringe upon the claim of A's patent, since in order to have four legs, a chair would have to have three (a plurality), thereby coming within the provision of A's patent claim.

232 THE SCIENCE OF ENGINEERING DESIGN

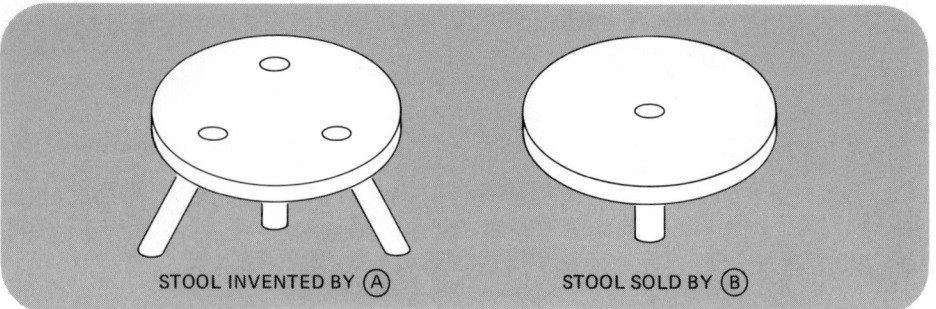

Figure 9-6 **PATENTED STOOL**

Consequently, even though A did not invent a stool or chair having four legs, his patent would prevent one from making, using, or selling such a device. One conclusion might be that A had a very good patent.

Mr. B makes and sells a stool having only a single leg (see Fig. 9-6) for the purpose of milking cows.

Inventor A immediately files an action of infringement against B alleging infringement of his patent claim: "A horizontal platform having a plurality of supporting vertical members."

In court, B asserts that his stool does not infringe since it has only a single vertical support while A's claim calls for a plurality of vertical supports.

The court decides in favor of B since A can only enforce his patent in what he has claimed.

PROPRIETARY INFORMATION AND TRADE SECRETS

Proprietary information may be considered confidential technical information, not generally available to the public, and giving one who uses it an advantage over competitors who do not know or use it. Such information includes:

TECHNICAL INFORMATION NOT PROTECTED BY PATENTS;

NEW ENGINEERING AND RESEARCH DEVELOPMENTS;

ENGINEERING PLANS RELATED TO TECHNOLOGICAL CHANGES;

NEW-PRODUCT INFORMATION PRIOR TO RELEASE;

UNIQUE MANUFACTURING PROCESSES;

PROCESSES OF MANUFACTURE USING UNIQUE MATERIALS;

ANALYSES, ESTIMATES, EVALUATIONS, PROPOSALS, AND SURVEYS FOR PRODUCTS;

CUSTOMER PROPOSALS DESCRIBING ITEMS OR FUNCTIONS NOT GENERALLY AVAILABLE TO THE PUBLIC;

PRICES, COSTS, ORDERS RECEIVED AND MISCELLANEOUS FINANCIAL DATA;

BUSINESS PLANS AND DATA ON THOSE PLANS.

Trade secrets are a type of proprietary information more frequently concerned with formulas, devices, or processes, sometimes known only by a group of individuals, each member of which knows only one particular facet of the information.

Proprietary information can be bought, sold, leased, or licensed. Normally this is brought about by a contract setting forth the conditions under which the information is to be used. Contracts of this type are fiduciary in nature, and violations may be far-reaching in their effects. Consequently, a breach of contract involving proprietary information is not regarded lightly. The unique characteristic of this type of contract, when breached, is that irreparable damage is done; that is, no retrieval of the information is practicable.

Even though most companies require an agreement to be signed by the employee legally in which he promises to refrain from revealing the organization's proprietary information to unauthorized persons, this information is sold under cover to competing companies every day. Most leading companies will pirate away individuals from their competitors with the bait of high salaries and liberal benefits to obtain information on special processes or new-product ideas. Industrial espionage is practiced among automotive firms prior to the unveiling of new models to the point that it is considered a farce today, and the "Big Three" are beginning to profit through clever advertising of the cloud of secrecy protecting their proprietary secrets. Soft-drink bottlers have been trying for years to learn the ingredients of Coca-Cola. Competition for outright searching of proprietary information and trade secrets is most predominant among cosmetics and the ladies' clothing industries.

The primary precaution to be taken in protecting proprietary information or trade secrets is to *give notice* that the information is confidential.

Once proprietary information is given improperly it is practically impossible to retrieve, and once proprietary information is received it is equally impossible to divest oneself of the stigma of having so obtained it.

PATENT INFRINGEMENT

Infringement of a patent consists of the unauthorized making, using, or selling of the patented invention within the territory of the United States during the term of the patent. If a patent is infringed upon, the patentee may sue for relief in the appropriate Federal Court. He may ask the court for an injunction to prevent the continuation of the infringement, and he may also ask the court for an award of damages. In such an infringement suit the defendant may raise the question of the validity of the patent, which is then decided by the court. The defendant may also declare that his actions do not constitute infringement. Infringement is determined primarily by the language of the claims of the patent, and if the defendant's product does not fall within the language of any of the claims, he is not infringing.

PROMOTING THE INVENTION

Assuming that the patent is obtained and one is an independent inventor, his next step is to profit from the invention. As stated earlier, there is no guarantee an inventor will receive a favorable market just because the Patent Office has agreed to recognize patentability. The inventor or his agent must persuade others to use the invention by pointing out its advantages. The Patent Office and other government agencies, however, can be of assistance in many ways, for example:

> Reliable promotional organizations may be obtained from the Better Business Bureau of the city in which the company is located, or from the Bureau of Commerce and Industry in the appropriate state.

> The inventor may obtain assistance in developing and marketing his invention from such local organizations as the Chamber of Commerce, banks, power companies, and railroads.

> Assistance may be obtained from one of the field offices of the U.S. Department of Commerce or the Small Business Administration.

> In nearly all states there are planning and development agencies or departments of commerce and industry that are seeking new products and new process ideas for manufacturers and communities within the state.

> Even though the Patent Office cannot advise in the development and marketing of an invention, it will publish a notice in the *Official Gazette* that the patent is available for licensing or sale. There is a small charge for this service.

The Business and Defense Services Administration of the U.S. Department of Commerce can often assist the inventor with information and advice, as its various industry divisions maintain close contact with all branches of American industry.

Addresses of the U.S. Department of Commerce field offices may be found in the publication *Patents & Inventions: An Information Aid for Inventors*, which may be obtained from the Superintendent of Documents, U.S. Government Printing Office, Washington, D.C. 20402.

EXERCISES

1. How is progress advanced when the granting of a patent gives the inventor the exclusive right to distribute or prevent others from distributing his invention?
2. Design either an illustration or slogan that one could identify with ball-point pens.
3. Consider yourself the inventor of the ordinary wooden pencil with eraser attached. Write a series of claims you feel would protect this invention.
4. Suppose you were to advise a friend on the patentability of his idea consisting of a shotgun projectile in the form of a 12-gauge ball (sphere) which is uniformly dimpled like a golf ball to provide for true flight. Would this idea be patentable and why?
5. Reviewing only one-third of the 30 claims written by King Gillette on the safety razor, try to write specifications for a razor that could legally compete with Gillette's razor.
6. Write additional claims to A's patent on the three-legged stool to protect the basic idea against B's infringement. Write any others you can think of.
7. Give an example of a situation in which one company acquired proprietary information or trade secrets which may have resulted in a loss to another company.
8. As an executive in a small industry, how would you propose to protect proprietary information and trade secrets?
9. Considering yourself an independent inventor who has received a patent on a new product in the consumer line, outline a plan for developing and marketing the invention through licensing or independent sale.
10. In recent years there has been much criticism levied against the U.S. Patent Office in its interpretation of the laws governing patent matters. Some say it is becoming increasingly difficult to obtain a patent, to convince the Patent Office that the invention is really new. Patent protection is far more often given to intellectual concepts than to their reduction to practice, and there is little to no protection given to design specifications. With what knowledge you have gained in reading this chapter, write a brief critique of the U.S. Patent System and discuss what reform(s) is/are needed if any.

10 THE DESIGN CRITIQUE

next monday

"A primary function of engineers is to generate new ideas. But this is not a limit to their job. Although usually untrained in the arts of persuasion, engineers must convince the people who hold the purse strings."

Source: Eugene Raudsepp,
Director of Psychological Research,
Deutsch & Shea, Inc., New York.

Unlike the advertising executive, the banker, the average businessman, or the industrial manager, the engineer is ill-equipped to sell his ideas. More often than not the engineer is represented by company officials in dealings with the client. These officials are usually not well-versed in engineering principles and must relate the client's questions and remarks back to the engineer. There can be (and often is) valuable information lost through such a second-hand exchange.

Unfortunately the engineer receives little to no training in the art of persuasion. He is a problem solver, a good listener, not active in community affairs, ill-at-ease when asked to speak before a group, and generally a loner. Organizations have been established to provide what might be considered the art of customer contact. This contact eventually results in the sale of new-product ideas, whether it be a new airplane design for the military or a self-sharpening lawn mower blade. Few engineers break into this inner circle, and those that do have abandoned engineering work for management positions.

This chapter is written for the practicing engineer who has novel ideas and the desire to have them accepted. It presents suggestions on how ideas may be brought to the attention of management or the client and how they may be persuaded to take a receptive point of view.

CRITICISM

Harsh criticism levied on good ideas is a most effective way of crushing them, as illustrated in Figure 10-1. All ideas must be verified in some way, and the

Figure 10-1 **CRITICISM**

engineer needs to discuss them with his colleagues in order to get information feedback. As illustrated in Figure 10-1, the individual with ideas usually feels hurt when his thoughts are squashed through a few well-known expressions. He immediately takes on a defensive attitude, and usually becomes silent when new ideas are requested. Unfortunately it is much easier to criticize than to praise. This is not to say that all ideas must be praised, but the critic has a responsibility to the originator of the idea to give credit for good ideas and constructive criticism for those that are not so good.

There are two types of critics in our society that the engineering designer must insulate himself against. The type 1 critic automatically says the idea is no good before the originator has finished explaining it. The type 2 critic tries to "one-up" all ideas. He goes it "one better". Although he seldom comes up with new ideas of his own, there seems to be no idea that he cannot improve upon. Unfortunately most of us at one time or another fit both categories. It is easier to think down than to think up.

Like the creative process discussed earlier, mental effort of a high order and some self-discipline are required to give just and constructive criticism. On the other hand, individuals who wish objective idea-feedback should seek out critics who will give reliable information and stay away from type 1 and type 2 critics.

DESIGN REVIEW

The design review is a well-organized, well-conducted session in which disciplines and specific groups involved with the evolution of a new product meet to exchange ideas related to optimizing the product. Reviews should be scheduled at regular intervals during phases of new-product design and development. Figure 10-2 shows the stages of new-product evolution from its beginning in an aid to its realization as a product with appropriately scheduled review sessions.

It is suggested that at least four scheduled design reviews be conducted at the end of each stage as discussed below.

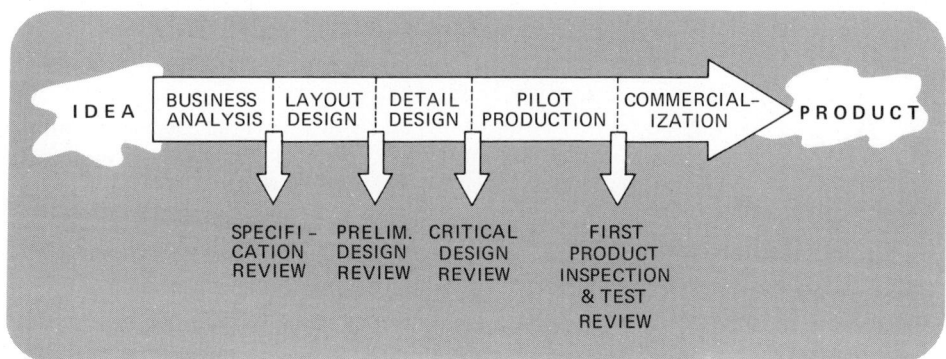

Figure 10-2 **CRITICAL STAGES FOR DESIGN REVIEW**

specification review

This review follows business analysis and draws upon the company's research and financial commitment to react to specifications written by the design team.

preliminary design review

Following conceptualization and the narrowing down of the design to one or two alternatives with a first-order analysis performed, the review at this stage can aid marketing with a feeling for the product, manufacturing will have additional lead time to begin to think about production, and so forth. This review is most important to design where decisions made earlier are verified in light of overall company requirements.

critical design review

Following the detail design stage, the product is usually in the form of a prototype or sophisticated mock-up and may now be reviewed critically by all disciplines. Here the client is invited (if there is a contractual account) and asked his reaction to the design.

first product inspection and test review

At the end of the first pilot production run, the product is critically inspected and tested for performance against earlier specifications. This information is presented in review before the final stage of commercialization is begun.

The design review is the responsibility of engineering design and should include no more than from 10 to 12 invited persons. Any more prevents a free exchange of ideas and often results in a nonproductive education session. To provide continuity and proper coverage, representatives of each group shown in Figure 10-3 should attend all design reviews. It is also important that the same person within a group try to attend all sessions concerned with that product. A review session generally consumes the greater part of a working day—from about four to six hours. For this reason, design reviews are usually budgeted from one to two percent of engineering costs on a project.

The design review should be chaired by someone from the engineering department who has a broad understanding of the overall technical problem but is not in a direct line of authority to the designer, since they would be reviewing themselves. He should be skilled in leading a technical meeting and have a high level of tact and discretion. Above all, it must be understood that the meeting is not a board of inquiry where the designer is on trial. A free exchange of ideas for the purpose of optimizing the design from the standpoint of function, cost, reliability, production, and appearance can reduce the risk associated with the introduction of a new or improved product.

Details of steps for planning a meeting, presentation techniques, as well as tips in managing the critique will be discussed later. It is important here, however, to consider the following items as effective lead-ins to the review:

1. Provide background information (often a history).
2. Define design requirements.

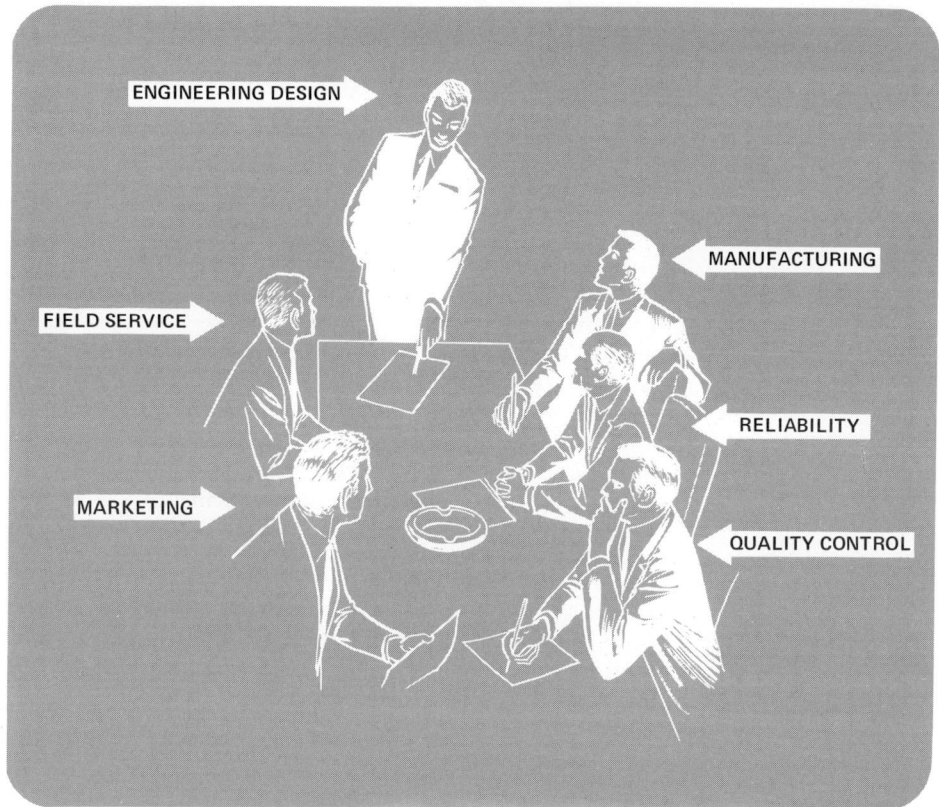

Figure 10-3   **DESIGN REVIEW**

3. Discuss the design approach (decision matrix).
4. Identify problems encountered or expected.
5. Show how the design meets the specifications.

The design review should proceed in an orderly manner under complete direction (but not control) of the chairman. The following events should develop either during or after the meeting:

1. Introduction (background information)
2. Review of previous design review
3. Progress report (CPM/PERT)
4. Discussion of problems encountered
5. Critique (free exchange of ideas)

6. Follow up. At the conclusion of the review, recorded minutes are assembled and circulated among participants and active items to be resolved are assigned by the chairman to specific members.

THE ENGINEER AS AN ENTREPRENEUR

"Throughout the innovative process one finds an entrepreneur who, either alone or as an inspirer and leader of a group effort, combats the conservative nature of man and drives forward with new untried ideas until they are finally introduced in the market place. . . . Technology transfer is basically a 'people transfer' process. It is people who take ideas and do things with them. An effective entrepreneur, who has the ability to exploit commercial applications of science and technology, is far more useful to a profit-oriented corporation than many millions spent upon research and development, which for lack of entrepreneurial skill, bad timing, or lack of market acceptance, may never result in salable, profitable products." *

The engineer, now more than at any other time in history, has the opportunity for entrepreneural activity. There is an awareness today of engineering contributions to the development of our standard of living and to the maintenance of national safety. Society is beginning to realize that the engineer's ability to translate the findings of the physical sciences into practical machines and processes forms the very basis of today's unparalleled standard of living.

Many engineers shy away from entrepreneural activity because they do not wish to be associated with a failure. In the success-oriented career of the entrepreneur, failure looms large. Yet failure is part and parcel of implementing a new creative idea. One can become an engineering entrepreneur if he can overcome this fear and is willing to pick himself up after a failure and try again. The effective entrepreneur has the following traits:

1. Straight, fast thinker.
2. Easy, off-hand speaker.
3. Can persuade others to his point of view.
4. Enthusiastic about his ideas and the creative ideas of others.
5. Able to stimulate others to creative effort.
6. Dedicated to what he believes in.
7. Highly individualistic.
8. A generalist, in that he has a broad knowledge encompassing many fields.

---

* Richard S. Morse, *Innovation and Entrepreneurship, Education for Innovation*, Pergamon Press.

9. His interests include those of the engineer, the inventor, the project manager, manufacturer, salesman, and the financier.

10. Immense faith in himself and the necessary drive to prove his "can't-be-done" critics wrong.

Above all things, the art of entrepreneurship involves the discipline of communications—communication on a one-to-one basis, before a committee or board of directors, or before a relatively large audience. The following eight steps concerning oral reporting can greatly assist the engineer toward entrepreneural activities, giving a technical paper, or making a speech before the local Rotary Club:

1. Decide on subject and approach.
2. Make notes of what you know.
3. Fill in knowledge with research.
4. Tie information together with an outline.
5. Write out the talk.
6. Convert it to speaking notes.
7. Study the notes.
8. Give the talk.

HOW TO SELL IDEAS

Realizing that he is apt to encounter judicial thinking, conservatism, resistance to change, and false security when attempting to sell a new idea, the successful designer has the staying power to prosecute the idea and find acceptance for it. Thomas A. Edison once stated: "Society is never prepared to receive any invention. Every new thing is resisted, and it takes years for the inventor to get people to listen to him and years more before it can be introduced." Although the situation has improved somewhat since Edison's time, the task of selling new ideas to others is still difficult.

Selling an idea is akin to selling an article in a store. The article is either needed by the buyer or is beneficial to him in some way. Recognizing this similarity can be a great aid in selling ideas.

Considering that an idea, in the opinion of the originator, is a good one and that he really believes in it, the following factors in idea selling can assist in its eventual acceptance:

Try the idea out on fellow engineers or immediate superiors before bringing it to top management.

Have all of the facts at hand. How much will it cost? Will it sell? Can we make it? Does it fit our product line? What changes will it precipitate?

If the idea is to be presented to a group or a committee, sell it to one or two members ahead of time to get them on your side.

Take a positive attitude that the idea will be accepted. Don't show that you have even the remotest thought it might be rejected. Enthusiasm often breeds acceptance among others.

Idea presentation should be concise and to the point. Use a vocabulary understood by the people to whom you are trying to sell the idea.

Be well prepared to argue your point of view with background material, research, charts and graphs, statistics, and models.

Don't assume an air of superiority when presenting the idea. At the meeting it is a good practice to recognize through credit the contribution of others who helped develop the idea.

Know the people you are speaking to—their temperaments, attitudes, idiosyncrasies, and preferences.

Write an effective report covering the idea in detail, its advantages and disadvantages, and a summary. Leave copies with the group. More often than not ideas are accepted at a later date when people think about what was said at the presentation. Time has a way of leveling-off and even reducing criticism.

THE DESIGN REPORT

Whether the design results in a device, system, or process, a design report must be prepared and distributed to engineering and management personnel to begin the production cycle. Often the report is required to assist in selling the design idea. It is estimated that the population who read design and engineering reports are 65% industrial executives, 15% engineering colleagues, 15% students and assistants; and 5% public relations people and the general public.

In preparing a design report it might be well to think about the following theme proposed by the late William Strunk, Jr., Professor of English at Columbia University:

> "Vigorous writing is concise. A sentence should contain no unnecessary words, a paragraph no unnecessary sentences, for the same reason that a drawing should have no unnecessry lines and a machine no unnecessary parts. This requires that the writer not make all his sentences short, or that he avoid all detail and treat his subjects only in outline, but that every word tell."

The design report must explain fully the design solution: how the device, system, or process works; its advantages and disadvantages; cost breakdown; and an idea of its potential among competing products (every design idea has a competitor). The report will be read by many of varying backgrounds, from company executives to production specialists, so it must be composed to sell the design and get across the designer's intent. The following is a list of the elements likely to be found in the most comprehensive design report. Needless to say, all are not used in all reports and many are used only at the author's discretion or by customer direction.

1. FRONT COVER—Index paper in color with or without window or company name or trademark.
2. TITLE PAGE—Name of proposed device, system, or process; to whom submitted; name of designer; date, and so on.
3. LETTER OF TRANSMITTAL—An appropriate letter to the official or client whom the designer considers the originator of the design task, stating that the design has been complied with and the solution is being transmitted to him for review. Here proprietary information may be established if necessary.
4. FOREWORD OR PREFACE—Usually the identification of the need and statement of the design goal.
5. ABSTRACT—A brief, general description of the design in exact terms.
6. TABLE OF CONTENTS
7. LISTS—Tables, mathematical symbols, abbreviations, appendices.
8. MAIN BODY OF REPORT
   a. Introduction
   b. Discussion—Description of how the device, system, or process works, its advantages and disadvantages.
   c. Cost breakdown and market potential
   d. Illustrations—Design layout drawings executed in a manner which clearly describes the design and complements the solution. The drawings should also communicate workability and feasibility of the solution in sufficient detail that a draftsman could prepare detail drawings at a later date without elaborate instructions. It is a good idea to begin with a pictorial assembly or exploded view with key parts identified and critical and also overall dimensions shown. Linework should be of sufficient quality (dark and clear) for reproduction purposes and for transparencies to be made when the design is presented in critique.
9. BIBLIOGRAPHY
10. GLOSSARY OF TERMS
11. APPENDICES
12. INDEX
13. DISTRIBUTION LIST
14. BACK COVER

246   THE SCIENCE OF ENGINEERING DESIGN

Many different techniques, equipment, and materials are available for reproducing a report, each having a bearing on cost, quality, time, and quantity. In determining which process or materials to use, consideration should be first given to:

Quality desired, which is determined by whether a report is a contract item and/or will be eventually submitted to the customer or if it is being generated for internal use only.

Whether charges for reproduction will be borne by a shop order (in which case it will have been either quoted in the contract or otherwise requested by the customer) or performed as an overhead department charge.

KEY: □ NOT PRACTICAL   ▨ ACCEPTABLE   ▦ PREFERRED   ▩ REQUIRED

	REPRODUCTION PROCESSES	VERIFAX	OZALID	DITTO	ELECTRO-STATIC (XEROX)	DIRECT IMAGE MASTER	PHOTO OFFSET	NEGATIVES SCREEN	NEGATIVES LINE
CONTRACT/DELIVERABLE REPORT	TEXT (PG. SIZE 8½" × 11")				6 OR LESS	TYPE ON MASTER	TYPE ON REPRO		
	LINE ART (ORIGINAL 7" × 9" OR LESS)				6 OR LESS	DRAW ON MASTER	MOUNT ON PAGE		
	OVERSIZED ORIGINAL (REDUCTION NECESSARY)								▩
	PHOTOS (HALF TONES)							▩	
INTERNAL DISTRIBUTION REPORT	1-6 COPIES	▨	ORIG. VELLUM		▦				
	7-10 COPIES	▨	ORIG. VELLUM		VELLUM MASTER	▦			
	10-30 COPIES		ORIG. VELLUM		VELLUM MASTER	▦			
	OVER 30 COPIES		ORIG. VELLUM	▦	VELLUM MASTER				
BINDING	1-20 COPIES	STAPLED OR LOOSE LEAF				PLASTIC SPIRAL			
	OVER 20 COPIES	SCREW POSTS				STAPLED CLOTH			

OVERHEAD CHARGE ⏟   SHOP ORDER ⏟

Figure 10-4   **REPRODUCTION PROCESSES**

The urgency connected with making the information available.

The number of copies likely to be needed for distribution and filing.

The table shown in Figure 10-4 summarizes popular reproduction processes in a form for easy reference and choice by the author of the report.

PLANNING AND PREPARING A PRESENTATION

Once the design report has been completed it is often presented in critique before a jury. The jury is a critical panel of experts in the fields of the design, finance, marketing, promotion, and sometimes includes the customer. Members of the panel are often government officials, consultants, management personnel, engineers, manufacturers, or the client. Even though the written report covers the design in detail, it can never answer all the questions in the reader's mind. The design critique, therefore, is an arena in which the design is formally presented to a jury for criticism—acceptance or rejection. The design must be sold at this time. Many of the techniques discussed earlier in this chapter—design review, entrepreneurship, how to sell ideas, and the design report—will greatly assist the designer in planning a successful presentation. In addition, the designer might wish to consider the following rules for planning and giving a presentation:

Know the design solution well and show enthusiasm for it.

Point out its advantages as well as its disadvantages.

Be prepared to accept criticism but at the same time stand up for the original concept.

Arrange the format in a manner that will make it easy for an aide to project illustrations and photographs.

Construct a set of prompter cards (index size).

Start on time, break on time, stop on time.

Keep firm control during the critical examination period but do not be a dictator.

The most important part of any critique is the selling and defending of your ideas.

Figure 10-5 compares a number of common visual display media to assist the designer in choosing the one best suited to the condition of the presentation. Once the visual media has been selected, the designer must undertake the task of preparing visuals or direct others in their preparation. The following tips may assist the designer in making his visuals more acceptable.

MEDIA	DESCRIPTION	ROOM CONDITION	AUDIENCE SIZE	WORDS PER VISUAL OR FRAME	ADVANTAGE	DISADVANTAGE
FLIP CHARTS	30" x 40" POSTERS	LIGHTED	10 – 15	20 – 25 MAX.	EASILY PREPARED. MAY BE WRITTEN ON DURING PRESENTATION.	DIFFICULT TO USE BEFORE LARGE GROUP. DIFFICULT TO TRANSPORT.
VIEWGRAPH	OVERHEAD PROJECTOR 8" x 10" TRANSPARENCY	LIGHTED	10 – 30	20 – 25 MAX.	EASILY PREPARED. EASY TO TRANSPORT AND FILE.	DIFFICULT TO ACHIEVE EFFECTIVE COLOR AND A VARIETY OF ART WORK.
LANTERN SLIDES	35 MM 2¼" x 2¼" 3" x 4¼"	DARKENED	20 – 60	20 – 25 MAX.	COLOR AND ART WORK EASY TO ACHIEVE. EASY TO TRANSPORT AND FILE.	TIME CONSUMING TO PREPARE. MUST BE SHOWN IN A DARKENED ROOM.
MOVIE PROJECTOR	8 MM or 16 MM SILENT OR SOUND	DARKENED	20 – 60	20 – 25 MAX.	A HIGHLY SOPHISTICATED PRESENTATION CAN BE MADE. IS USUALLY IMPRESSIVE.	EXPENSIVE. PRESENTOR DOES NOT TAKE AN ACTIVE ROLE. TOO PERMANANT.
MOCK UP	CARDBOARD BALSA PLASTIC CLAY	LIGHTED	5 – 10	20 – 25 MAX.	SHOWS WORKINGS AND FUNCTION OF PARTS. EASY TO VISUALIZE.	DIFFICULT TO SHOW TO A LARGE GROUP. DIFFICULT TO TRANSPORT.

Figure 10-5 **COMPARISON OF VISUAL DISPLAY MEDIA**

get the attention of the audience.

It is a good idea to first present something "light" to set the audience at ease (as well as yourself). Usually a cartoon, quote, or photograph relating to the design is good for the first visual, be it a flip chart, viewgraph, or lantern-slide projection.

the first technical slide

This slide should be a pictorial assembly or exploded view drawing or photograph showing clearly the workability of the device. Too often the designer

THE DESIGN CRITIQUE   249

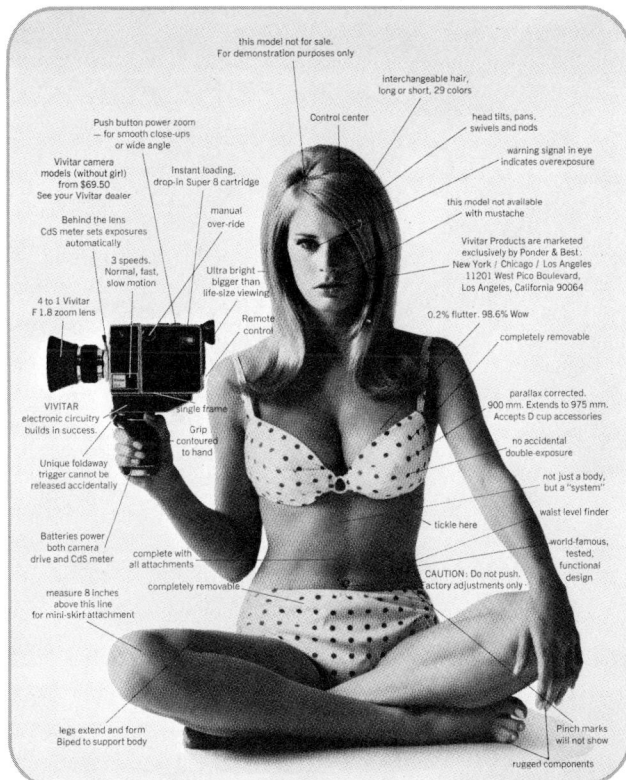

Figure 10-6   **GET THEIR ATTENTION**

will present a detail drawing as his first visual and be half way through the presentation before the audience understands how the product works.

An effective technique of animating an overhead projection transparency is through the use of *overlays*. The base transparency is prepared first. Through critical registration, up to three overlays can be easily hinged to this base and folded into place on the projection stage as the visual is presented.

An excellent method of showing the internal workings and relationships of moving parts of a device is through a two-dimensional plastic mock-up placed directly on the projection stage of the overhead projector. A pointer may be used to identify parts. Functional parts may be moved while the fixed member is either taped or held firm on the projection stage (glass plate).

A common problem in presenting projection materials stems from trying to put too much information in the illustration. If it is a table, an outline, or a chart, the maximum number of words that can be seen and understood at one time is from 20 to 25. This means a chart of four lines of five to six words each. A dra-

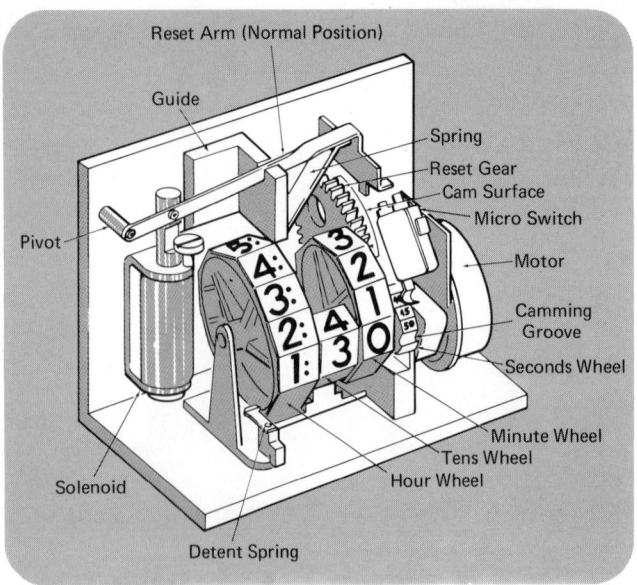

Figure 10-7 **FIRST TECHNICAL SLIDE**

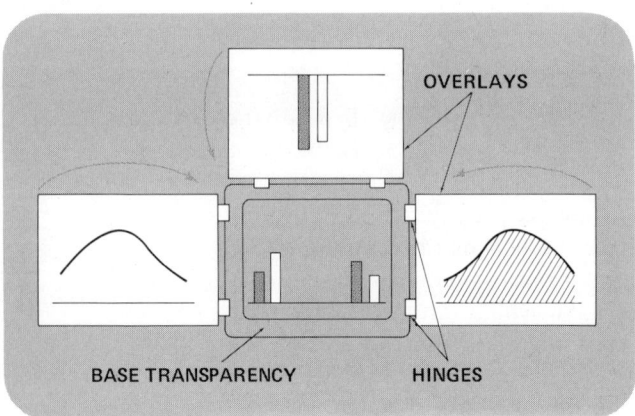

Figure 10-8 **USE OF OVERLAYS**

matic method of revealing each line in a chart or outline can be accomplished through the use of *fold-outs* in which the sentence is allowed to appear on the screen at the appropriate time with all other lines blacked out.

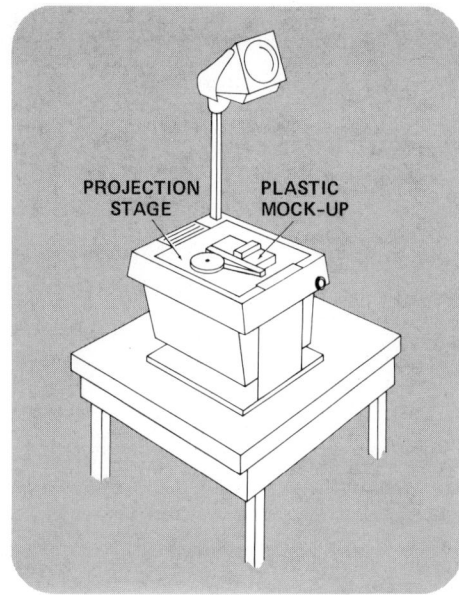

Figure 10-9   **PLASTIC MOCK-UP ON PROJECTION STAGE**

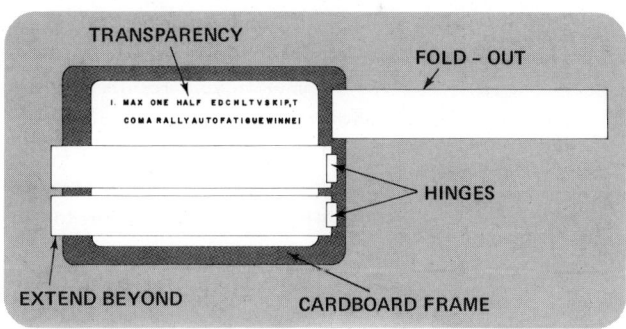

Figure 10-10   **USE OF FOLD-OUTS**

PRESENTATION CHECK LIST

Once the presentation has been prepared and the time and place of meeting set, the designer's attention must be given to physical arrangements of the actual meeting. Although this responsibility is usually delegated to an assistant, it might be a good idea for him to consult the following check list once arrangements have been made.

Sufficient chairs available and seating arrangement checked.

Display material available.

Chalk and eraser available for chalkboard.

Projection equipment and screen available (including spare bulbs).

Trained projectionist available.

Arrangements made for coffee breaks.

Pencils and paper available for note-taking.

Person assigned to take minutes (record action items, questions, and significant comments).

Ash trays available.

Podium, pointers, public-address system available if needed.

EXERCISES

1. To test the validity of the statement "it is easier to think down than to think up," perform the following experiment. Ask three friends, colleagues, or fellow students to list the things they feel are good about and the things they feel are wrong with the idea: *A scheme has been devised to microencapsulate an active chemical ingredient in roofing shingles that will cause the surface to turn light on bright sunny days and dark on cloudy days.* Summarize your results and draw a conclusion.

2. Show an effective organizational structure for design reviews in the form of a flow diagram or organization chart for a company designing and producing lawn mowers (both electric and gasoline). The lawn mowers are sold through distributors. In addition to listing the personnel invited to attend the review, include information about the frequency and content of meetings.

3. Why is the engineer ideally suited for entrepreneural activity?

4. Consider that you are employed by a container company and have thought of a novel idea for opening cans. Outline a plan by which you will attempt to sell this idea to your boss and then to top-level management.

5. Assuming that your idea for opening cans mentioned in question 4 has been accepted by top management and the design has now been completed, list the necessary elements of a design report to present the design to a fruit juice company. Also write specifications on how the report is to be reproduced and bound, considering 15 copies are to be made.

6. Design the first visual (slide, transparency) to be used in a presentation of the can opening idea of question 4. Remember that the visual is to be used to relax the audience and get their attention.

7. What visual medium would you select to assist you in giving a technical paper on a limited budget before a mixed audience of from 30 to 40 people in a high school classroom? Why did you choose this medium?

8. What size of print (when projected on a screen) would be required for an audience to read without difficulty at a distance of from 30 to 35 feet? (Try various sizes to determine the optimum).

9. In your opinion, what is the most effective medium or technique of visual communication and why?

10. What item or items would you add to the check list preceding the exercises in setting up a design presentation?

B BIBLIOGRAPHY WITH ABSTRACTS

The purpose of this bibliography is to make available additional information and a different slant on the topics of creativity, inventivity, innovation, and engineering design. The abstract is intended to serve as a meaningful preview of the contents of the publication so that the reader may decide whether the book provides material of interest to him before he troubles himself with locating the reference.

ENGINEERING DESIGN

Asimow, Morris, *Introduction to Design.* Englewood Cliffs, New Jersey, Prentice-Hall, Inc., 1962.

As the title suggests, this short book (135 pages) develops an inclusive discipline of design. Categories of problems typical of the design process are identified and related to analytical techniques which can resolve them. The author cites problems in making design decisions, allocating available resources, estimating risks of design failure, optimizing deterministic processes, optimizing stochastic processes, and predicting the behavior of tentatively-formulated systems.

Each design project has a history which is particularly its own. However, a common pattern becomes evident as the various problems are initiated and developed. The author examines this pattern and exposes the methodology of design by which ideas about needs are projected creatively into engineering prescriptions for transforming suitable resources into useful, physical objects.

Buhl, Harold R., *Creative Engineering Design.* Ames, Iowa, The Iowa State University Press, 1960.

A cleverly written book in a unique format suggesting methods to develop latent creativity and decision-making ability and to obtain insight into the creative process. The book outlines and discusses ways to "get off the beaten path," ways to think imaginatively and design creatively. There are concrete suggestions and techniques intended to enable one to gain confidence and proficiency in the application of tools; keep a flexible approach, use trial-and-error, intuition, association, application, and the like; and change one's thought habits to admit imagination with reason.

Dixon, John R., *Design Engineering: Inventiveness, Analysis, and Decision-Making.* New York, McGraw-Hill Book Company, 1966.

The author states in the preface that engineering design is not essentially an "art," but rather an activity that can be researched, analyzed, and taught. "I believe that problem solving (engineering design is a particular kind of problem solving) is a high-level intellectual activity." This book discusses the topics of inventiveness, engineering analysis, and decision-making and gives examples of each through well chosen case studies illustrating the techniques involved.

These techniques include the creative process, brainstorming, inversion, fantasy, and systematic search for new combinations when dealing with the topic of inventiveness; problem formulation, physical principles, checking, computation, and evaluation when considering engineering analysis; optimization, probability, statistics, reliability, and management methods (decision theory, game theory, CPM, PERT) in the area of decision-making.

At the end of each chapter the author summarizes past material and gives a preview of what is to be discussed later through an effective technique called "Summary and Preview".

Edel, D. Henry, Jr., *Introduction to Creative Design.* Englewood Cliffs, New Jersey, Prentice-Hall, Inc., 1967.

This text presents a unified introduction to various mental procedures and associated application systems that are involved in creating new designs. Social, economic, production, and aesthetic aspects are considered along with related engineering aspects. Definite mental synthesis procedures are described to aid the reader in applying them at different phases of the design process.

Treating the topics of basic design concepts, the text describes design process developments, design phases, and communication. An evaluation of external factors influencing design treats the environmental and socioeconomic aspects, the material factors, and the human element. In the development of design procedures, detailed sections examine creative thinking processes, analysis procedures, and the practical aspects of invention and patents. The multifaceted design process is delineated through discussions of exploration and formulation, preliminary design, and detailed design for production.

Gibson, John E., *Introduction to Engineering Design.* New York, Holt, Rinehart and Winston, Inc., 1968.

The purpose of this text is to introduce students to the broad spectrum of engineering. By presenting the various major areas of engineering, the author provides a perspective that will enable the student to choose more wisely among the different career opportunities available to him in the field. More importantly, the author attempts to motivate and sustain student interest in engineering and makes the following course work more rewarding.

The concepts of preliminary design are used to introduce the macro-techniques of engineering—energy exchange, information, thermodynamics, human factors, and the fundamental aspects of economics. Such concepts can be treated at an exciting, conceptual, and mathematical level for the beginning student and, at the same time, provide an introduction to the fields in which he will take subsequent courses. The material on data reduction and statistics will be especially useful in this regard.

Harrisberger, Lee, *Engineersmanship: A Philosophy of Design.* California, Brooks/Cole Publishing Company, 1966.

A short, light-hearted, easy-to-read guide to the vital consideration of ideation, cost, styling, and communication in engineering. Designed to dramatize the importance of being a complete engineer. The book deals with the real-life domain of design engineering that exceeds the traditional boundaries of computation and analysis. The text is written in a readable and refreshing style and cleverly illustrated with meaningful cartoons throughout. Techniques are presented to aid the engineer in getting ahead in his profession and include such topics as blackboxmanship, bigmanship, me-too-ism, first-classmanship, ideation, marketology, value engineering, kludgemanship, colorsmanship, and clinchesmanship.

Johnson, Ray C., *Optimum Design of Mechanical Elements.* New York, John Wiley & Sons, Inc., 1961.

Probably the only book of its kind devoted to optimum design of mechanical elements, the author presents a method of design which is explicit in nature, can be applied to innumerable original design problems, minimizes "cut-and-dried" practice, and takes into account the inherently unavoidable limitations confronting design engineers. The thirteen chapters in this 535-page book are easy to follow and contain numerous examples and worked-out design histories. Material covered includes approximations for explicit design; effect of manufacturing errors on product performance; optimum choice for method of analysis; mechanical properties of materials; statistical consideration for fact or of safety; and principles of optimum design and application to axially loaded members, torsion shafts, beams, shaft with combined loading and gears.

Jones, J. Christopher and Thornley, D. G., eds., "Conference on Design Methods," held in London, England, 1962. New York, The Macmillan Company, 1963.

This volume contains the collected papers presented at the Conference on Design Methods at Imperial College, London. The full texts of all papers are included, together with transcription of the recorded instruction to the group discussions. The Conference, believed to be the first to be concerned with the methods, processes, and psychology of the design act was focused on seeking out and establishing systematic methods of problem solving, especially those problems associated with design. The Conference sought a means by which design could be taught as a creative process that could be aided by an experience with academic knowledge while at the same time keeping the imagination free from inhibitions.

Contributions to the Conference include the following papers: Determination of the Components of an Indian Village, Creative Methods in Painting, The Creative Process, Problems of the Design of a Design System, The Relevance of System Engineering, A Systematic Approach to the Problems of Town and Regional Planning, A Method of Systematic Design, Communication in Problem-Solving Groups, Some Experiences of Structural Analysis with the Aid of an Electronic Digital Computer, The Morphological Approach to Engineering Design, Psychological Aspects of the Creative Act, The Conception of Shape and the Evolution of a Design, A Systematic Method for the Teaching of Architecture, A Methodology for the Design of Instruments.

Krick, Edward V., *An Introduction to Engineering and Engineering Design.* New York, John Wiley & Sons, Inc., 1965.

This book presents a clear, concrete, and realistic description of engineering practice, an extensive introduction to important abilities of the competent engineer, and a motivating description of the fields in which he can profitably apply his talent. The book is divided into three parts, each of which illuminates an important facet of engineering.

Part I describes the nature of engineering by means of (1) a series of case studies showing problems an engineer must solve, how they are defined, and what is involved in their solution; (2) a discussion of the evolution of engineering and its relationship to science; (3) a description of the knowledge, attitudes, and skills that today's professional engineer must possess.

Part II introduces in detail three basic engineering skills: representation (modeling), optimization, and design. Five of the book's eleven chapters are concerned with engineering design.

Part III points out the opportunities awaiting the future engineer, with special emphasis on such fields as space technology, medicine, computer technology, resource development and conservation, aquaculture, and transportation.

Liston, Joseph and Stanley, Paul E., *Creative Product Evolvement.* West Lafayette, Indiana, Balt Publishers, 1964.

This book summarizes an extensive exploration of the methods found to be most effective for evolving new products, and it presents a step-by-step procedure for conceiving, describing, proving, and communicating new-product ideas. The first part is mainly devoted to synthesis and the known methods of stimulating imaginative thinking from the preparatory steps in evolving a complex product, to conceptual techniques, to spatial visualization. The later part treats useful analytical and experimental methods of proving the feasibility and soundness of proposals for satisfying recognized needs. These methods include feasibility and optimization studies using analog and digital computers, experimental confirmation of feasibility, and planning the feasibility test facility. The book has many illustrations and examples to complement the text material and contains an appendix devoted to several excellent practice problems in creative synthesis intended to stimulate and develop one's ability to deliberately conceive and describe ideas for new or improved products.

Middendorf, William H., *Engineering Design*. Boston, Allyn and Bacon, Inc., 1969.

One useful technique developed by the author in this book is the flow of information which is motivated throughout by a design procedure presented in the first chapter. This gives the book the structure of a unified story. As one chapter ends, the reasons for discussing the subject of the following chapter should be apparent. On the other hand, skipping part or all of a chapter which may be of no interest to a reader or a group of students will usually not impair the study of succeeding material. This gives an unusual degree of flexibility to the use of the book.

The book has four major goals: (1) To help the reader understand that skill in analysis is a necessary prerequisite for the activity of design. (2) To present topics that the designer needs but which are normally not covered elsewhere in a typical engineering curriculum. (3) To make the reader aware that design requires a high degree of creativity in the development of alternative designs and in the development of design procedures. (4) To offer at least some help by using the modern techniques of synthesis, optimization, decision theory, computer-aided design, and reliability theory, toward elevating design from the intuitive art practiced by many designers to the fine blend of art and science it should be.

Mischke, Charles R., *An Introduction to Computer-Aided Design*. Englewood Cliffs, New Jersey, Prentice-Hall, Inc., 1968.

This volume presents a basic approach to the problem of engineering use of the computer for engineering design purposes. Fortran is the language used and examples were processed on the IBM 360 although the approach employed is independent of the language or the machine. The primary purpose of the book is to give the reader a clearer conception of how to approach engineering problems and to enable him to enlist the aid of the computer in design much more effectively than at present. Supporting the text are useful appendices devoted to problems, documentation, subroutine listings, and a bibliography. Among the book's special features are concentration on engineering design problems rather than on numerical analysis techniques, providing information suitable to all engineering specialties, and introducing the Iowa CADET (Computer-Aided Design Engineering Technique) algorithm.

Roe, P. H., Soulis, G. N., and Handa, V. K., *The Discipline of Design* (experimental edition). Boston, Allyn and Bacon, Inc., 1967.

The content of this book represents an identification of those individual areas of "studious enquiry" most generally relevant to the problems of modern technological design. Moreover, the text attempts to show that the particular subjects, together with their interaction in the process of design, have a meaningful form which constitutes a true "discipline of design."

In the past, design has often been studied in a fragmentary fashion. There has been a tendency merely to examine the actions, methods, and procedures associated with particular classes of objects. Thus there exists machine design, graphic design, control system design, architectural design, all of which are regarded as legitimate fields of endeavor in their own right. This book suggests that the methods and procedures used in design can be studied as a unified subject, not dependent upon the nature of the object being designed. Thus, while primarily using engineering examples, the book takes the point of view that design is a systematic process which can be applied to a broad range of problems. It discusses orderly procedures starting from the initial expression of a design problem and ending with its final physical realization.

Spotts, M. F., *Design Engineering Projects*. Englewood Cliffs, New Jersey, Prentice-Hall, Inc., 1968.

This book contains design projects that pertain to actual machines and devices now in production. These actual devices provide the criteria determining whether or not an acceptable solution has been achieved, according to the author. Over 100 design projects are intended to develop engineering judgment and practice in decision-making. The final portion of the book describes the qualities which a design must have if the component parts are to be feasible within a particular process. The treatment is from the standpoint of the designer who wishes to become acquainted with the capabilities of the manufacturing process. Contents of the book include: Creativity and Development of Designs; Projects; Iron and Steel Castings; Die Castings; Stamping, Drawing, and Related Methods; Plastic Moldings; Die Forgings; Cold and Hot Heating; Screw Machine Parts; Machining; Miscellaneous Production Methods; Fluid Power Systems; Dimensions and Tolerances; Properties of Materials; Strength of Materials; and Mechanical Components.

Woodson, Thomas T., *Introduction to Engineering Design*. New York, McGraw-Hill Book Company, 1966.

The broad approach of this complete textbook makes it suitable for the engineering student's first course in design. The text contains unique and professionally-oriented material on design methodology, engineering innovation, order-of-magnitude analysis, and modeling. It also presents a fresh approach to mathematical modeling by means of the criterion function equation.

A full interdisciplinary array of possible projects is suggested, and ample appendix material is supplied to help both student and instructor. The tone realistically gives the student a preview of his professional future. The exercises at the end of each chapter provide further development of engineering concepts illustrated in the text by including worked-out examples.

CREATIVITY AND INNOVATION

Alger, John R. M. and Hays, Carl V., *Creative Synthesis in Design*. Englewood Cliffs, New Jersey, Prentice-Hall, 1964.

This monograph discusses the purpose of engineering, the nature of engineering design, and the design and creative processes. The design process is explored through the morphology of problem-solving and decision-making with several well-chosen case histories to illustrate techniques. Creativity is related to engineering design and methods of enhancing it, establishing the proper attitude, and methods of discovery are discussed in relation to individual effort as well as group activity. A series of problems is included to test the reader's creative ability. A chapter entitled Scheduling Innovation outlines a simplified PERT (Program Evaluation Reveiw Technique) as a method of project planning and control and shows how the critical path is determined.

De Simone, Daniel V., ed. and introduction, *Education for Innovation*. New York, Pergamon Press, 1968.

This book contains the published results of a conference on creativity and innovation in engineering education held at Woods Hole, Cape Cod, Massachusetts in 1966. The conference was sponsored by the National Academy of Engineering, the National Science Foundation, and the U.S. Department of Commerce.
 The articles of the contributors, preceded by an extensive introduction by the editor, concern the problem of keeping engineering education alive and relevant. It explores the environment for engineering education and ways in which to make it a live and exciting experience. It discusses the kinds of emphasis for making education relevant to the intensive problems of our time—and of any nation, whether it be one struggling to enter the twentieth century or a post-industrial civilization, as is the United States. New insights into the processes of invention and innovation are developed and analyzed in terms of their significance to education. While recognizing that engineering education should encourage students to strive for the mastery of fundamentals, the discovery of the relatedness of things, and the cultivation of excellence, this book argues that these aims are not enough. Engineering education to be creative must stimulate the imagination of students. It should encourage them to believe they can do the impossible from time to time.

Ghiselin, Brewster, ed., *The Creative Process.* New York, Mentor Books, The New American Library of World Literature, Inc., 1960.

In this fascinating symposium, 38 of the world's outstanding men and women reveal how they actually begin and complete creative work in such fields as art, literature, and science.

These brilliant thinkers tell in their own words how they chose their particular forms of expression, acquired knowledge, and mastered techniques necessary in their work. The difficulties artists have in recognizing their own talent, and the roles inspiration, direction, and patience play in producing creative work are also discussed in this provocative volume. Here, too, is a fascinating analysis of how new ideas are born and developed.

Gordon, William J. J., *Synectics.* New York, Collier Books, 1961.

This book is the result of some 15 years of experimentation in training creative capacity. It not only identifies the psychological processes involved in creativity but also explores the methods by which creative potential in individuals and groups can be stimulated and directed to the solution of specific problems.

Synectics theory, which holds that creative efficiency in people can be markedly increased if they understand the psychological process by which they operate, is effectively demonstrated in dozens of case studies of groups in actual problem-solving sessions. With broad implications for the sciences, the arts, and education, Synectics has proven particularly meaningful and provocative to businessmen and executives whose livelihood depends on the development of new ideas.

Hutchinson, Eliot D., *How to Think Creatively.* New York, Abingdon Press, 1949.

This book is concerned primarily with an exposition of the process of creative thought as seen in the experience of contemporary thinkers and with the bearing of these processes upon education, aesthetics, religion, and research. The author substantiates whatever theory is necessary by concrete evidence carefully selected from the most authentic instances of actual production.

Material to document creative techniques is drawn from about 250 of the most famous contemporary thinkers of both England and America, men and women from every profession whose achievements entitle them without question to be called creative. The author states that "in no way do I undertake to explain creative ability. . . . The problem can be dealt with only descriptively; explanation can be only approximate." Instead, the book studies the objective experiences of those who are productive. It notes their characteristics, observes their mental habits, and hangs their intellectual wardrobes on the line. Through this technique, one may venture a guess as to the function and the pattern of the creative process.

Koestler, Arthur, *The Act of Creation.* New York, The Macmillan Company, 1964.

The first part of this book proposes a theory of the act of creation—of the conscious and unconscious process underlying scientific discovery, artistic originality, and comic inspiration. It endeavors to show that all creative activities have a basic pattern in common and to outline that pattern.

The aim of the second part (book two) is to show that certain principles operate throughout the whole organic hierarchy—from the fertilized egg to the fertile brain of the creative individual—and that phenomena analogous to creative originality can be found on all levels.

Osborn, Alex F., *Applied Imagination.* New York, Charles Scribner's Sons, 1963.

This book brings together practically all that is known of the principles and procedures of creative thinking. Step by step, it formulates the practical techniques by which the creative imagination can be more productively utilized. Its primary function is to teach the reader to understand and to apply his own innate creativity to all aspects of his personal and vocational life.

The book is unique in that it presents applied imagination as a formal systematized discipline. The discipline is presented through the development of functional methods with an emphasis on actual problem-solving techniques.

Samson, Richard W., *The Mind Builder.* New York, E. P. Dutton & Co., Inc., 1965.

This unique book is designed for anyone who wants to improve his power of thinking creatively: scientists, engineers, artists, those in business or the professions. Both text and exercises have been especially prepared for individuals working alone. The author explains how language works, what the different phases of analytical thought are: (1) thing-making—how we use words for things we observe; (2) qualification—how we separate or isolate qualities from objects; (3) classification—how we proceed to sort things into classes, types or families; (4) Structural Analysis—how we are able to think more clearly about an object by breaking it up into its parts; (5) Operation Analysis—how we divide happenings into the stages in which they occur; (6) analogy—how we learn to recognize relationships in seemingly unconnected situations.

The second half of the book is devoted to the exercises, 288 in all, which enable the reader to train himself in the above methods. Arranged into seven groups and varying in degrees of difficulty, each group contains exercises building each kind of analytical thinking.

Taylor, Calvin W. and Williams, Frank E. eds., "Instructional Media and Creativity," *Proceedings of the Sixth Utah Creativity Research Conference.* New York, John Wiley & Sons, Inc., 1966.

Some 15 of this country's leading psychologists, researchers, and educators gathered for four days to discuss how instructional media might help foster and encourage creativity in our primary and secondary schools in America.

Papers were delivered on the following topics: Creativity Through Instructional Media; Basic Problems in Teaching Creativity; Can Existing Instructional Media Be Used for Creativity?; Some Personal Observations on Creativity; Implications of Creativity Research Findings for Instructional Media; Instructional Media in the Nurturing of Creativity; Use of Films and Television for Creative Teaching; Imagination: Developed and Disciplined; Creation and Instructional Media; Instructional Media for Creativity in the Arts; Stocktaking Between the Areas of Creativity and Instructional Media.

Von Fange, Eugene, *Professional Creativity.* Englewood Cliffs, New Jersey, Prentice-Hall, Inc., 1959.

Designed to be equally valuable for those of all interests, this book shows distinctly what creativeness is and how to contribute creatively through the effective interplay of aptitudes, resources, and planning. It is written to be of benefit both to the casual reader seeking stimulation through example and technique as well as to those who wish to study in depth the path to creative accomplishment.

The first portion of this work is designed to convince the reader that he has creative talent and to show him how he may more fully use his mental powers and resources to create. The latter portion discusses in detail the planning of creative work to insure (1) that worthwhile objectives are established; (2) that the creative work itself is conducted efficiently; (3) that the results, when achieved, will be accepted by others.

DESIGN METHODS

Bartee, Edwin M., *Engineering Experimental Design Fundamentals.* Englewood Cliffs, New Jersey, Prentice-Hall, Inc., 1968.

The author states in the preface, "The purpose of this book is to provide the fundamental concepts that are necessary to accomplish effective planning, design, and analysis of engineering experiments, so that the use of experimental equipment and resources may be optimized." The effective planning of engineering experiments is emphasized as a problem involving the application of the scientific method and the methodology of engineering design. Engineering modeling techniques are emphasized through two steps in the model-building method (1) design of the experimental model (both structural and functional) and (2) design of the analytical model.

The general outline followed in presenting the methodology of experimental design contains (1) formulating the experimental problem, (2) analysis of the experiment, (3) design of the experimental model, (4) design of the analytical model, (5) conducting the experiment, and (6) deriving a solution from the model.

Crede, Charles E., *Shock and Vibration Concepts in Engineering Design.* Englewood Cliffs, New Jersey, Prentice-Hall, Inc., 1965.

This monograph is intended as a reference for use by students and practitioners of engineering design. The material is divided into four chapters, each covering a discrete phase of the design of a machine or structure. The first chapter discusses the formulation of mathematical models; that is, the transformation of the actual system into an idealized system to be used directly in the writing of applicable equations. The second chapter discusses the solution of the equations, and particularly the presentation of the results in a form that is useful for design purposes. The last two chapters are somewhat more applied in nature. They discuss (1) methods of limiting the source of shock or vibration and (2) methods for controlling the transmission of or the response to the excitation. In each instance the discussion of the practical aspects of design is integrated with the preceding theoretical discussion so the former serve as examples of the latter.

Greenwood, Douglas C., *Manual of Electromechanical Devices.* New York, McGraw-Hill Book Company, 1965.

Over 100 specific types of electromechanical devices are discussed and analyzed in this easy-to-follow book. It brings to the design engineer an understanding of and information to choose and design the best device for a particular product system. The book was written to fulfill the need of the design engineer, who has experienced in recent years the linking of two main disciplines; that of mechanical and electrical engineering, to learn from one source design information on electromechanical devices. Design aids contained in the book include tips on what can and cannot be done with certain alloy substances, standards for determining component cost/efficiency ratios, tested ways to estimate long-range mechanical life, and 14 checkpoints for specifying a switch. Specific devices discussed in minute detail include magnets, solenoids and solenoid valves, relays, fractional-horsepower and instrument motors, synchros, electromechanical power transmission components, switches, and ancillary control devices.

Greenwood, Douglas C., ed., *Engineering Data for Product Design.* New York, McGraw-Hill Book Company, 1961.

The material contained in this book is selected articles from *Product Engineering*, and edited and classified to facilitate easy reference. All material has been chosen by an experienced design engineer on the basis of direct interest and usefulness to the product engineer. The information contained in 12 chapters (430 pages) can be quickly and easily adapted to almost any design needs as they relate to such areas as metals and alloys, non-metallic materials and finishes, and design analysis involving beams, torque, moment of inertia, bearings and shafts, springs and vibration, and mechanical control. The material is exceptionally broad in scope and contains a large number of tables in the form of charts and nomograms for the selection and checking of almost all widely used mechanical components. Several modern techniques are included, such as dielectric heating, nameplates, metal whiskers, radiation characteristics of metals, high temperature spring materials, polyamide-cured epoxy, and applications of radioisotopes.

Greenwood, Douglas C., ed., *Mechanical Details for Product Design*. New York, McGraw-Hill Book Company, 1964.

    This book is intended to be used "right at the desk" of a design engineer to furnish quick, workable ideas and answers to complex design problems that might otherwise take much longer to solve. The book is profusely illustrated with methods for solving problems that design engineers face in their everyday work. These methods have been drawn from recent issues of *Product Engineering*. The book's ten chapters contain detailed coverage of Accessories; Basic and General Design; Control and Materials Handling; Fastening and Joining; Hydraulics and Pneumatics; Mechanical Movements and Linkages; Mechanical Power Transmissions; Spring Devices; and Welding and Brazing.

Greenwood, Douglas C., ed., *Product Engineering Design Manual*. New York, McGraw-Hill Book Company, 1959.

    This manual contains articles from *Product Engineering*, presented in a practical "how-to" format. Careful editing and arrangement makes it possible to locate easily over 100 product design topics from information on adhesives to ultrasonic applications. This material is freely supplemented with illustrations, formulas, tables, and reference charts. It gives the reader the "feel" for a possible design alternative or solution to a sticky problem by showing what is good, what has been done, what is accepted, and what is new. Included are detailed sections on accessories, assembly, clutches, couplings, bearings and mounts, control and measuring devices, dimensions and design, drives, electrical-electronic-and-magnetic components, mechanical movements and linkages, miscellaneous design aids, shaft seals, springs, and welding and brazing.

Hix, C. Frank, Jr. and Alley, Robert P., *Physical Laws and Effects*. New York, John Wiley & Sons, Inc., 1958.

    Realizing that new products, new inventions, and whole new industries often originate from an idea inspired by an unusual or unapplied law or effect, the authors have painstakingly combed not only the current literature but also almost forgotten first editions of early science books and journals. Their effort has resulted in the first convenient, centralized source of information on the subject, and as such represents a major step in making the consideration of both familiar and unfamiliar laws and effects a practical part of the engineer's approach to problems.

    The book is comprised of three different cross-reference systems, indexed so that pertinent information can be located rapidly. The systems include (1) description of laws and effects, (2) cross-reference by fields of science, and (3) cross-reference by physical quantities.

Lipson, Charles, *Wear Considerations in Design*. Englewood Cliffs, New Jersey, Prentice-Hall, Inc., 1967.

   This monograph presents the current state of the art of wear considerations in design by summarizing the various forms of wear as related to design. It cites current experimental data on forms of wear with references for more detailed information. Wear is classified as adhesive, abrasive, pitting and spalling, cavitation, galvanic corrosion, fretting corrosion, stress corrosion. The text also discusses selective subjects such as friction and lubrication and high temperature surface phenomena. One important feature is the discussion of surface protection against wear and corrosion which includes electroplating, anodizing, metal spraying, electromechanical smoothing, hard-facing, carburizing, induction hardening, nitriding, tufftriding, cyaniding and carbonitriding, and flame-hardening.

Ray, William S., *An Introduction to Experimental Design*. New York, The Macmillan Company, 1960.

   The author starts with the simplest ideas and principles of design[1] and analysis, takes up in sequential order a few central developments while avoiding many peripheral ones, and ends with certain interesting and advanced topics and issues. The early chapters are intended to stimulate the reader to reason about the basic problems of planning and evaluation. At the same time some elementary statistical topics are reviewed. As the discussion proceeds, the sophistication of the statistical approach increases, but repeated attempts are made to encourage the reader to integrate common sense and statistics and to strive for a useful and rewarding level of understanding.
   Certain standard designs which occur repeatedly in actual practice are emphasized. Designs which are the exception in practice or which involve the user in more than the normal number of uncertainties either were not selected or are not emphasized.

Robenstein, Allen B., Rathbone, Robert R., and Schneerer, William F., *Engineering Communications*. Englewood Cliffs, New Jersey, Prentice-Hall, 1964.

   This monograph presents a unified approach to engineering communication. It states that communication is an indispensible activity in the engineering design process. Based on the conviction that the engineer must communicate with himself as well as others, the book carefully develops the basic concepts of information theory and relates it as an integral part of the modern theory of engineering design. This provides the logical basis to which the rules for effective reading, writing, and drawing are related.
   Part one discusses the area of communication involving the role of communication in engineering design, the mathematical theory of communication, and communication systems. Part two discusses the reader: the writer and the report; oral reporting; graphics in engineering design; how to sketch; techniques of sketching; charts, graphs, and mathematical constructions: and presentation.

---

[1] Design here does not refer to innovation but is used in the context of the design experiment.

Ruskin, Arnold M., *Materials Considerations in Design*. Englewood Cliffs, New Jersey, Prentice-Hall, Inc., 1967.

This book deals with the application of knowledge of materials to engineering problems. Single phase materials are studied in light of the functions all materials perform; that is, interaction with energy and reacting to forces. Characterizing the entire book is the unifying point of view that, unlike energy which can be thermal, mechanical, electrical, magnetic, or chemical, materials can react in only a limited number of ways and can resist or yield forces in a similarly limited number of ways.

Multiphase materials are described by comparing and contrasting their behavior with single phase materials. The knowledge of manufacturing feasibility and the use of engineering data illustrate the significance of applying materials considerations throughout the design process. Contents include: Perfect Crystals; Imperfections in Single Phase Materials; Properties of Single Phase Materials; Generation of Multiphase Structures; Properties of Multiphase Materials Compared with Properties of Single Phase Materials; Selecting Suitable Materials—Design Combinations.

Ruskin, Arnold M., *Selection of Materials and Design,* Selected Papers. New York, American Elsevier Publishing Company, Inc., 1968.

This volume consists of five lectures delivered at the autumn course of the Institute of Metallurgists which was held jointly with the Institution of Mechanical Engineers at Eastbourne, England in 1966.

The first paper by P. H. W. Wolff sets the perspective for the whole volume in presenting the concept of engineering design and surveying, with examples, the stages of investigation and decision in a project design involving problems of material selection.

The second and third papers are complementary. The second, by A. S. Kennedy, in stressing fundamental scientific aspects deals also with the limits within which those designing the materials must work. This, in turn, leads to the third paper by N. P. Inglis, a discussion of the effects the properties of the selected material must have on the technology of manufacture of the product. The engineer's difficulties in fabrication and inspection are analyzed and the problems arising from unnecessarily fine specification are discussed.

In the last two papers by T. Broom and W. J. Arrol, a review is given of trends for the future both in the development of materials better suited to meet likely future engineering demands and in the increased understanding of the more effective use of materials at present available.

Starr, Martin K., *Product Design and Decision Theory*. Englewood Cliffs, New Jersey, Prentice-Hall Inc., 1963.

This book presents the nature of decision theory and how it can be utilized to improve product design decisions. Various kinds of design situations are discussed, extensive detail having been foregone. Mathematical and logical methods are employed, but the reader need have only a surface acquaintance with these subjects. However, an ample supply of references has been included to direct the more advanced reader to further study.

HUMAN FACTORS IN DESIGN

Damon, Albert, Stoudt, Howard W., and McFarland, Ross. A., *The Human Body in Equipment Design.* Cambridge, Massachusetts, Harvard University Press, 1966.

This book makes available, for the first time, extensive quantitative data on human body size and biomechanics, placed in a broad biological context and applied, following clearly stated principles, to the design of equipment for human use. Both kinds of information—the body measurements and the biomechanical data on the range, strength, and speed of the body's movements and tolerance to various forces—are prerequisite for evaluating current designs or planning new ones in terms of the physical characteristics of the users.

The data and principles presented comprise the most complete and integrated treatment to date of anthropometry in equipment design. All measurements were selected for their relevance to human engineering problems in the new field of biotechnology.

Dreyfuss, Henry, *The Measure of Man.* New York, Whitney Library of Design, 1960.

In the offices of the industrial design firms of Henry Dreyfuss in both New York and Pasadena hangs a creed which reads: "We bear in mind that the object being worked on is going to be ridden, sat upon, looked at, talked into, activated, operated, or in some other way used by people. When the point of contact between the product and the people becomes a point of friction, then the industrial designer has failed. On the other hand, if people are made safer, more efficient, more comfortable—or just plain happier—by contact with the product, then the designer has succeeded."

The information made available in this portfolio is presented in only 16 diagrams that could save money and hours of research by the designer. The diagrams contain anthropometric data for the adult male and female as well as male and female children, foot measurements and basic foot controls, human strength, body clearances, climbing data, ingress and egress, basic control data, and basic display data.

The diagrams are supplemented by a check list for designers containing manual controls, pedals, visual displays, auditory signals, sensory signals, anthropometric conformity, safety, illumination, environment, and maintenance.

Two life-size figure charts suitable for mounting (25 × 76 in.) of the 50th percentile adult male and female containing all essential measurements are also contained in the portfolio for designers who prefer to work in full-size scale.

Gagne, Robert M., ed., *Psychological Principles in System Development.* New York, Holt, Rinehart and Winston, Inc., 1965.

This book attempts for the first time to present a theory of psychotechnology of man-machine systems. It achieves the integration of what has heretofore been variously called human engineering, human factors engineering, or engineering psychology on the one hand and personnel psychology or personnel and training research on the other. Each chapter was written by men in the fields of research, systems science, and psychology, and include the following: Psychology and System Development; Human Functions in Systems; Men and Computers; Human Capabilities and Limitations; Human Tasks and Equipment Design; Task Description and Analysis; The Logic of Personnel Selection and Classification; Aids to Job Performance; Concepts of Training; Training Programs and Devices; Team Functions and Training; Proficiency Measurement; Assessing Human Performance; Evaluating System Performance in Simulated Environments; The System Concept as a Principle of Methodological Decision.

McCormick, Ernest J., *Human Factors Engineering.* New York, McGraw-Hill Book Company, 1964.

This well-illustrated and fact-filled book deals with some of the problems and processes involved in efforts to achieve simple and clear-cut objectives—"designing things so people can use them effectively and creating environments that are suitable for human living and work." Material to form the basis of the book was drawn from a variety of disciplines, including physical anthropology, physiology, climatology, engineering, and psychology. Much of the material deals with the results and implications of research investigations of some of the practical problems of design of man-machine systems and environments. The book is divided into six parts: The Human Aspects of Man-Machine Systems, The Human Organism, Human Processes in Man-Machine Systems, Space and Arrangement, Environment, and Personnel and System Integration.

Martin, W. Edgar, "Children's Body Measurements for Planning and Equipping Schools," Special Publication No. 4, U.S. Dept. of H.E.W., U.S. Government Printing Office, Washington, D.C., 1955.

This study was conducted to secure reliable and current information on the body measurements of children in working positions which are characteristic of the varied learning experiences of pupils in our schools today. Examples of some of these positions are sitting erectly and writing and reading positions at tables and desks; standing, reaching, and working at laboratory and shop benches; and reaching for stored materials in bookcases and on shelves.

Morgan, Clifford T., Chapanis, Alphonse, Cook, Jesse S., III, and Lund, Max W., eds., *Human Engineering Guide to Equipment Design,* Sponsored by Joint Army-

Navy-Air Force Steering Committee. New York, McGraw-Hill Book Company, 1963.

This book provides a guide in human engineering which the designer can use in the same manner as handbooks in other areas to assist in solving design problems as they arise. The primary emphasis in the guide is on recommended design principles and practices in relation to general design problems rather than on the compilation of research data. Topics covered include: the Man-Machine System; Visual Presentation of Information; Auditory Presentation of Information; Speech Communication; Man-Machine Dynamics; Design of Controls; Layout of Workspace; Arrangement of Groups of Men and Machines; Design for Ease of Maintenance; Effects of Environment on Human Performance; Anthropometry.

Murrell, K. F. H., *Human Performance in Industry*. New York, Reinhold Publishing Corporation, 1965.

This book covers the latest advances in human engineering in a way which suggests many fresh ideas for raising employee performance and simultaneously enhancing employee moral. Emphasis is placed on the study of the relationship between man and his working environment, or ergonomics. The book attempts to answer such questions as: What are the specific work operations best performed by man—and which are those better allocated to a machine? How are the functions of man and equipment scientifically balanced for optimum results? What are the conditions under which workers are now known to perform at their best? All four major aspects of human engineering are covered in detail: (1) the capabilities and limitations of the human body; (2) factors to be considered in the design or redesign of all types of equipment, from seats and machines to controls and instrumental displays; (3) environmental factors such as illumination, temperature, noise, and vibration; (4) the all-important organizational factors that help determine where change is needed and how to take action.

White, William J. and Schneyer, Solomon, *Pocket Data for Human Factor Engineering*. Buffalo, New York, Cornell Aeronautical Laboratory, Inc. of Cornell University, 1964.

This pocket guide, written in handbook form for ease of reference, is intended to provide the designer with basic human-engineering data and establish design recommendations for maximizing the efficiency of man-machine systems. Areas covered with tables of data include: Physical Dimensions of Operator and Workspace; Visual Displays; Auditory Displays; Controls; Instrument Panel Design; and Environmental Considerations.

Woodson, Wesley E. and Conover, Donald W., *Human Engineering Guide for Equipment Designers*, 2d ed. Berkeley, California, University of California Press, 1964.

The first edition of this book broke new ground in an effort to induce industrial engineers to "start with the man" and to base their designs on him rather than taking him into account only at the end of the design process. To be practical, man-centered designs must draw on man's physical and psychological inventory. The original edition attempted to provide such information simply and graphically. It was sufficiently successful to require several printings, was translated into Asian and European languages, and was imitated by other books that appeared afterward.

The present edition is much enlarged, for it embraces a wider range of subjects and examines each of them in greater depth. More important perhaps, it draws on a far more extensive body of experience than was available when the original edition appeared 10 years ago. Most of the material is the product of nearly 20 years of research and practical experimentation by the authors in the fields of industrial engineering, equipment design, systems engineering, and industrial psychology.

BACKGROUND MATERIAL

Clarke, Arthur C., *Profiles of the Future.* New York, Bantam Science and Mathematics, Bantam Books, 1967.

It is impossible to predict the future, and all attempts to do so in any detail appear ludicrous within a very few years. This book has a more realistic yet at the same time more ambitious aim. It does not try to describe the future, but to define the boundaries within which possible futures must lie. If we regard the ages which stretch ahead of us as unmapped and unexplored country, the author attempts to survey its frontiers and to get some idea of its extent. The detailed geography of the interior must remain unknown—until we reach it.

With a few exceptions, notably Chapter 8, the author limits himself to a single aspect of the future—its technology, not the society that will be based upon it. This is not such a limitation as it may seem, for science will dominate the future even more than it dominates the present. Moreover, it is only in this field that prediction is at all possible; there are some general laws governing scientific extrapolation, as there are not in the case of politics or economics.

Klemm, Friedrich, *A History of Western Technology,* translated by Dorothea W. Singer, Cambridge, Massachusetts, The M.I.T. Press, 1964.

This history of technology is in the form of contemporary writings—from technologists, churchmen, naturalists, poets, economists, and statesmen—which reveal how historical circumstance altered the direction of technical development, and how the intellectual forces of a period influenced and were in turn modified by technical progress. The documents selected begin with Greco-Roman times, span the industrial revolution, and continue into our own day.

Attention is directed less to individual aspects of technology than to its problems in general. Historical details have therefore been omitted, and an effort has been made, by extensive use of contemporary documents, to reveal the forces which guided the development of technical advance in this or that direction.

Mason, Otis T., *The Origins of Invention: A Study of Industry Amoung Primitive Peoples.* Cambridge, Massachusetts, The M.I.T. Press, 1966.

"The devices of primitive man are the forms out of which all subsequent inventions arise. The fire sticks of savages are the earliest form of illumination by friction. The tribulum is the modern thresher with stone teeth. The kaiak furnishes the lines of the swiftest racing boats. The sewing machine makes no new loops. Warfare is still cutting, bruising, or piercing, and our most precious maxims antedate literature. The whole earth is full of monuments to nameless inventors, and the history of the development of the inventive faculty is the history of humanity."

This classic study, originally published in 1895, traces some of our modern industries to their origins and shows how the genius of man, working upon and influenced by the resources and the forces of nature, learned the art of invention. The term *invention* is used in a fundamental and logical sense, that is, discovering how to perform specific acts by some new implement or improvement or substance or method. Basically, each invention is a change in one or all of these aspects.

MANAGEMENT TECHNIQUES

Banning, Douglas, *Techniques for Marketing New Products.* New York, McGraw-Hill Book Company, 1957.

This book focuses its attention on the role of small business in the introduction of new products as vital to the country's expanding economy. The purpose of the book is to combine under one cover all salient aspects of (1) marketing, (2) advertising, and (3) managing the introduction of a new product by a small business. The book can be divided into the three parts enumerated. Chapters 2 through 6 are concerned with the problems of merchandising. Here is reviewed aspects to product selection, evaluation, and choice of distribution channels. Chapters 7 through 9 are concerned with the broad field of advertising. Here is discussed in detail elements of sales psychology, copywriting techniques, direct mail, advertising production, and media selection. Chapters 12 through 15 deal with the problems of general management, reviewing significant administerial, legal, and financial aspects of small business organizations.

O'Brien, James J., *CPM in Construction Management* (*Scheduling by the Critical Path Method*). New York, McGraw-Hill Book Company, 1965.

This book explains the function of the Critical Path Method (CPM)—the technique developed for the purpose of using electronic computers to schedule construction.

First, the reader is given a practical picture of what CPM is. Second, he is shown how to compute it. Third, the operations of the electronic computer are described. Fourth, the role of the computer in CPM is explained. (PERT is also discussed since one may find it a requirement of some operations.) Finally, the reader is shown how to use CPM most effectively on specific projects.

Value Engineering in Manufacturing, American Society of Tool and Manufacturing Engineers. Englewood Cliffs, New Jersey, Prentice-Hall, Inc., 1967.

The concept embraced by this book is that value engineering includes the total value program, of which manufacturing engineering, from the product drawing to the shipping platform, is an integral part. From this point of view, the book provides a thorough coverage of the theory, principles, application, and administration of value engineering and analysis. Outstanding features of the book include: (1) Meaning and Analysis of Functions: the general concept of function; meaning of use, esteem, and exchange values; basic vs. secondary vs. unnecessary functions; using and evaluating functions. (2) Principles of Value Engineering: problem recognition and definition; the role of creativity and ways to achieve it; the criteria for comparison; the element of choice. (3) Value and Decision: the decision process; theory of the decision matrix (linear programming); concept of utility; make-or-buy; elements of queuing theory and Monte Carlo method. (4) Scheduling of VE Activity: manual system; Gantt charts; PERT charts and technique; network logic; critical path method (CPM); use of control charts. (5) Role of Management in VE: orientation and responsibilities; centralization vs. decentralization; level of VE in the organizations; small-plant activity; budgeting and auditing; merit recognition.

PATENTS AND INVENTIONS

Tuska, C. D., *An Introduction to Patents for Inventors and Engineers* (formerly titled *Patent Notes for Engineers*). New York, Dover Publications, Inc., 1964.

The author, formerly Director of the Patent Department of RCA, presents a fundamental picture of what constitutes statutory invention—examining negative definitions and essential requirements as interpreted by the courts. This book gives sound, up-to-date guidance on proper protection of individual rights when the possibility of a patentable invention arises, indicating the danger signals along the path from the original moment of discovery to the actual patent ownership. This practical, easy-to-follow guide contains the following chapter headings: Invention in Popular Sense; Statutory Invention; Nature of Statutory Invention; Statutory Invention as a Practical Matter; Records of Invention; Prosecution of Patent Applications; Interferences; Patent Approval; Ownership and Use of Patents.

U.S. Department of Commerce/Patent Office. Superintendent of Documents, U.S. Government Printing Office, Washington, D.C., "General Information Concerning Patents," 1966.

The purpose of this booklet is to give some general information about patents and the workings of the Patent Office. It answers many questions commonly asked of the Patent Office. It is not intended for the patent lawyer, nor is it intended to be a textbook on patent law. Consequently, many details are omitted and complications have been avoided as much as possible. It is intended that the booklet will be useful to inventors and prospective applicants for patents, to students, and to others who may be interested in its subject matter, to give them a brief general introduction to the subject.

U.S. Department of Commerce/Patent Office. Superintendent of Documents, U.S. Government Printing Office, Washington, D.C., *Patent Laws*, 1965.

This book contains the legal aspects that govern the granting of a patent. The book is divided into three parts. Part I concerns itself with the Patent Office, Part II with patentability of inventions and grant of patents, and Part III with patents and protection of patent rights. In case of a legal problem or question related to a patent or an invention, before one consults an attorney this book is highly recommended.

U.S. Department of Commerce/Patent Office. Superintendent of Documents, U.S. Government Printing Office, Washington, D.C., "Patents and Inventions: An Information Aid for Inventors," 1966.

The purpose of this pamphlet is to help inventors in deciding whether to apply for patents, in obtaining patent protection, and in promoting their inventions. The importance of patents is discussed and six basic steps are outlined to serve as a practical guide in obtaining a patent. Also discussed is marketing and developing the invention and answers to questions frequently asked. Appendices include Addresses of U.S. Department of Commerce Field Offices and Address of Small Business Administration Field Offices.

U.S. Department of Commerce/Patent Office. Superintendent of Documents, U.S. Government Printing Office, Washington, D.C. *Rules of Practice of the United States Patent Office in Patent Cases*, 5th ed., 1965.

This book states the rules governing the practice and procedure in the Patent Office with respect to application for patent and related matters, and the recognition of attorneys. The rules covered in the book range from Fees and Payment of Money; Prosecution of Application and Appointment of Attorney or Agent; Who May Apply; the Application; the Petition, Oath or Declaration; Specifications; the Drawings; Models; Exhibits, Specimens; to Claims. Also included are conventional symbols for draftsmen including Material Symbols, Electrical Symbols, and Mechanical Symbols.

U. S. Department of Commerce/Patent Office. Superintendent of Documents, U.S. Government Printing Office, Washington, D.C., "The Story of the United States Patent Office," 4th ed., 1965.

    This most readable and interesting booklet traces the history of the Patent Office from 1790 when President George Washington signed the bill which laid the foundation of the modern Patent System to the year 1965 and a Patent for Dispensing Container for Tablets. The story is told through the patents that have been issued, the most noteworthy of which are described in the booklet. It is most interesting to read and contemplate how many of the inventions described contributed to the progress of the United States and the general benefit of mankind. For anyone preparing a lecture on the subject of Patents and Invention, this booklet is a must.

A1 DESIGN PROBLEMS

The design problems contained in this section are intended to introduce the new designer to a number of real problems and sharpen his abilities to undertake more difficult projects and cases included later in sections A–2 and A–3. These problems identify the need for a device or system that is not presently available. Their solution requires the application of the design process with emphasis on the conceptual phase and some analysis. Primary emphasis is placed on imagination and invention with special attention given to the development of a series of alternative solutions (concepts). In attempting to solve the problems presented here, the designer is advised to apply the orderly process of design beginning with the definition of goal and continuing through the research, task specification, ideation, and conceptualization phases into analysis, if required, until a solution description is reached.

PROB. 001   SPILL-PROOF FUNNEL

When filling the gas tank of an outboard motor, power lawn mower, or power saw, with the use of a funnel, the fuel often overflows onto the engine, due to the difficulty in predicting the level of gas in the tank. This is extremely dangerous, especially when the engine is hot. One way to correct this overspilling would be through the use of a funnel designed to automatically close once the tank was 95 percent full.

Design a spill-proof funnel that will automatically close when the level of fuel (liquid) reaches the full or almost full level. The partially-filled funnel might (in some way) serve as an indicator as to when this level is reached. An added feature that could enhance the design would be a lever or float so that the over-poured fuel in the funnel could be emptied back into the tank. This device, in a very inexpensive way, would prevent the spilling of any toxic or inflammable fluid.

PROB. 002   SPEED RECORDER

Legislation is presently being considered to require all motor vehicles to have installed a device that would indicate the vehicle's speed for at least the last mile traveled prior to any required inspection. Should a vehicle be stopped for speeding, a record of the speeds attained during the last mile would be available to the inspecting officer. It is felt that such a device will reduce greatly the increasing toll of accidents which are attributed to speeding and in addition eliminate argument between inspecting officer and violater.

Design a continuous reading velocity device that will record for a period of one mile the vehicle's speed. This device should, in some way, be connected to the vehicle's speedometer and/or odometer. The recorded velocity should be visible to the vehicle operator, but should be so constructed that the device cannot be tampered with to reveal a false or incorrect reading.

PROB. 003   AUTOMATIC CHANGE DISPENSER

The modern trend in industry is to serve lunches, coffee, as well as certain articles of clothing through automatic vending machines. While some machines will accept change in any denomination, most vending machines require the exact change (cigarette machines, laundramat, etc.). At the present time many large industries must employ a teller in lunchrooms for the purpose of making change for paper money. Therefore there is a demand for an automatic change dispenser that will accept paper money in any denomination (say 1 to 10) and return the correct change.

Design an automatic change dispenser that will accept one, five, and ten dollar bills and return the correct change in the denomination of two quarters, four dimes, and two nickels and remaining bills. The dispenser should also be able to detect counterfeit currency and sound an alarm, such as the ringing of a loud bell.

PROB. 004   PARKING METER ZEROING DEVICE

The revenue lost from second occupant parking is so sizable that public officials have requested a meter be designed so that excess time left by a previous occupant would be automatically removed upon departure of that occupant. The new user of the meter would then drive into a parking space while the meter is reading zero time.

Design an attachment for the standard parking meter using some vandal-proof scheme that will zero the meter upon departure of a vehicle presently using it. Extreme care must be exercised so that (1) all makes and sizes of cars will be affected, and (2) the meter cannot be disrupted by playful children or commonplace occurrences such as someone standing in front of the meter and the like.

PROB. 005 AUTOMATE MANUFACTURING PROCESS

The manufacturer of ladies' summer shoes has an attractive shoe idea in which he has encountered certain problems. One feature of the shoe is an elasticized fabric sewed to both sides of the shoe across the instep. The design of the shoe requires this instep strap to be relaxed (material gathered in pleats) when the shoe is displayed and to be stretched tightly across the foot (material elongated) when the shoe is worn.

The shoe manufacturer is primarily interested in producing the elasticized fabric for the instep strap rather than paying a premium for this material through a jobber. The production shop has automatic sewing machines capable of sewing continuously and having a table 24" in length. The manufacturer wishes to feed two materials into this machine. The elastic material is to be fed in a relaxed state. As it enters the machine it must be stretched 75% constantly over 24 inches, and it is to leave the machine in a relaxed state. Simultaneously an equal width of fabric must be fed into the machine to be sewed continuously to the elastic over the 24" length of sewing table.

Design a mechanism to be attached to the sewing machine that will accomplish the automatic positioning of fabric to elastic for purposes of sewing one to another as described above. In addition, include a device to cut straps (relaxed state) 6" in length as the elasticized fabric leaves the 24" sewing table.

PROB. 006 NEW-PRODUCT DESIGN

You have drawn a card listing the name of an electric appliance or article that has become a standard "best seller" to the average family. Consider that you are employed by a company in the position of design engineer; the boss walks into your office on Monday morning and announces that the product you have drawn is beginning to lose sales appeal due to innovations in a similar device produced by your competitor.

You are required to drop all previous product work and concentrate on the design innovation of this product in order to give it new life on the competitive market. Your task is one of coming up with a new twist or improvement to this product that will enhance its sales appeal. The boss has reminded you that any improvement in the product must not increase the selling price more than 10%.

Card A. Electric iron
Card B. Bread toaster
Card C. Electric mixer
Card D. Electric fry pan
Card E. Electric can opener
Card F. Roaster oven
Card G. Electric knife sharpener
Card H. Kitchen radio

PROB. 007   AUTOMATIC TOLL GATES

Many toll roads, such as the Connecticut Turnpike and the Garden State Parkway, collect tolls by having traffic pass through toll booths at periodic intervals to pay a fixed fee. Others, such as the Massachusetts Turnpike, the New Jersey Turnpike, and the Pennsylvania Turnpike, charge a rate based on where the vehicle enters and leaves the highway.

The process of collecting tolls in either case is expensive, it delays the motorist, and often causes traffic jams. The periodic toll systems have solved this problem by installing automatic toll collecting devices. This is a relatively simple problem where the amount to be paid is fixed. The "pay when you get off" stations have not yet devised a system which works.

This identifies the need for a system which is as efficient and effective as the fixed fee automatic systems to be used in the "pay when you get off" highways. This design requires a complete study of the problem so that the proposed system will improve the present situation (cost of collector and inconvenience to the motorist).

PROB. 008   REDESIGN OF UTILITY POLE

Nature has endowed America with great natural beauty, but this beauty will soon be obscured unless we plan our constructions more carefully with some attention to form as well as function. The utility pole is just one example of our neglect and indifference.

Walk or drive down any street in any community in the U.S.A. and observe the utility pole that carries electrical and telephone service to our homes, stores, and institutions. Compare the buildings, trees, sign posts adjacent to the pole and conclude as to your feelings their related forms and functions. Have you taken a photograph of a scene and later when viewing the developed print found that a utility pole was more prominent than the scene you wished to record?

Consider the average utility pole found outside any door as to form and function. Draw a sketch of it. Design an aesthetically pleasing pole, paying particular attention to form as well as function than will blend into the casual observer's scene or will enhance the spacial environment in which it exists. Consider also the relative costs of the new pole as well as the load carrying and erection capabilities.

PROB. 009   COLLAPSIBLE RADAR REFLECTOR

Design a lightweight, compact, collapsible radar reflector and tripod stand that will erect itself upon impact. The designer should concern himself with the linkage structure, methods of attracting links, spring loaded devices, etc., rather than the radar power unit. The solution should include a discussion of possible usage for such a device.

PROB. 010   ARTICULATED LINK TOWER

There is a basic need for a compact, unfolding type tower that can extend itself to a height of about 50 ft. while lifting a 1100 lb. payload. The device should be such that it could be installed on the rear of a truck, possibly on a horizontally mounted drum, that would lift a workman and equipment for purposes of repair of high tension cables, use by telephone linesmen, building repair, sign installation, window washing, etc.

PROB. 011   VARIABLE GEOMETRY STRUCTURES

"Variable Geometry" structures in the form of compact (tubular or flat) packages are required to fit the dimensions of a nose-cone or the limited volume available in a space-craft. Once the structure reaches its destination, it is required to unfold (spring-loaded joints, expanded by an air bag, centrifugal force if ejected from a spinning vehicle, etc.) for purposes of satellites, antennas, reentry vehicles, space stations, space housing, etc. The basic design solution could consist of rings with articulated semicircular or triangular arches. These could be combined to produce a variety of structures. A combination of triangular arches with rigid rings at different diameters would provide a structure that could fold within itself to form a condensed conical shape. A combination of two rigid rings of the same diameter and circular arches could be made to form a spherical shape of maximum volume and minimum surface area. Design such a structure and show its workability through a model.

PROB. 012   TENSION MEASURING SYSTEM

It is desired to determine forces in permanent lightweight structures such as antenna towers, power line towers, etc., so that creep in the material or shifting of soil can be determined. A method for determining forces carried by cables, in place, would make it possible to analyze and keep a recorded case history on such structures.

The desired system should have the following characteristics:

1. Measure tensions of 10,000 lbs. range in a cable. It would be desirable if the system could be adapted to a variety of ranges.
2. Measurements should be accurate within 5%.
3. System must be capable of reading without changing the load in the cable.
4. System or device should introduce no additional flexibility or instability to the structure.
5. Device should be as simple and inexpensive as possible, but it can be assumed that technicians are available to operate it.

PROB. 013   PORTABLE (TAKE HOME) DESIGN KIT

Due to the increasing number of students enrolled in design courses in the College of Engineering, it is impossible to provide adequate facilities for construction of design mock-ups and prototypes. The realization and testing of paper designs in the form of physical models plays an essential role in the education of an engineer.

This problem of laboratory space could be solved if an effective and inexpensive take-home design kit was invented for students enrolled in introductory design courses. The kit should be reasonably light in weight, easily carried, self-contained, and should contain adequate tools, materials, and components for the student to construct devices involved in design projects but at the same time should not suggest how the design solution is to be accomplished or limit it in any way.

PROB. 014   AUTOMATED POSITIONING AND INSPECTION

Random positioned transistors are to be prepared for the final process of attaching leads to the emitter, collector, and base terminals. Design the necessary handling equipment to automatically orientate the transistors face up and the

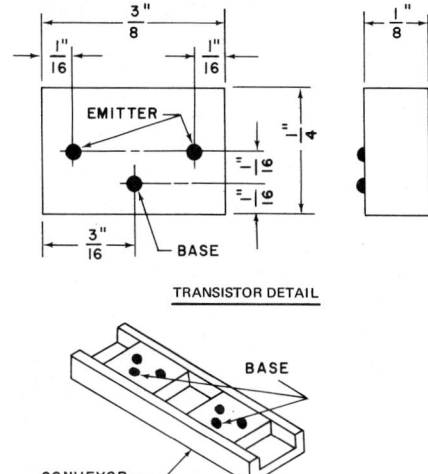

TRANSISTOR DETAIL

CONVEYOR

base terminal facing front, as shown in the conveyer chute. At some point along the system, inspect the emitter to base resistance and reject all transistors whose value of resistance falls outside the range of 9.2 to 9.3 KΩ.

PROB. 015   AUTOMATED ASSEMBLY AND INSPECTION

Design a fully automated system to assemble and test the Ball Bearing Assembly as shown. The inner race, outer race, and separator ring are fed to the assembly area on conveyor belts. The ball bearings are fed to this area along gravity chutes. The bearing unit may be redesigned to accommodate ease in assembly. The design solution should consider the following factors:

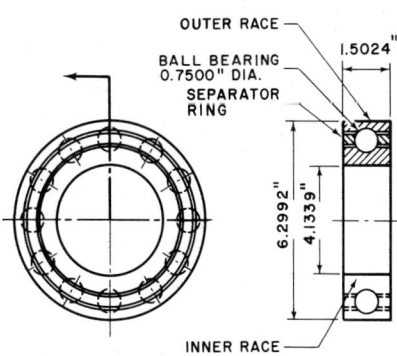

BALL BEARING ASSEMBLY

1. Show each part entering the assembly area (transports).
2. Design the assembly area to be completely automated. Include mechanical, electrical, pneumatic, and/or hydraulic hardware to accomplish the assembly. Include control equipment to meter and index hardware involved in the automated assembly.
3. Design a device to inspect each assembled unit for a minimum frictional torque of 3 inch-ounces and reject all bearings under this minimum.
4. Transport bearings from the assembly area to that packaging of units can easily be handled.

PROB. 016 AUTOMATED MANUFACTURING PROCESS

Design a fully automated system that will receive the housing of a hydraulic cylinder, as shown, in the form of a casting, and perform the following machining operations:

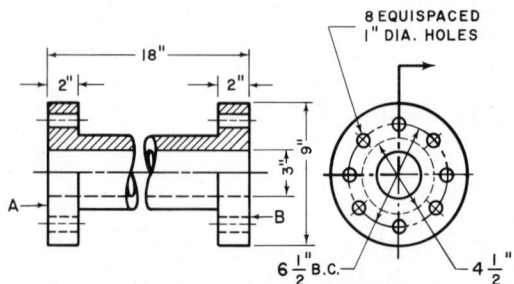

HYDRAULIC CYLINDER HOUSING

1. Face surfaces A and B.
2. Ream the 3″ dia. hole to size.
3. Drill eight 1″ dia. holes on the $6\frac{1}{2}$″ dia. bolt circle.

The design solution should consider the following factors:

1. Orientation of the casting for machining operations.
2. Types of machining equipment necessary to handle task.
3. Jigs and fixtures necessary for the positioning and clamping of the casting. Include control equipment to index, transport, and position castings in the automated area.
4. Transport finished castings from the machining area in a manner that can be adapted to the next phase, that of assembly.

PROB. 017   SELF-PROPELLED WATER SKIS

The production of fiberglass boats with reliable high speed motors (outboard and inboard-outboard drives) has in recent years opened up a whole new industry of water sports. Waterskiing and aquaplaning, which include skis, tow ropes, life belts and vests, wet suits; skin diving; sport fishing; marsh boating with shallow water drive, are new industries dependent upon the small power boat industry.

One method of "breaking into" this industry is to design a product that is already accepted by water enthusiasts but does not require a large investment to enjoy yet is safe in its operation. Such a product could very well be self-propelled water skis. If a pair of skis could be devised requiring no boat for assistance and give the same maneuverability and thrill as present regular and slalom skis, it could command a large part of the water sports market. Here is an excellent opportunity for the young designer with imaginative ideas.

PROB. 018   SELF-CLOSING WINDOWS WHEN IT RAINS

Present manufacturers of windows for the home have shown few if any innovations in window design in recent years. This industry seems to rely on quality and ease of installation to maintain good customer relationships. There have been preservatives added to wood sash, toxic treatment against termites and decay, aluminum storm and screen units, and opening and closing hardware, but these could hardly be classified as real innovations in design.

Consider yourself employed by a small window manufacturer who wishes to increase sales but is prevented from doing so by the image of the big four in windows, Anderson, Morgan, Brosco, and Reuten. One method of entering this competition is through a new product which would be completely novel to the window industry. Such a product could be a device that is either built in upon installation or attached to the standard window (casement or double hung) that would close the window when it rains. You are instructed, as a design engineer with this company, to research the need for such a device and present a series of alternative concepts to management so that a decision may be made as to the feasibility of marketing such a product.

PROB. 019   SPEEDOMETER FOR SMALL SAILBOATS

Design a speedometer that may be easily installed on the average small sailboat (sunfish, sailfish, penguin, lightning, comet, 110, 210, snipe, star, etc.). This instrument should in no way interfere with the sailability and speed of the boat. There is a genuine need for such a device since there are no boat speedometers available that will record speed accurately below 10 mph. Such a device would be effective as an instrument in "tuning" a sailboat for optimum speed and to compare hull speed at each point of sailing (reach, run, tack, and with spinnaker, etc.).

PROB. 020   BOAT PIER FOR FRESH WATER LAKES

Design a boat and swimming pier to be used in fresh water lakes in the northern part of the country. The pier must be relatively inexpensive, light in weight, durable, and able to be set in place by one person. It should be adjustable to uneven bottom conditions (sudden drop-off, rocky, muddy). Remembering that lakes freeze in the north (from 12″ to 36″), the pier must be removed and stored each fall and reassembled at the lake in the spring.

A2 DESIGN PROJECTS

Design Projects like the problems in Section A-1 represent realistic engineering situations requiring imagination and creativity on the part of the designer to arrive at a reasonable solution. These projects, however, are a bit more involved and require the application of the entire design process for their solution. More time must be allowed for the designer to examine each phase of the design process as it is applied to the project. While many of the problems of Section A-1 were completed at the conceptual phase, projects in this section are expected to be presented in complete solution form involving analysis, feasibility, prototype whenever applicable, manufacturing capabilities, cost breakdown, and consideration of distribution and consumption.

PROJ. 021   ANALOG COMPUTER

design specification

Assume you are a mechanical designer employed by a company engaged in research and development work for the U.S. Navy on defense contracts. You have a general background education, during your undergraduate years, in Kinematics of Machinery and an introductory knowledge of Analog Computer Systems. On Friday, March 31, 19–, while you were thinking about recreational activities over the coming weekend (that last chance at spring skiing, a date with the intelligent blond, two days with the kids, two days of loafing and sleep, etc.), the following problem was dropped on your desk with a solution required to be presented to the top brass on April 21, 19–.

"The diving officer and the navigator of an atomic submarine must be provided with a continuous display of information similar to that supplied by a destroyer's D.R.T. (dead reckoning tracer) as follows:

288   THE SCIENCE OF ENGINEERING DESIGN

SHIP'S GYRO HEADING (o)   DISPLACEMENT N-S
SHIP'S SPEED (Knots)   DISPLACEMENT E-W
ANGLE OF DIVE (o)   TOTAL HOR. DISP.

RATE OF DIVE
ACCELERATION OF DIVE

The quantities in the first column are to be considered as input signals and all other quantities are to be computed. The quantities RATE OF DIVE and ACCELERATION OF DIVE are the vertical components of ship's velocity and acceleration.

Present the design of a mechanical analog computer complete with interconnectors, scale constants, and direct output displays in console form as required."

Once you have survived the initial shock of having your pleasant dreams on this Friday morning interrupted and realize that you want to keep this job, you begin to remember the following:

In working out the design for an analog computer system it is first necessary to obtain the relation between the input and output quantities in mathematical form. Once this is accomplished, the designer proceeds to mechanize the equations using standard computer elements (first in the form of a block diagram and then as a schematic showing computer elements in symbolic form). Finally, the system is detailed and means of interconnecting elements and handling of constants are devised and the console and display are finally designed.

PROJ. 022 REMOTE MECHANICAL HAND FOR NUCLEAR REACTOR

An engineering college is involved in the building of a small reactor for laboratory experiments in conjunction with its nuclear engineering program. Your company has been engaged to supply a remote mechanical hand for the purpose of inserting carbon control rods in the reactor.

Minimum Requirements: Design a mechanical linkage for removing rod A from the storage rack and inserting it into hole A of the reactor as shown in Figure 1. Key dimensions between storage rack and reactor face are shown in Figure 2. All controls (mechanical, manual, electrical, or pneumatic) must terminate on the "safe" side of the shield.

Additional Requirements: Design the device to accomplish the same function for any other rod-hole combination as illustrated in Figure 1.

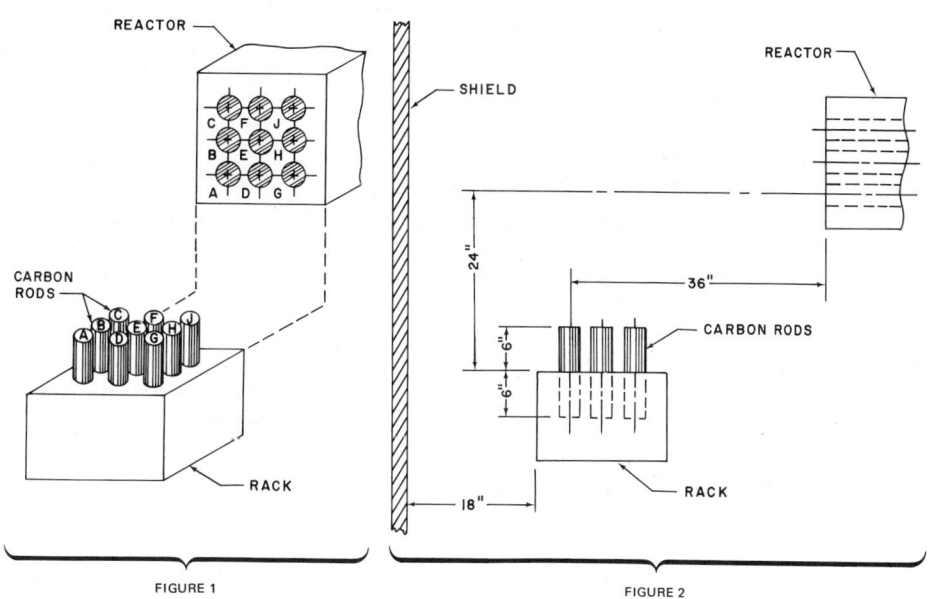

FIGURE 1

FIGURE 2

PROJ. 023   TAIL GATE ASSIST FOR STATION WAGON

design specification

THE NEED

During the last decade the American automotive industry has brought into popular use the combination shopping cart, school bus, pick-up truck, mobile home, and family car known as the Station Wagon. This vehicle has had more effect on the driving habits of the country than any of the so-called new developments in autos in the past twenty years. Its popularity continues to grow and along with this, the uses to which it can be put seem to extend wider and wider. Recent innovations in Station Wagon design among manufacturers are centered on convenience, as evidenced by the 1962 Studebaker with the sliding roof panel to allow articles to be carried and extend beyond the roof line, the 1965 Buick with glass panels providing the "sky-view," and the 1966 Ford line with the "Magic Doorgate" (swings open like a door and folds down like a tail gate).

One of the difficulties which a station wagon owner faces is that his wagon can easily transport objects which he cannot easily lift into it. Assuming a standard passenger car suspension, it would not be unreasonable to expect to place a weight equivalent to three adults at the rear of the vehicle (about 600 lbs.). Objects which one might be expected to transport but have difficulty in lifting into the rear of the wagon include outboard motors, lawn mowers, snow blowers, small trees for planting, etc.

THE GOAL

Design a system which will permit the use of the station wagon tail gate as a platform and elevator to raise heavy loads from the ground to the level of the floor. The system should be hand operated if possible and capable of increasing the weight-lifting capacity of the average man by one hundred and fifty pounds (150#). The system should be contained in the existing body work and should not increase the cost of the car more than $150.

EXHIBIT 1   Tail Gate Assembly
EXHIBIT 2   Body Assembly
EXHIBIT 3   Side Body Assembly

1967 Ford Station Wagon
Courtesy Ford Motor Company

PROJ. 023 (CONT) 291

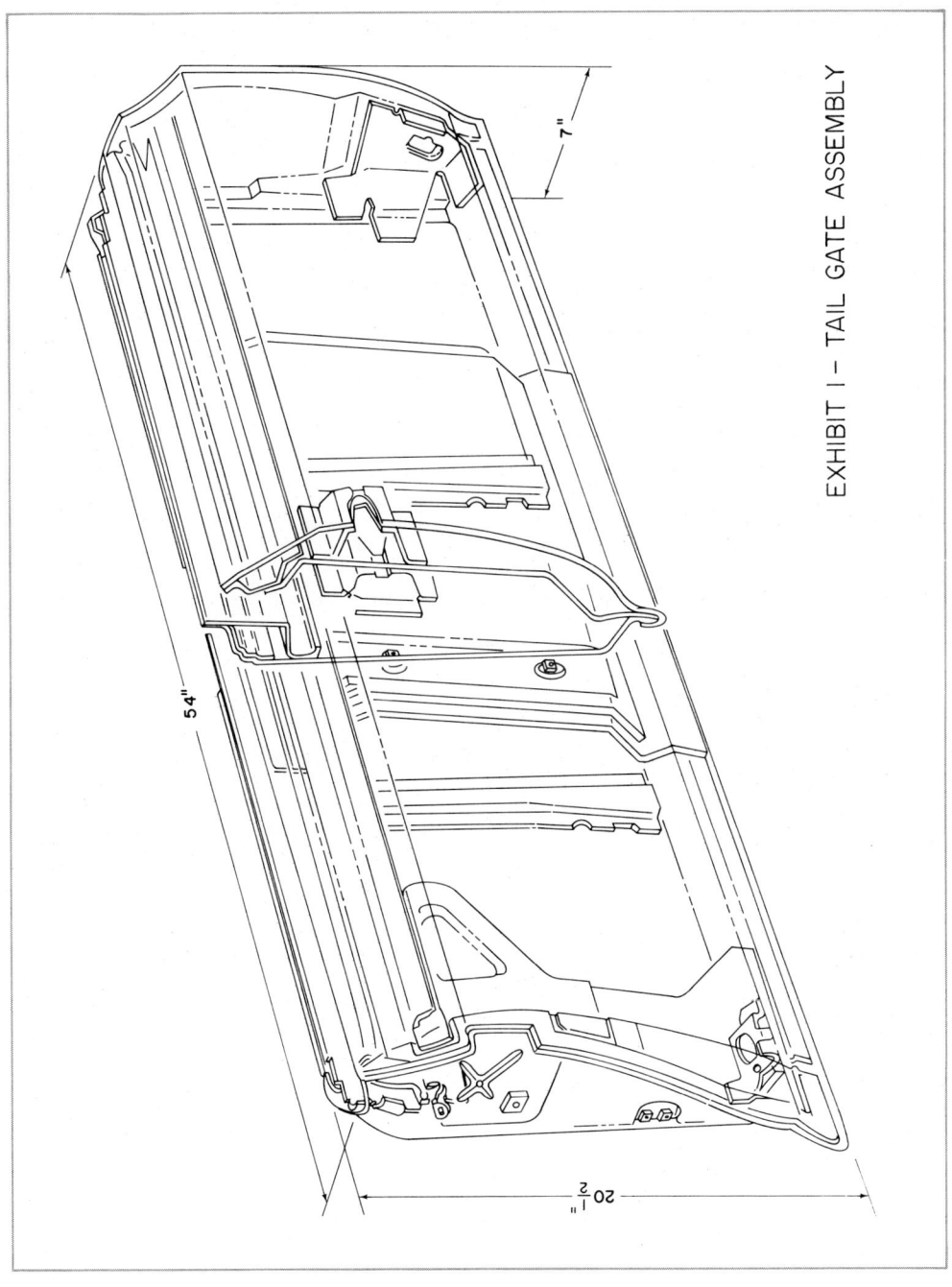

EXHIBIT I – TAIL GATE ASSEMBLY

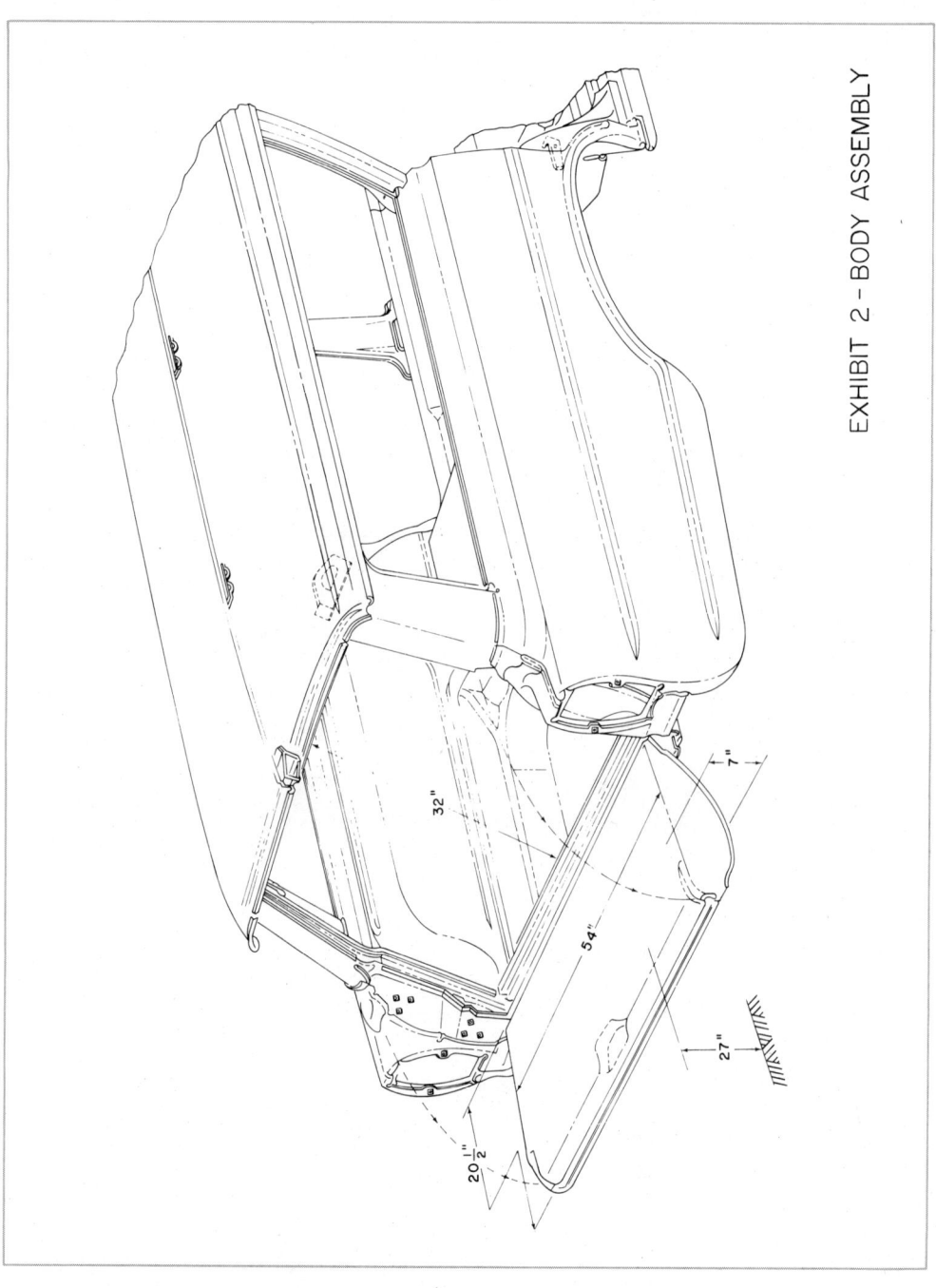

EXHIBIT 2 - BODY ASSEMBLY

PROJ. 023 (CONT)   293

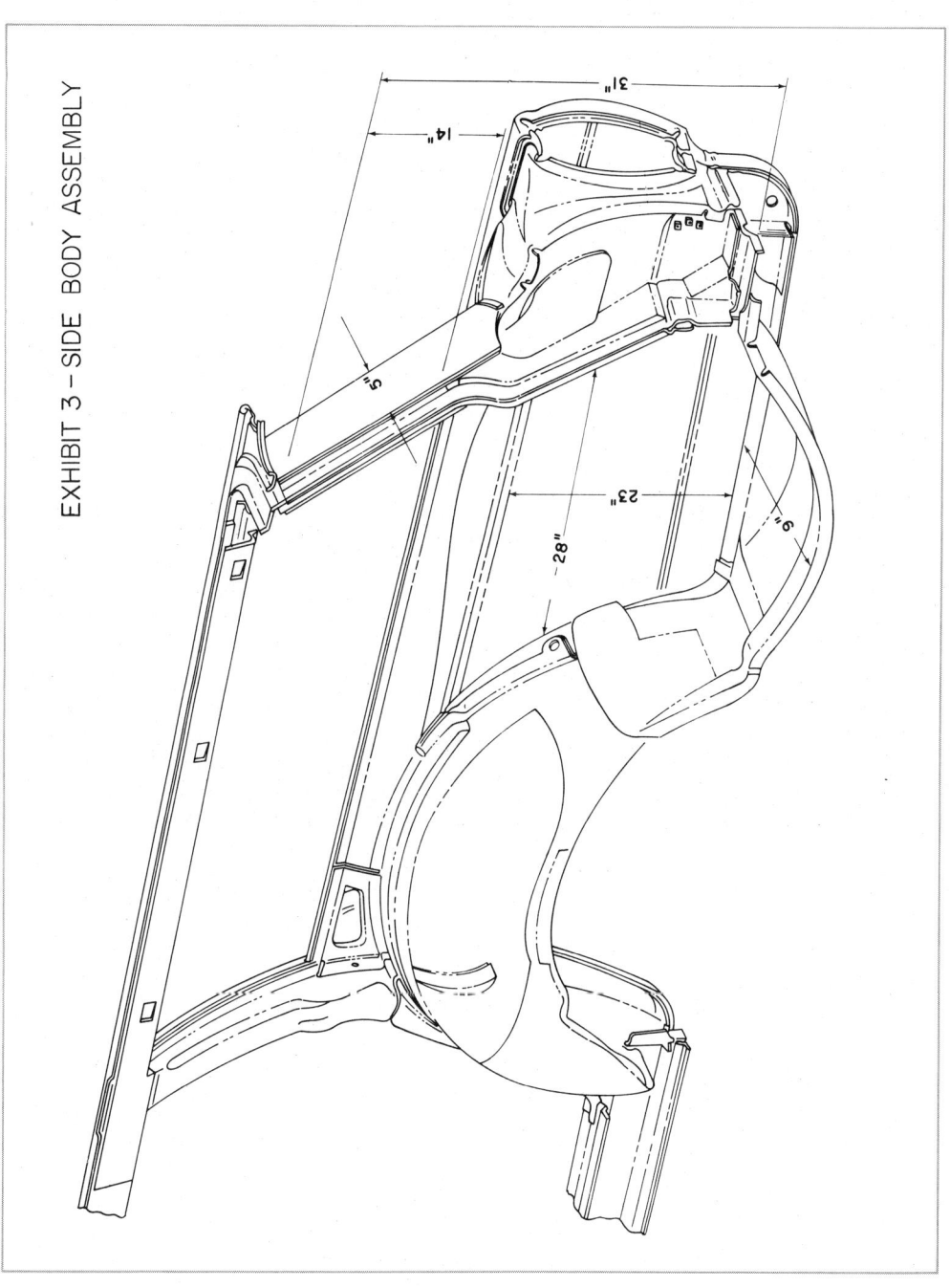

EXHIBIT 3 – SIDE BODY ASSEMBLY

PROJ. 024   PEACE CORPS REQUEST

March 22, 19--

Peace Corps

Sincelejo, Bolivar, Columbia

South America

Vita
230 State Street
Schenectady, New York   12305

Dear Sirs:

    We are now having a baseball tournament in Sincelejo and we would like to be able to mark the baseball field in the proper manner. Will you please send me a set of working plans for a simple machine that can be used for marking the baseball field.

    We have plenty of sheet metal and other steel items that we can use for the construction of the apparatus. We will be marking the field with lime.

Thank you very much,

Nicolas Mansito, Jr.
NICOLAS MANSITO, JR.
Peace Corps Volunteer

Nicolas Mansito Jr.
Voluntario-Cuerpo de Paz
Sincelejo, (Bolivar)
Columbia, South America

PROJ. 025   TRANSITIONAL TECHNOLOGY

Dec. 1, 19--

Dear Sirs —

    Your service is indeed a valuable and necessary one. I was very happy to learn of it and hope that you can be of service to us, as I have a number of requests and/or problems, which I hope you can help me solve.

    (A) Designs for a Seed Cleaning Machine Operated by Hand — I have seen such a machine that was operated by an electric motor, but am unable to remember the details. It would be utilized for cleaning cereal grains such as rice, wheat and corn. We will have to manufacture the screens ourselves. At our disposal we have a good assortment of tools including an arc welder.

    (B) Plans for a Garden Tractor with a Seed Drill Attachment — Garden tractors are available here only through importing. This one would have to be operated by a 3 to 5 h.p. gasoline motor. Plans for attachments such as a tiller, rotary hoe, plow and mower would also be useful.

    (C) Designs for a Jute Fiber Extraction Machine — It should be hand operated and designed to leave the pith of the jute intact, as it is utilized in building.

    (D) Designs for a Row Seed Drill Powered by Hand.

    (E) Plans for a Hand Powered Corn Husking Machine — for removing the kernels from the cob.

    (F) Lastly a personal request that would insure my toleration in the task of pursuing the realization of these plans. How do you make vodka from potatoes? How do you convert barley into malt? And do you have any recipes whereby molasses could be utilized in place of sugar for beer making?

    Sincerely,

P.C.V. Ron Gillespie

P.O. Doyabari
Dist. Nadia
West Bengal
INDIA

PROJ. 026   POWER ASSIST SHOVEL

MORE – POWER MOWER CORPORATION

To: Advanced Design Group   Date: April 25, 19--

From: F.A. Mitchell, President

Subject: Power Assist Shovel

     The present inclination of the American public towards motorized do-it-yourself gadgets has, as you know, put our company in a very favorable position in the market place. Our sales have increased on an average of 50% per year over the last four years and the profit figures have followed suit. Most of this increase is due to our line of powered garden tools and related items (power mowers, hedge clippers, roto tillers, snow blowers, etc.). We know that we have excellent equipment and our reputation is quite good in this area, but the need for expansion is indicated.

     At my request, our sales department with some help from outside consultants has made a study to attempt to predict the future behavior of this line, as well as trying to gain some insight into our customers' needs and desires. The results were startling. The study indicates that although at this moment sales are doing very well, we might be rapidly approaching a saturation point. If this happens we might find ourselves in a difficult position. The most likely solution seems to be a new product. With this in mind it has been decided to seek a product which might fit in with our present marketing program, fill a need which presently exists and pick up the anticipated slack in our production. If our Marketing Staff can pin saturation point figures down, we just might be able to get a real jump on the competition.

     One idea has occurred to us which we feel has a great deal of potential, a POWER SHOVEL. We have discussed this idea with the marketing group and they concur with management's opinion.

     The idea is to make a shovel with some sort of power assist which will aid the user in performing several or many of the functions for which a shovel can be used. It seems to be certain that this is the one tool which is hard to use and which no one has attempted to improve.

     We would like to see proposed designs for this device on <u>May 13 for Critique on May 16, 19--</u>. We expect these designs to be thoroughly analyzed and detailed by this date for we hope to have prototypes built and tested by June.

PROJ. 027   TRANSPORTATION OF VENUS DE MILO

TRAFALGAR MARINE INSURANCE LIMITED
71 - LOMBARD STREET - EC 3
LONDON

February 1, 19--

B.C. Big Corp.
1100 College Avenue
Medford, Massachusetts   02155   U.S.A.

Gentlemen:

    The Louvre Museum in Paris is considering lending the statue of Venus de Milo for temporary display at the Boston Museum of Fine Arts. This company has been requested to provide insurance for the sculpture during her journey. After careful consideration, our board of directors has decided to submit a proposal for the tranportation and insurance of this priceless and delicate work of art. We have taken the position that we will not consider insuring unless full control of the operation is in our hands.

    It has been decided to request your group among others to examine the problem and propose methods for accomplishing this task. We feel that none of the usual methods for handling a shipment of this type will be satisfactory. We would like to see a system which is capable of bringing the statue through any credible accident.

    Several of our directors will be in Medford on April 1, and they would like to examine your proposals at that time. Pertinent data on the statue is on the accompanying drawing.

                                                   Sincerely yours,

                                                  Alec S. Wilson

ASW/ehh   Alec S. Wilson
Enclosure   Vice-President

298   PROJ. 027 (CONT)

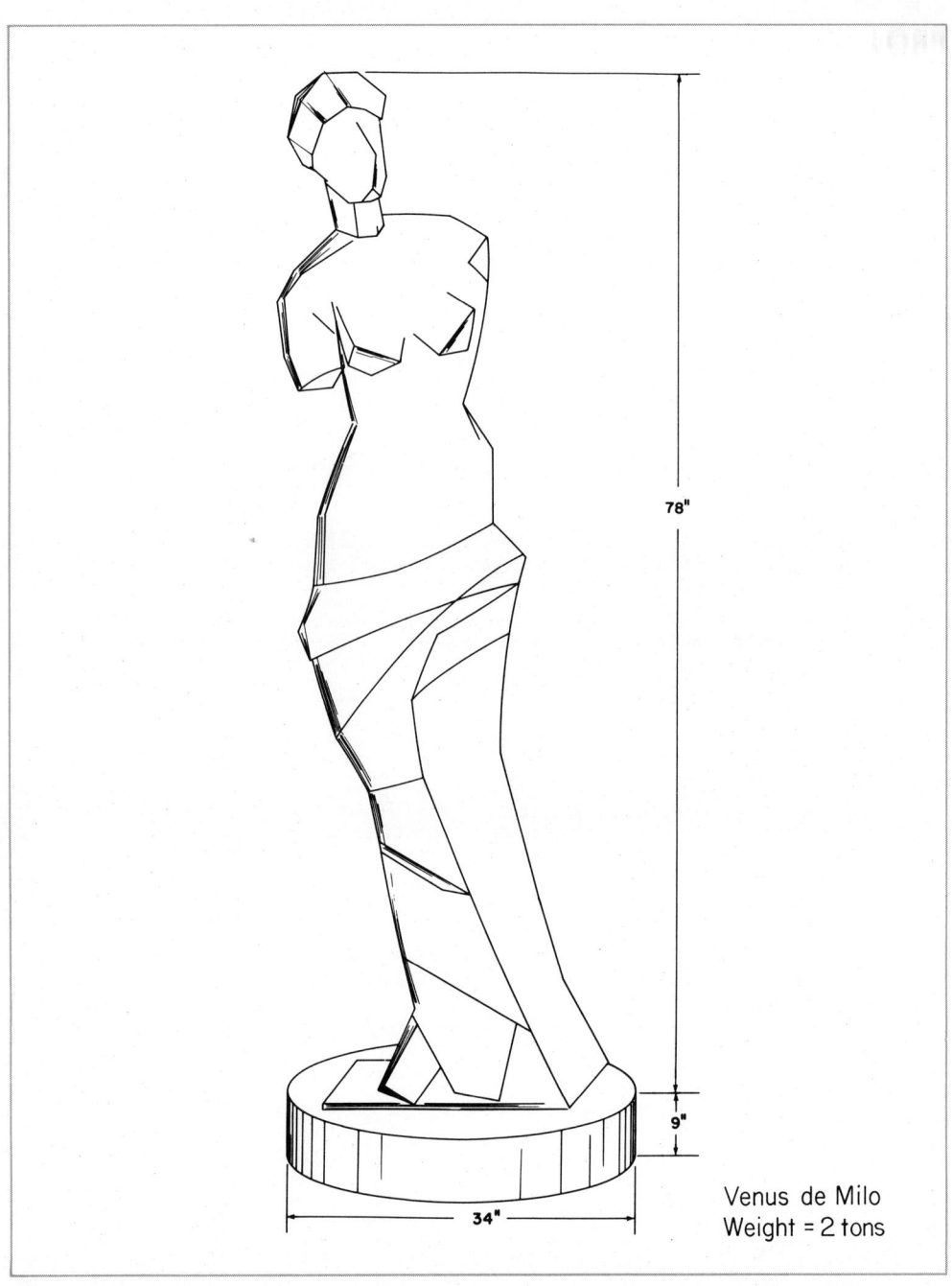

Venus de Milo
Weight = 2 tons

PROJ. 028   SPACE TOOLS

POWERMORE TOOL COMPANY, INC.

1123 COLLEGE AVENUE

MEDFORD, MASSACHUSETTS 01255

February 14, 19--

M E M O R A N D U M

From: Chief Project Engineer
To: Advanced Design Group
Subject: Initial Design of Space Tools as pilot program for NASA contact.

    Our company has been fortunate in being awarded a pilot design contract by the Air Force under sponsorship by NASA to develop initial designs and prototypes of <u>SPACE TOOLS</u> to be used by man to perform assembly tasks in space in connection with the Manned Orbital Laboratory Project. Project MOL, scheduled for launching in mid-19--, will make use of Gemini 8 as a vehicle to require an astronaut to climb out of his orbiting capsule and tighten four bolts with a special tool designed for the Air Force, hopefully by this company.

    This memorandum assigns the Advanced Design Group the task of designing a minimum reation power tool to exert a torque of from 15-40 ft. lbs. for tightening a bolt, yet impart only a 0.0116 ft. lb. **reaction** on the dangling astronaut. All power must be self-contained in the unit and the tool must be equipped with screwdriver, hammer, thread tap or drill attachment, as well as a socket wrench. Our competition has proposed a device that generates a short-duration, high amplitude torque in one direction for tightening through an instrument that is quite bulky to handle.

    It should be remembered that the acceptance of this contract by management places our company in a competitive position with other leading tool companies. If our designs are considered superior by Air Force officials, this could mean a sizable and continued contract. All design solutions and prototypes are to be presented before Air Force officials on March 11, 19--.

PROJ. 029   MOBILENAUT

POWERMORE TOOL COMPANY, INC.

1123 COLLEGE AVENUE

MEDFORD, MASSACHUSETTS 01255

February 25, 19--

M E M O R A N D U M

From: Chief Project Engineer
To: Kinematic Design Group
Subject: Initial Design of Space Simulator as pilot program for NASA Contract.

    Our company has been fortunate in being awarded a pilot design contract by the Air Force under sponsorship by NASA, to develop initial designs and prototypes of <u>SPACE TOOLS</u> to be used by man to perform assembly tasks in space in connection with the Manned Orbital Laboratory Project hereafter referred to as project MOL. The advanced Design Group has been assigned this project with a scheduled date of completion on March 11, 19--.

    In conjunction with this project a simulator device must be designed and constructed so that the effect of the proposed space tools may be tested in the laboratory under conditions as they exist in space. This laboratory device should take the form of a six-degree-of-freedom mechanical simulator which reproduces the mobility problems an astronaut will experience and endure under reduced gravity.

    This memorandum assigns the Kinematic Design Group the task of designing an experimental <u>MOBILENAUT</u> in which an astronaut will be strapped into the cradle, his legs and torso are then immobile, but his arms are free to provide torque by pushing away from a spacecraft, pulling a tethering line, or using a tool. The astronaut must be unlimited in freedom in the pitch, roll, and yaw directions as well as translational movements along the X, Y, and Z axes. Ventilation and breathing lines will be provided, allowing the subject to work wearing a pressurized (3.7 psi) space suit. A 12-cfm vented flow cools the suit.

    If this device can be sucessfully developed by our company for internal testing of space tools, it may lead to additional contracts for the testing and study of small thrusters, hand held air hoses omitting air under 100 psi pressure producing a jet effect of about 5 lb. thrust, the changing of counterweights to simulate movement on the moon's surface of five-sixths of subject's weight, and experiments of assembling a space station. Needless to say, the successful completion of this design study means a great deal to the company.

    All design solutions and prototypes are to be presented before management and advanced design groups on March 4, 19--.

PROJ. 030 VITA REQUEST

COOPERATIVE FOR AMERICAN RELIEF EVERYWHERE, INC.

KATIGSAK BUILDING
KALAW AND MABINI STREETS
MIDWEST REGION
NIGERIA

CABLE CAREINC., NIGERIA
TELS 482-6
     483-6

VITA, INC.
230 State Street
Schenectady, New York 12305

Gentlemen:

    I am a Peace Corps Volunteer in the Midwestern Region of Nigeria. Recently, I have been given the responsibility of expanding the primary school facilities in the region. In order to do this I will need detailed plans and materials estimates for simple, 3 room cement block school buildings.

    Can you create for me, or put me in touch with, someone who can supply me with this sort of information? My needs are for plans which will not take a very high level of skill to carry out. Thus, for example, the rafter patterning and construction should be as basic as possible.

    Questions that I'd be grateful if you'd answer in detail are:

1. Will I need to use reinforcing rod as an integral part of the wall or should that be confined only to the supporting pillars which will offer the main support to the wall?

2. Will the foundation have to be tied in to the cement slab floor, or should the floor be poured last?

3. Should I use any thing other than a wooden beam across the top of my window holes in order to support the weight of the wall above? What size beam (or piece of steel) is necessary to bridge what size gap with what amount of wall above it?

4. How much eave space is required to provide adequate shelter from hard tropical rain, thus allowing windows to remain open during rain storms?

5. What is the simplest way to build a two-child primary school classroom bench and desk?

    If you need any more detailed information in order to answer my questions, please write me as soon as possible.

    Thank you for your assistance.

                              Yours,

                              Alan Crew

PROJ. 031 AUTOMATED ASSEMBLY

M E M O R A N D U M

March 12, 19--

From: Chief Engineer
To: Automation Design Group
Subject: Automated assembly of fluorescent light starters

    The Surestart Corporation has engaged our company to present the design of an automated assembly system for one of its major products. This product is a small "can type" starter for fluorescent lamps which is presently being hand-assembled. Due to the rising cost of labor in its union shop, the company finds they must raise the price on this item, placing them in a poor competitive position. Management is considering the possibility of completely automating the assembly process as a means of achieving a reduction in unit price on these starters. Detail drawings of the components and assembled starter are attached for your information.

    There are seven unit parts in the assembly; the case, an insulating tube, the base, a capacitor, a thermal relay, and two contact prongs. Of these parts, all but the base are delivered, ready for assembly, to the assembly area. The base is stamped from fiberboard on the assembly floor. The capacitor and the relay are assembled "in house" at another location and shipped in bulk. The contact prongs are shipped in bulk from an outside supplier. The case and insulator are also purchased but these are shipped in an orderly arrangement.

    The present assembly process consists of two sub assembly steps and the final joining. The first sub assembly is the electronic module. This involves one worker laying the relays on a conveyor, another placing the capacitors with them, a third worker twists the leads of these two units together. Three others are meanwhile sorting and assembling the contacts and the base pieces. Another uses a press to peen the rivet like end of the contact. The prepared base is threaded with the twisted wires. This operation is done by several workers in parallel. At the next station these contacts are heated and then a small drop of solder is placed on the outer face. The solder "sweats" up the tube and a good contact is usually formed.

    The second sub-assembly consists of the case and insulator. These are hand assembled by two workers who are supplied with parts by a third.

    Final assembly consists of placing the electronic unit in the case and bending over the tabs. This is done with a tool similar to a pliers but built for this operation. The excess solder is then ground off the face of the contacts and the unit proceeds to testing.

    A unit is considered acceptable if an application of 50 volts will cause the relay to open in not less than 1.5 seconds nor more than 3 seconds, and remain open for 10 seconds.

    We will meet on March 18th and at that time finalize the overall methods to be used. At that time the specific designs of stages will be assigned to the various members of the group. These designs should be prepared for presentation on April 23.

PROJ. 031 (CONT)   303

DETAIL OF COMPONENTS

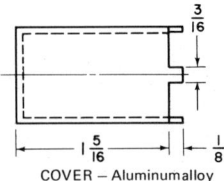
COVER — Aluminum alloy

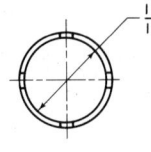

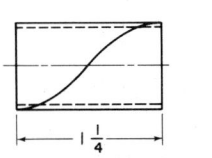
INSULATOR — Cardboard

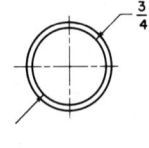

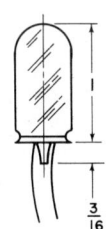
THERMAL RELAY — Assembled (glass)

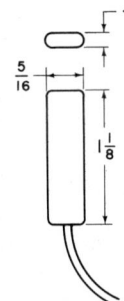
CONDENSER — Assembled
(Wire and paper)

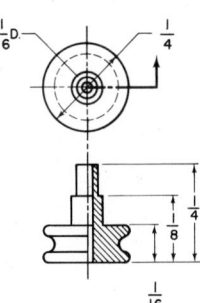
CONTACT — Brass
(two req'd)
Scale: 3X

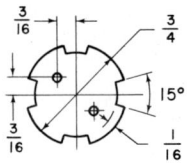

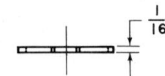
BASEPLATE — Fiberboard

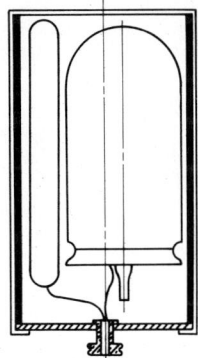
SECTION THROUGH STARTER
Scale: 2X

PROJ. 032   TUBE FLIGHT

April 9, 19--

M E M O R A N D U M

To: Department Design Group
From: S.E. Hall, Director of Research
Subject: High Speed transportation system for the Boston to Washington Corridor.

There is a great need at the present time and an urgent need in the future for an all-weather, ground-level passenger transportation system from Boston to Washington following a route to be known as the Northeast Corridor. This is a 460-mile route along the Eastern seaboard between Boston and Washington which embraces the nation's most heavily unbroken populated area, coupled with a virtually consistent corridor of manufacturing and industrial activities. The Department of Commerce is engaged in a program to upgrade and realize passenger transportation service along this corridor, or as Senator Claiborne Pell (D.-R.I.), the godfather of the plan calls it: the "megalopolis."

This proposal was bought to light in President Johnson's State of the Union message. The President's plans calls for a three-year research and test program, an initial $20 million outlay the first year. The long range research program could extend over the next decade and a half. However, the demonstration project could come into being next year.

As currently visualized, some $10 million would be spent for research and development of new ground transportation systems, about $8 million for demonstration projects, and $2 million for statistical studies.

The Department Design Group is hereby assigned the task of performing basic research studies for such a transportation system to meet the demands stated in this memorandum, to undertake feasibility studies on available systems, and to produce an initial design study of a basic system to form the basis for a proposal to the government for a contract to produce a workable design of the transportation system for the Northeast Corridor.

Our operating plan is to work as a group during the initial phases of investigation and then divide into sub-groups for concentrated design of sub-systems. Each sub-group leader will report directly to me as coordinator of the project. Our task is one of proving the feasibility of a specific system of transportation and producing preliminary design studies that can be defended in critique before invited government officials, the latter part of May 19 -- .

With my present knowledge of tranportation systems that are capable of producing safe high-speed passenger service, I have concluded that the system involving a vehicle in a tube, invented by Dr. Joseph V. Foa, Head of the Department of Aeronautics and Astronautics, Rensselaer Polytechnic Institute, is the most feasible. This conclusion was reached through a knowledge of the Tokaido Line, operated by the Japanese National Railways between Tokyo and Osaka at speeds in excess of 150 MPH; monorail systems as used in Seattle, Washington, and Houston, Texas; and the magnetically floated vehicles being investigated by Westinghouse Electric Corp.

Our first task, therefore, will be the investigation of the possibility of transporting a vehicle in a tube, which hereafter will be known as: TUBE FLIGHT. Additional data and research studies will be distributed to you as the design investigation develops.

PROJ. 033 STAIR-CLIMBING WHEEL CHAIR

NATIONAL INVENTORS COUNCIL
U. S. DEPARTMENT OF COMMERCE
WASHINGTON, D. C. 20550

There is a great need for a STAIR-CLIMBING WHEEL CHAIR. The objective of such a chair is to give an active handicapped individual an additional range of mobility. The desired design should enable him to cope with the usual problems he might encounter traveling to and from work, moving about industrial buildings, and the like. It must be remembered that the chair will perform the usual wheel chair function approximately 95% of the time; therefore, not too much of the conventional wheel chair's versatility and convenience should be sacrificed in providing the climbing function. In this connection it might be well to point out that many active handicapped are able to fold a conventional wheel chair, put it in an automobile and drive to work. If this cannot be accomplished with a climbing chair, the overall objectives of providing the handicapped with independent mobility will not be achieved.

The following factors are to be taken into consideration:

1. WEIGHT OF OCCUPANT - Assumed to be 200 pounds.

2. WEIGHT OF CHAIR - 50 to 75 pounds, maximum.

3. COLLAPSIBILITY - Capable of being folded by user and stowed in the interior of standard automobile or taxi cab. Special loading ramps which are not carried as a part of the chair system or special modifications to automobile to accomodate chairs will not be considered.

4. WIDTH OF CHAIR - 25 inches maximum. The American Standards Association utilized this width, as the maximum for standard wheel chairs, in establishing architectural standards which will provide the physical handicapped with freedom of movement in buildings and other structures. However, a lesser width for a chair or a means of temporarily reducing the full open width for passage through narrow doorways would be a definite asset in negotiating door openings in private homes or other structures which do not conform to the new standards.

5. TURNING ABILITY - The turning radius should be held to a bare minimum. Turning should be accomplished without damage to floors or carpets. Chair must be able to negotiate an L - or a U - type stair landing. Ability to negotiate an L-type landing as small as 3 feet by 3 feet would be considered an asset.

6. CLIMBING ABILITY - Chair should be able to negotiate street curbings and any stairs with average height risers and depth of tread as found in office buildings and homes. Ascent and descent should be performed in presence of litter and moisture without damaging stair treads and risers. Any transition or adjustments required between normal and ascending and descending functions must be accomplished in a minimum time to eliminate delays at street curbings and stairways.

7. PROPULSION SYSTEM - Chair may be propelled by occupant or by a motorized unit, however total cost, weight, and callapsibility requirements are the same for either type of drive. If self-driven, a minimum arm strength of 10 pounds may be used in devising drive mechanism.

8. CHAIR SAFETY - The presence of an attendant, though undesirable, is permissable during ascent or descent. However, the chair occupant should always be in a reasonable state of balance so no more than 15 to 25 pounds of weight will be transferred to attendant. The chair must be "fail safe" to prevent uncontrolled descent in the event something happens to occupant, attendant, or mechanism.

9. COST - In order for the maximum number of handicapped to be able to purchase a stair-climbing wheel chair, every effort should be made to keep its cost at a bare minimum. The retail cost should be no more than $200. Current models of standard tubular frame folding wheel chairs are priced at approximately $150 retail.

10. OTHER CONSIDERATIONS - Chair should be capable of performing all its normal functions without undue jostling or jouncing of the occupant. Since most paraplegics have little control of the lower limbs, a method of retaining them in position is often required. The foot rests should be movable to enable the occupant to raise himself to a standing position on the floor.

11. The chair should not require the installation of special ramps, mechanical contrivances, or electrical outlets in buildings.

STANDARD TUBULAR FRAME WHEEL CHAIR

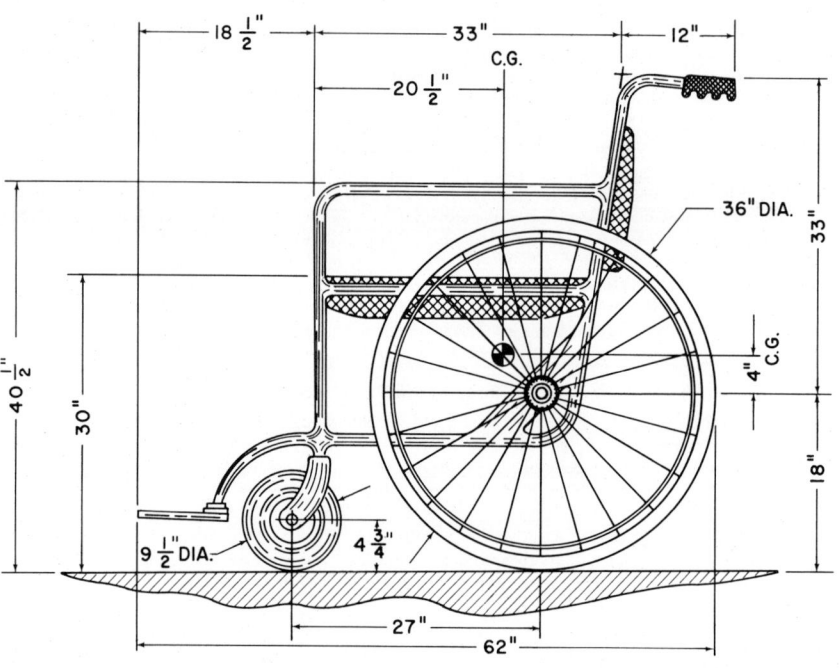

PROJ. 034 COMPUTER CONTROLLED PARKING GARAGE

DESIGN SPECIFICATION

Design a fully automatic computer-controlled parking garage system accommodating at least several hundred automobiles and operated by a single cashier-attendant.

NEED: There is a basic need for an automatic short-term parking garage designed for busy locations with high land costs and applicable to a large variety of layouts, including plots too narrow for other types of parking. Such a system must be an improvement upon conventional elevator or ramp parking systems.

SPECIFICATIONS: All operations must be performed by one attendant from his control console. The system should be capable of parking about 27 cars in a ten-minute period. To park a car, the attendant selects a key for an empty stall, inserts it in the control console's keyhole. The key acts as "punch-hole card" for the computer and is carried by motorist as a claim check. Computer should not only "remember" and "direct" parking and unparking operations but also calculate the parking fee.

SCOPE: Due to the large scope of this problem, designers are to be divided into teams in order to undertake the study of separate aspects of the design solution. Each group should meet regularly with other teams in order that their solution will fit into the overall scheme.

TEAM 1-STRUCTURE AND STALL DESIGN	TEAM 2-LIFT AND CONVEYOR MECHANISM DESIGN	TEAM 3-COMPUTER PROGRAM AND DESIGN
_____	_____	_____
_____	_____	_____
_____	_____	_____
_____	_____	_____

PROJ. 035   SIMULATOR FOR AUTOMOTIVE DRIVER TEST

M E M O R A N D U M

April 13, 19--

To: Advanced Engineer Design Group
From: J.R. Smart, Chief Engineer
Subject: Design of Simulator for the Automotive Driver Test

The increasing number of automobiles using the nation's highways, lack of standardization of inspection procedures, and the inaccurate search of information in the operator's test have resulted in record fatilities in automobile accidents in recent years. The highway engineer states that the rate of accidents could be reduced through improved highway design. Improved highways exist, but fatalities on these highways are on the increase. Many states have excellent inspection procedures, Michigan has none. Standardization of inspection procedures as well as hand signals, roadside signs, etc., may be a problem of states rights. The automotive manufacturer is engaged in accident prevention research and has recently claimed that fatal accidents will eventually level off between 25,000 and 30,000 per year through safety devices engineered into the automobile, but sees no hope of reducing this figure.

The motoring public will continue to have accidents, many of them fatal, no matter what preventive measures are taken, but the rate of accidents can be greatly reduced through an intelligent solution to the problem. One of the greatest and possibly the most important area in accident prevention needing continued study and research is the operator (driver) himself. No automobile of the future, optimum highway design, efficient inspection procedures, policing of the highways, platoon system of travel, etc., is the answer to safe driving as long as the instrument of travel (automobile) is in the hands of a human operator.

It is about time the Design Engineer turned his talent to the problem of automotive accident prevention and attempted to advance the state of the art in this area.

The Advanced Engineering Design Group is hereby assigned the task of studying **the operator of motor vehicles with an attempt to measure the kind and type of in**dividual that should be allowed by law to operate an automobile. This study is to be a group project in which designers will work together in teams **studying and attempting** to solve different phases of the problem. Each **team's** activities will be under the supervision of a Project Engineer who will report regularly to the Chief Engineer who is responsible for the overall coordination of this research. Areas of responsibility will be assigned when the Group can define specific areas through meetings in joint session.

The first phase of the project will concern itself with a study of the present driver test used in the state of _____ and certain recommended changes in this test. Secondly, the operator will be studied in detail so that the Group can conclude his limitations, factors concerning judgement, and what information must be taught about safe driving. Third will be the design of a <u>simulator</u> for purposes of determining the operator's ability to operate a motor vehicle.

The results of this assignment will be presented for critique by the Advanced Engineering Design Group during the week of May 25, 19--, to representatives from the Registry of Motor Vehicles, American Automobile Association, Automotive Legal Association, and Highway Safety Engineers.

GROUP 1 - DISPLAY SIMULATION

Proj. Engr._____
_____
_____
_____
_____

GROUP 2 - VEHICLE SIMULATION

Proj. Engr._____
_____
_____
_____
_____

GROUP 3 - SYSTEMS ANALYSIS

Proj. Engr._____
_____
_____
_____
_____

GROUP 4 - HUMAN FACTORS

Proj. Engr._____
_____
_____
_____
_____

Group 1 - Visual and audio display design including hardware and system analysis.

Group 2 - Simulator Hardware - Vehicle and interior components.

Group 3 - Integration of visual and audio display with simulator hardware and sub-systems and output display.

Group 4 - Human engineering for purposes **of suggesting input information** and measurement of results.

A3 DESIGN CASES

Unlike the design problems and projects of sections A-1 and A-2, the cases included here are written descriptions of actual events or circumstances confronting the design engineer. The facts presented are based on real-life situations. Company and individual names have been changed so that the case cannot be related to its source. This in no way affects the value of the design experience.

The case method of design requires the same effort of creativity, imagination, and design skill on the part of the designer, but in addition requires that decisions be made in the context of the case situation. In other words, the design solution must fit the case. This requires the designer to sift the data in order to select and develop useful information. Like real-life situations, occasionally crucial information is missing. In this instance the designer may utilize his own experience and knowledge or seek additional information upon which to base his design. More often than not he must make assumptions which eventually will be defended in light of the case situation.

CASE 036   3C-COPY CENTER, CORP

3–C — COPY CENTER, CORP.

525 STATE STREET
Boston, Massachusetts 02180

PHONE: 834-3311
TWX: Holy - 188

March 4, 19 – –

Mr. James R. Prince, President
THINK, INC.
32 Boston Avenue
Medford, Massachusetts 02155

Dear Mr. Prince:

  The 3C – Corp. has seen the copy business grow from carbon paper ten years ago to the "copy anything" industry of today. Copying is now a way of life in the office, among students, and in the home. Its impact has recently been felt in a revision of the copyright laws.

  Demands placed upon the copy industry are, in order of importance: convenience, permanence, contrast of copy, and unit price. We have been able to meet all of these among our competitors in the line of vending machine copiers. These machines are located in commercial airline terminals, libraries, and stationery stores. Although competition is difficult, we have been able to hold our own in this market.

  It is our wish now to diversify our product by introducing to the consumer field a small portable copier about 12" x 12" x 1" in size. This copier should appeal to students, salesmen, and possibly architects and engineers. It would be good if this device could copy all colors, all sheet sizes, as well as pages from bound books and magazines. Its price must be such that it will sell to the market previously suggested and its per copy price must be competitive with vending machines.

  Due to the splendid reputation of THINK, INC., in producing successful designs, almost upon demand, we at 3C would like to contract your services to offer possible solutions to the copy problem which I have outlined. Please advise me as to your group's reaction to our request.

Sincerely yours,

James P. Corbitt

James P. Corbitt, Manager
Marketing Division

JPC:m

March 7, 19 - -

Mr. James P. Corbitt, Manager
Marketing Division
3C - Copy Center, Corp.
525 State Street
Boston, Massachusetts 02180

Dear Mr. Corbitt:

Thank you for the letter of March 4 and of your confidence in THINK, INC.

We have made a quick initial study of the copying field and would like to attempt proposing a solution to your problem request. We believe a compact and portable MINI-COPIER can be designed to measure up to the requirements set forth in your letter. The design will require a series of laboratory experiments as well as a complete market survey on our part in order to advise you on its potential.

Our design group will begin work immediately on this project and will be prepared to present alternate solutions to 3C management on April 4, 19 - - in our offices.

Terms of our contract will be negotiated at a later date. Shall we get together for lunch, say March 17, to establish these terms?

Cordially yours,

James R. Prince, President

JCP:v

CC: P.H. Hill
    Chief-Idea Group

EXHIBIT "A"

314   CASE 036 (CONT)

The photos below will give the designer an example of the four major copiers presently available.

LARGEST VOLUME USER

PORTABLE DESK COPIER

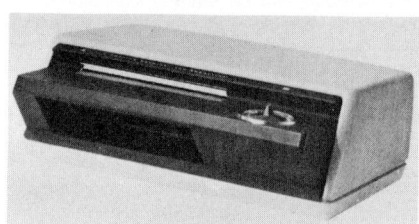
DESK TOP COPIER

PORTABLE COPIER

INDUSTRIAL COPIER

EXHIBIT "B"

On the next page are samples of six popular copy processes. Each has an application and advantage in a particular field of copying. The following descriptions will give the designer some idea of the process involved:

PHOTO OFFSET - Probably the oldest and best quality copying process available. It is expensive for short runs but inexpensive for a large number of copies (100 or more). Here a negative is made by photographing the original. The negative is then offsetted to a master plate in which the print surface is left bare and the remaining surface is oiled. Ink will stick to the bare surfaces and not to the oiled. In this way the master when inked will provide copies. Expense is in photography and machine to operate master.

THERMOFAX - This process uses a specially coated paper that turns dark upon being exposed to concentrated heat. Any original whose print contains a heavy carbon deposit will allow heat to bounce off it thereby producing the image on the copy paper. Thermal process which is inexpensive for a small number of copies will not reproduce color, ball point pen, half tones, etc.

HECTOFAX - Same as Thermofax but uses a standard duplicator special carbon master that will reproduce about 30 copies of the quality of the included sample. Has same copy limitations as thermofax.

OZALID – Original must be either transparent or semi-transparent (tracing vellum). When projected on to a special coated paper through exposure to light (contact print) and fixed in ammonia (fumes) a print will result. Inexpensive process for small numbers of large engineering and architectural drawings. Will fade in time and rather quickly in sunlight.

VERIFAX – A photographic process in which original is transferred to a plate (matrix) upon exposure to light. The matrix will produce from 3 to 6 copies when rolled against special (high gloss) copy paper. Good copies but an involved process.

XEROX – A popular and convenient 1 to 10 copy process in which graphite particles are attracted to the print of the original through electrostatic transfer to the copy paper. Once the paper is fixed (electrically) a copy results. This process has difficulty copying dark areas. Also copy device is rather large (floor model type).

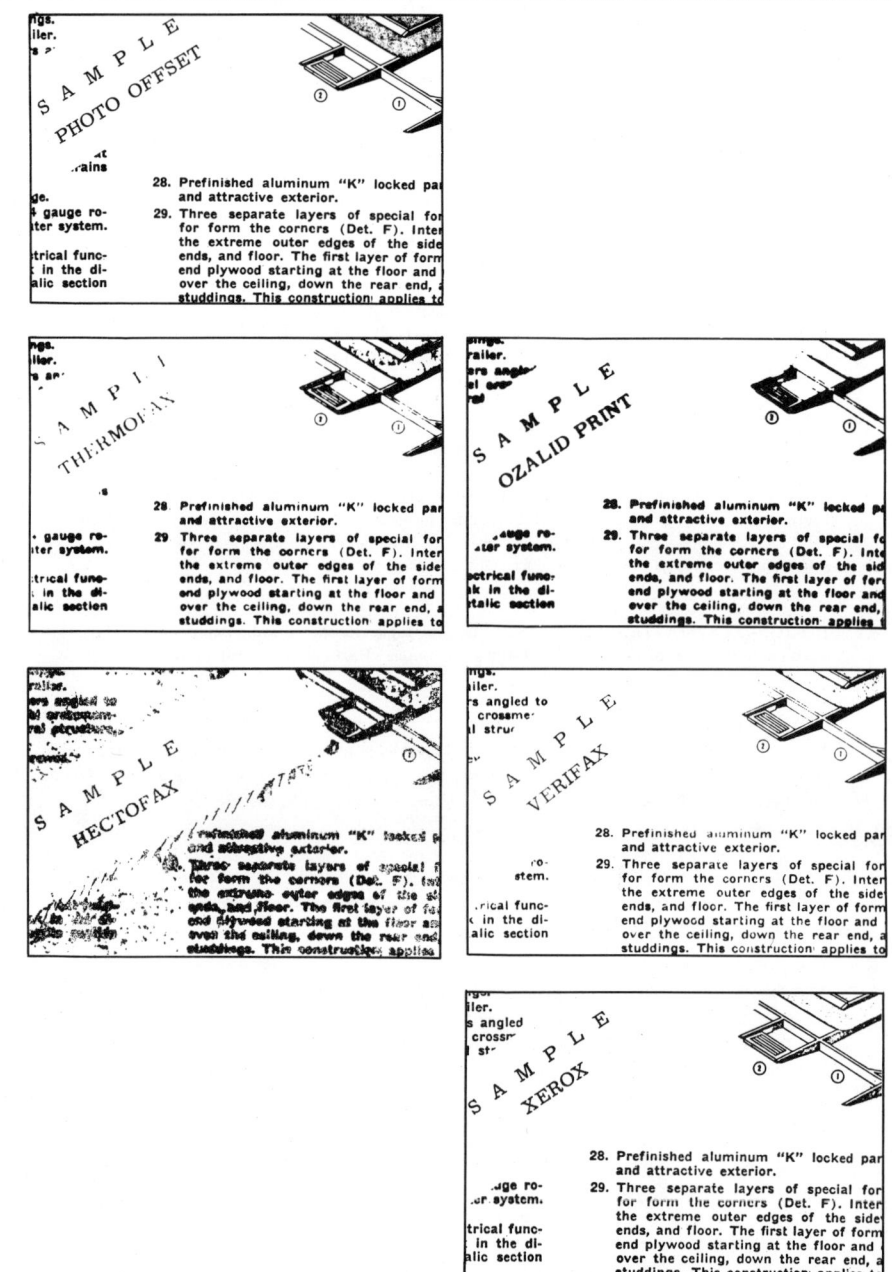

CASE 037   GANSON MANUFACTURING CO.

GANSON MANUFACTURING CO.
186 LANSING STREET
NORTH PENNINGTON, CONNECTICUT 12183

March 1, 19--

Mr. Percy H. Hill, Director
DESIGN CONSULTANTS, INC.
1286 Boston Avenue
Medford, Massachusetts 02155

Dear Mr. Hill:

    The Ganson Manufacturing Company is interested in engaging the services of DESIGN CONSULTANTS for the design of a new product line. Knowing of your company's reputation in the innovation of new products and its success in diversified areas of invention, we feel that your team approach to problems can benefit our company. A recent market survey indicates that manufactured items in the area of devices to teach scientific principles to secondary school children will be a rapidly expanding field in the near future, and we wish to become a pioneer from a manufacturing point of view. Management has consented to invest a sizeable capital expenditure for this product line if unique and effective devices are acceptable. Needless to say, a product line in this area will put Ganson in a favorable competitive position with small firms in neighboring areas.

    I have permission from the board at Ganson to offer DESIGN a sizeable consulting fee (in the order of six figures) to design and present working prototypes of educational instructional devices (scientific if possible) for secondary school children that can be manufactured and marketed by our plant.

    We are prepared to pay for your services through initial design presentations and an additional (sizeable) fee upon acceptance of device or devices by our company. We are engaging the services of two additional firms to propose designs on a competitive basis.

Mr. Percy H. Hill, Director
March 1, 19--

Page 2

    If your company is interested in this proposal, I would suggest that you instruct design teams to begin work at once for time is of the utmost importance. We will expect to review your company's proposed designs on April 8, 19--.

    Please find enclosed a brief description of the Ganson Manufacturing Company as background material for your consideration. I would expect that you will treat this material in a confidental nature within your firm.

JCC/ch

Very truly yours,

John C. Carling

John C. Carling
General Manager

DESIGN CONSULTANTS, INC.

1286 BOSTON AVENUE

MEDFORD, MASSACHUSETTS 02155

March 7, 19--

Mr. John C. Carling, General Manager
GANSON MANUFACTURING COMPANY
186 Lansing Street
North Pennington, Connecticut 12183

Dear Mr. Carling:

    Thank you for your complimentary letter of March 1, and for the proposal to design educational devices for manufacture by GANSON. We have had some experience in this area and have the talent on hand to innovate such a product. I hereby accept the contract for DESIGN to initiate design studies for the invention of products of the type you have in mind. I would like to meet with you on March 11, to negotiate fees associated with this work.

    Design teams will be briefed on March 9, and will begin work immediately following the session on the innovation of the educational devices. We feel sure that our group can meet the scheduled presentation date of April 8, 19-- , and I can promise you that we will have several interesting designs and prototypes for your consideration.

    I am taking the liberty of copying your enclosed description of GANSON for distribution to our design group. You can be assured that it will be kept in the strictest confidence.

PHH/mv

                                                  Sincerely,

                                                  Percy H. Hill
                                                  Director

GANSON MANUFACTURING COMPANY

In February, 1965, Mr. John C. Carling was appointed general manager by Mr. Peter J. Ganson, President of the Ganson Manufacturing Company. Mr. Carling, age fifty-six, had wide executive experience in manufacturing products similar to those of the Ganson Company. The appointment of Mr. Carling resulted from management problems arising from the death of Mr. Richard Ganson, founder and, until his death, in early 1964, president of the Ganson Company. Mr. Peter Ganson had only four years experience with the company, and in early 1965 was thirty-four years old. His father had hoped to train him over a ten-year period, but his untimely death had cut this seasoning period short. The younger Ganson became president when his father died and had exercised full control until he hired Mr. Carling.

Mr. Peter Ganson knew that during 1964 he had made several poor decisions and noted that the morale of the organization had suffered, apparently through lack of confidence in him. When he received the profit and loss statement for 1964, the net loss of over $170,000 during a good business year convinced him that he needed help. He attracted Mr. Carling from a competitor by offering a stock option incentive in addition to salary, knowing that Mr. Carling wanted to acquire a financial competence for his retirement. The two men came to a clear understanding that Mr. Carling, as general manager, had full authority to execute any changes he desired. In addition, Mr. Carling would explain the reasons for his decisions to Mr. Ganson and thereby train him for successful leadership upon Mr. Carling's retirement.

The Ganson Manufacturing Company made only three industrial products, quality gears, precision ground shafts, and quality bearings. These were sold by company salesmen for use in the processes of other manufacturers. All of the salesmen, on a salary basis, sold the three products but in varying proportions. The Ganson Company sold throughout New England and was one of eight companies with similar products. Several of its competitors were larger and manufactured a larger variety of products than did the Ganson Company.

-2-

The dominant company was the Hamra Company, which operated a branch plant in the Ganson Company's market area. Customarily, the Hamra Company announced prices annually and the other producers followed suit.

Price cutting was rare, and the only variance from quoted selling prices took the form of cash discounts. In the past, attempts at price cutting had followed a consistent pattern; all competitors met the price reduction, and the industry as a whole sold about the same quantity but at the lower prices. This continued until the Hamra Company, with its strong financial position, again stabilized the situation following a general recognition of the failure of price cutting. Furthermore, because sales were to industrial buyers and because the products of different manufacturers were very similar, the Ganson Company was convinced it could not individually raise prices without suffering volume declines.

During 1964 the Ganson Company's share of industry sales was 12 per cent for type gears, 8 per cent for shafts, and 10 per cent for bearings. The industry-wide quoted selling prices were $2.45, $2.58, and $2.75, respectively.

Mr. Carling upon taking office in February, 1965, decided against immediate major changes. Rather he chose to analyze 1964 operations and to wait for results of the first half of 1965. He instructed the accounting department to provide detailed expenses and earning statements by products for 1964. In addition he requested an explanation of the nature of the costs including their expected future behavior.

To familiarize Mr. Peter Ganson with his methods, Mr. Carling sent copies of these items to Mr. Ganson, and they discussed them. Mr. Ganson stated that he thought precision shafts should be dropped immediately as it would be impossible to lower expenses on shafts as much as 30 cents per cwt. In addition, he stressed the need for economies on bearings.

-3-

Business seemed to stabilize at year ending 19-- , and Mr. Carling began to plan the launching of a new product line in the summer of 19 -- . His plans include a new product that could be constructed in the present plant but of a different nature than previous lines in order to present a challenge to the sales force in hopes of "getting them out of a rut."

In seeking out a new product line, Mr. Carling plans to invest a sizeable fee for consulting firms to compete for the design of the product.

GANSON MANUFACTURING COMPANY

PROFIT AND LOSS STATEMENT

FOR YEAR ENDING DECEMBER 31, 1964

Gross Sales.....................		$10,589,405
Cash discount...............		156,578
Net Sales.......................		$10,432,827
Cost of Manufacturing.........		7,529,758
Manufacturing profit.............		$2,903,069
Less: Selling Expense..........	$1,838,238	
General Administration...	653,020	
Depreciation............	458,440	$ 2,949,698
Operating Loss....................		$ 46,629
Less: Other income...............		$ 21,065
Net Loss before Bond Interest......		$ 25,564
ADD: Interest on bonds.........		145,283
Net Loss after All Charges........		$ 170,847

CASE 038　　HARDT TOOL AND HARDWARE CO.

THE HARDT TOOL AND HARDWARE COMPANY
11 State Street—Somerville, Massachusetts 02144 — 628-5000
"Tools of Quality for the Working Man"

February 21, 19--

Mr. Percy H. Hill, Manager
ENGINEERING DESIGN GROUP, INC.
1200 Boston Avenue
Medford, Massachusetts 02155

Dear Mr. Hill:

    I am enclosing a transcript of a meeting of the board of the Hardt Tool and Hardware Company regarding serious marketing problems involving our hand drill line---Model No. 6507. This transcript pretty well defines our problem here at Hardt. Since your group has a national reputation in solving problems of this nature, I have permission from the board to engage DESIGN GROUP in seeking a solution to this problem.

    We are interested in reducing the price of the unit approximately 20% without a great deal of loss in quality. Since we are competing with a foreign manufactured item that is almost identical to our model, we are very much interested in simplification of the basic unit and innovations in styling that will give it competitive sales appeal. The new design or design changes should fit within the present framework of manufacturing and assembly in our Somerville Plant, for we already have a rather large investment in machine shop equipment involved in the production of this model.

    Since time is an important factor concerning this problem, pressure is already being brought to bear on my division from the top. I have arranged a meeting on March 8, 19-- so that we may review the DESIGN GROUP's solution to this problem. I hope that you can comply with this date.

    If you need additonal information concerning Model No. 6507 please call my office.

Very truly yours,

Leonard Brenner
Vice-President, for Marketing

LB:v

THE HARDT TOOL AND HARDWARE COMPANY

Transcript from Executive Meeting

On February 4, 19--, the executive planning board of the Hardt Tool and Hardware Company met for the first of a series of important meetings. The company, one of the larger industrial and consumer tool manufacturers in the country, had enjoyed a large share of the consumer tool market for a number of years. Recently, however, increasing imports of foreign-produced tools had cut rather deeply into their market share, and this series of meetings had been called to discuss what action should be taken to meet this threat.

Leonard Brenner, vice-president for Marketing, opened the meeting with these remarks: "As we all know, many of our hand tools, particularly in the consumer market, are meeting stiff foreign competition. In the last year, I estimate that our market share has dropped at least five per cent, and I hardly think that it will stop there. Many of my district sales managers have interviewed hundreds of their wholesalers and retailers, and the consensus is this--there is no advantage in stocking our tools rather than imported stuff. Nearly every one of our sales features had been copied in one form or another by foreign manufacturers, and side by side there is little difference between competing products. The trade can purchase foreign stuff at better than twenty per cent below our prices and still net higher margins. Gentlemen, I feel that some drastic changes are in order."

Lou Silverstone, Advertising director, retorted quickly, "Len, I really can't see how this problem has developed. We offer the consumer a full line of tools, from nail sets all the way up through power tools, and much of our line is interchangable. Most of this import stuff is a one-shot proposition, which the buyer may never be able to match or replace. The full line is **what** we've stressed in our advertising, and I think more emphasis on it might do the trick."

"Lou, that's just the problem," replied Brenner. "It seems that the typical consumer tool customer, the home handyman, is only interested in one tool at a time. When he sees our product for, say $3.50, and an import for $2.98, they appear basically identical, he picks the lower priced tool. He doesn't care about a full line, nor is he concerned about replacement parts--all he's interested in, is the particular tool he's buying at the moment."

Jack Rickard, president of Hardt Tool, broke in, "Lou according to the last batch of sales figures you issued, some of our worst losses seemed to be in the hand-drill line. From what you've found in the interview data, do you feel the same problems exist here as with the rest of the consumer tools?"

"Jack, generally speaking, the hand drill situation is a typical example of our problems. We've always tried to produce a quality drill - one that would stand up well both in the market and also in use. And I think the production boys have done a darn good job on our present model. But, as I said before, I think we are

-2-

overpriced in the market--our costs have to come down so we can compete favorably with the imports. I don't mean that quality has to go out the window, but something has to give."

Richard broke in again quickly with, " O.K., let's not get hot over this. I think the design question deserves a lot of thought, but we're not meeting here to find a scapegoat. Personally, I think that George may have put us on the right track; perhaps design modification will provide the cost-cutting answer. Len?"

"Jack, while we're on the question of design, I think we should consider not only production cost but also marketability. As I said earlier, there is little to differentiate our product from the foreign competition when they're side by side on a shelf. If we worked at this, perhaps we could design in something - and I'm not sure what it might be - that would make our product stand out. I realize we're trying to cut costs, but couldn't the drill be made more saleable without a manufacturing price increase?"

"Len, I don't know. Your idea definitely has merit, and perhaps changes can be made that would cut our costs and also differentiate the product. I've had a design consultant in the office a few times lately and we've been talking over problems like this. Let me see what he has to say on this matter, and perhaps his firm can get us a report on the drill by our next meeting. I think we can agree that design changes which will facilitate manufacture and also increase the marketability of the product are what we're looking for here."

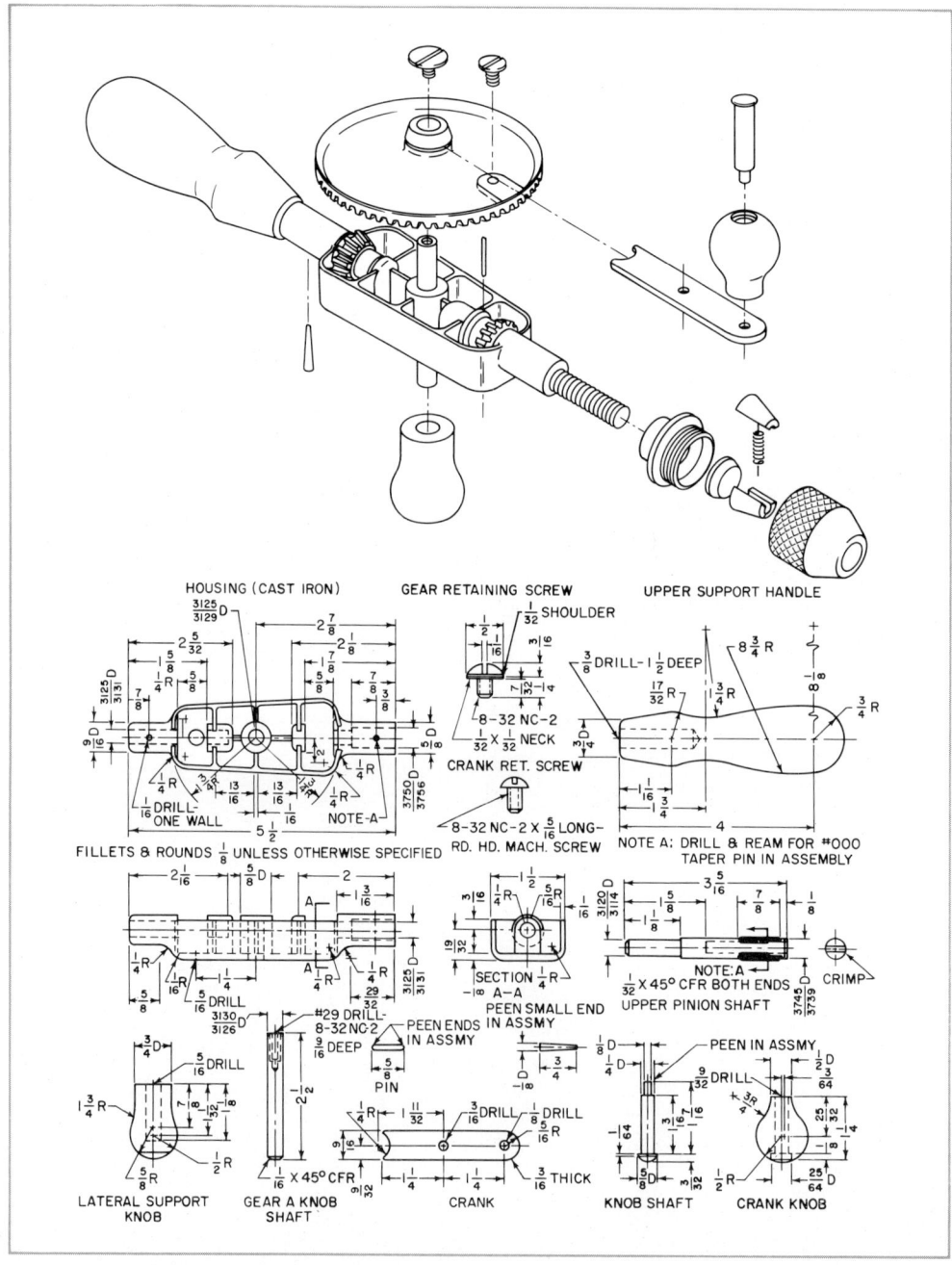

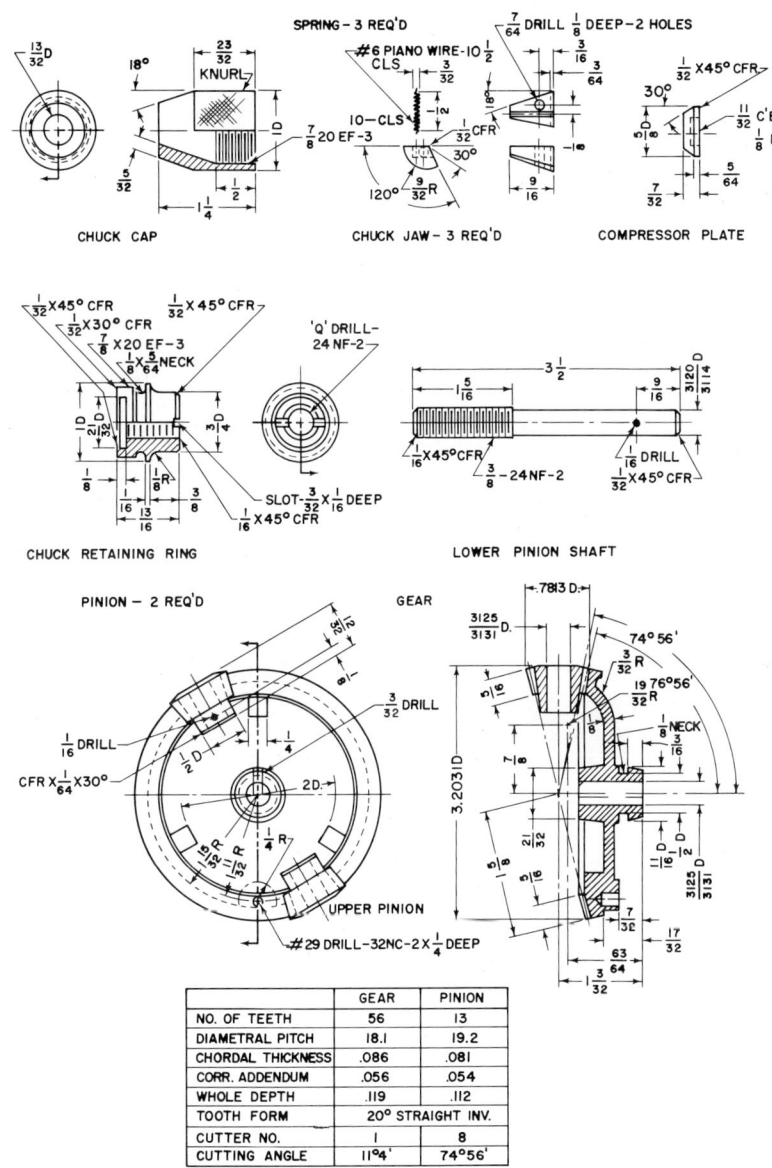

CASE 039   ZENITH CONSTRUCTION CORP.

ZENITH CONSTRUCTION CO.
ROUTE 128
WALTHAM, MASSACHUSETTS

January 22, 19--

Mr. Percy H. Hill, President
ENGINEERING DESIGN ASSOCIATES
125 College Avenue
Medford, Massachusetts

Dear Mr. Hill:

    Our corporation has a reputation in the heavy construction industry as a manufacturer of graders, earth moving equipment, cement mixers, erection hoists and cranes, etc. The present state of the economy dictates that we diversify our product line in order to maintain our corporate obligations. We would, therefore, like to market a new product line, not related directly to our present equipment, that will sell to the building industry.

    One idea that received a great deal of attention at a recent board of directors meeting is that of automation equipment related to the construction of buildings. We feel that a line of equipment that would reduce the labor and time factor involved with building erection and finishing would have an extensive impact on the construction market in this country as well as overseas and in underdeveloped countries that show a need for improved housing.

    Our present thoughts center around three possible machines, although your designers may have alternate suggestions.

    The machine that we wish to manufacture are:

1. FOUNDATION MACHINE - This machine must be able to excavate and pour a foundation for a one story building, waterproof the foundation, pour the basement slab as well as the floor slab. The machine should require minimum of supervision (as automatic as possible) and must be programmed from the construction drawings.

2. BLOCK LAYING MACHINE - This machine must lay conventional cement blocks on a prepared foundation that results in the walls of a building, following a predetermined program. The machine must accept blocks, mix mortar, lay the wall, point the blocks and grout when required and leave openings for doors and windows. The machine should require only one operator.

3. INTERIOR FINISHING MACHINE - This machine must be capable of painting an empty room with no prior programming of its dimensions. The machine must contain detection devices to sense the location of windows, doors, ceiling, and floors so as to not paint these areas. It should be completely automatic in that it will detect a predetermined starting point, move the necessary shielding devices and sense when it has returned to its original starting point and automatically shut-off.

Mr. Percy H. Hill
January 22, 19--
Page (2)

    Due to the reputation of Engineering Design Associates in the area of automated equipment and the known talent of its designers who specialize in design work in the area of creative functional devices, we are prepared to offer you a contract to present final designs on these three machines so that prototypes may be constructed for purposes of testing. Would you please let me know as soon as possible if your company wishes to undertake this contract for I must report back to our board of directors on February 25.

                                                  Very truly yours,

                                                 Frank R. Casely
                                               Executive Vice-President

FRC/eh

ENGINEERING DESIGN ASSOCIATES

125 COLLEGE AVENUE

MEDFORD, MASSACHUSETTS

January 29, 19--

Mr. Frank R. Casely
Executive Vice-President
Zenith Construction Corp.
Route 128
Waltham, Massachusetts

Dear Mr. Casely:

Thank you for your letter of January 22 and your confidence in my company to undertake the design for prototypes of the automated building machines. I met with my associates several times last week to discuss the feasibility of designing these machines in light of present materials and manufacturing facilities, and we have concluded that they can be constructed and produced to work efficiently. We are pleased to accept the design contract and will begin task specifications and the conception phase immediately. Would you arrange a meeting as soon as possible so that we may finalize the design contract terms. I will also wish to meet with your management group at regular intervals to discuss our progress on the design and to act on decisions related to feasibility and workability.

I want to thank you again for the opportunity you have given my company to produce this design. We are quite excited to be responsible for the design of automated machinery that may lead to a major breakthrough in the construction industry.

Sincerely,

Percy H. Hill
President

PHH/mah

M E M O R A N D U M

February 2, 19 --

TO: ACTION
        Design Group A (Ernesto E. Blanco), Design Group B (Allan H. Clemow),
        Design Group C (James P. O'Leary)

        INFORMATION - Chemical Engineering Group (Martin V. Sussman)

FROM: P.H. HILL

SUBJECT: Design Assignment of Automated Equipment for the
            Construction Industry.

    You will find attached correspondence related to our recent design contract with the Zenith Construction Corporation. This contract means a great deal to our company so every effort must be made in producing the optimum machines according to Zenith's request. The following design work is hereby assigned to Groups A, B, and C.

        Design Group A (Ernesto E. Blanco) - AUTOMATED FOUNDATION MACHINE

        Design Group B (Allan H. Clemow) - AUTOMATED BLOCK LAYING MACHINE

        Design Group C (James P. O'Leary) - AUTOMATED INTERIOR FINISHING MACHINE

    Since our Chemical Engineering Group is presently involved in the design of a processing plant in Turkey, they will not directly be involved with the initial design work but you should feel free (and encouraged) to use their members on a consultation basis.

    Each group leader should begin immediately to study his design assignment and to discuss it with his designers. Task specifications must be completed by February 19, conceptualization by March 10, and initial design and analysis of concept by April 14. I have set a target date on this project for discussion and possible acceptance of the design of prototype with Automated on May 14, and 17.

    The three groups will meet at regular intervals to discuss feasibility and later workability.

    All interoffice correspondence related to this project will be referred to as the AUTOMATED BUILDING BUILDER and subtitled group assignments. The code to be used is: ABB - FM, ABB - BL, ABB - IF.

ENGINEERING DESIGN ASSOCIATES
125 COLLEGE AVENUE
MEDFORD, MASSACHUSETTS

March 5, 19 --

M E M O R A N D U M

ABB-BL

To: Design Group (B)
From: A.H. Clemow, Project Engineer
Subject: Task Specifications on Automated Block Laying Machine for the Automated Building Builder.

T A S K: The machine must lay conventional cement blocks on a prepared foundation that results in the walls of a building, following a predetermined program. The machine must accept blocks, mix mortar, lay the wall, point the blocks and grout when required, and leave openings for doors and windows. This machine should require only one operator.

T A S K S P E C I F I C A T I O N S

1. MINIMUM BUILDING CODE SPECIFICATIONS.
   (a) Foundation.
       4 ft. below grade, 4 ft. - 4 inches above grade.
       8 in. thick reinforced poured wall.
       Basement floor slab = 3"
       Reinforced floor slab = 4"
   (b) Shell (one story building).
       Residence - six rooms - 1200 sq. ft.
       Small industrial building - 10,000 sq. ft.
       Door openings = 3'0" min.
       Window openings = 2'0" X 4' 0" min.
   (c) Roof.
       Prefab construction, wood rafters, wood sheathing, asphalt shingles, to be belted in place.

2. BLOCK DIMENSIONS.

   Weight = 60#
   1/4 " compacted mortar, lay loose 3/4".
   Blocks must be layed so that holes line up for wiring and heating.
   Each joint (vertical and horizontal) must be pointed inside and out.

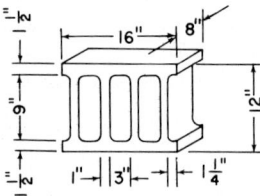 BLOCK SIZE

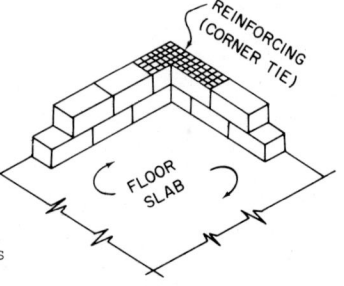

3. SPECIFICATIONS ON LAYING WALL.
   (a) Machine must lay straight wall, make right angle corner, and set reinforcing ties at each corner-each course. Machine can be guided by rollers on prepared floor slab.
   (b) Machine must be programmed to leave blocks out at window and door openings.
   (c) Controls are needed to level and plumb blocks and point joints.

4. FOUNDATION PLAN TO BE PREPARED BY GROUP "A" (ABB - FM).

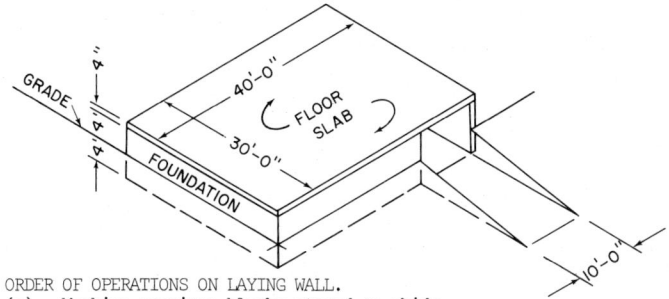

5. ORDER OF OPERATIONS ON LAYING WALL.
   (a) Machine receives blocks stored on skids.
   (b) Blocks must be wet before placement.
   (c) Mortar is mixed from stock (bagged mortar and water added from reserve).
   (d) Machine begins at one corner, spreads layer of mortar (3/4"), lifts block and guides it into place, presses plumb to 1/4" joint, points joint.
   (e) Machine lays straight wall turns corner through roller contact on floor slab. Machine must place reinforcing screen at each corner.
   (f) After one layer of block is placed, machine must raise one layer on staging to begin next course.

6. SIZE OF MACHINE.
   (a) Must be small enough and not too heavy to be moved over highways and through the city.
   (b) Should run on tracks to provide a solid platform.
   (c) Should be able to be transported on a low-bed trailer.
   (d) Machine should be constructed so that it will be in operation 30 min. after arriving at site.

7. MACHINE SPECIFICATIONS
   (a) Machine will automatically place door frames and lintles.
   (b) Machine must saw blocks through programming when ending an uneven course.
   (c) Heater and cover must be provided for severe weather conditions.
   (d) If electricity is to be used, it must have non-snag connections with source or carry its own generator.
   (e) If gasoline or diesel engine is used, must store fuel.
   (f) Should run all parts such as mixer, block layer, pointer, crawl and elevation mechanisms off of one power source.
   (g) Machine may be programmed through magnetic tape, punch tape, punched cards, indexing head, or pegged drum.

8. MISCELLANEOUS SPECIFICATIONS.
   (a) Machine noise level should be such as not to be objectionable.
   (b) Machine must be safe in operation.
   (c) Cost of renting machine and operation should be competitive with the trade.
   (d) Machine must construct the shell of a building with an appreciable saving in time.
   (e) Waste must be kept to a minimum.
   (f) Machine must measure setting time on mortar and blocks.

9. SYSTEM DIAGRAM

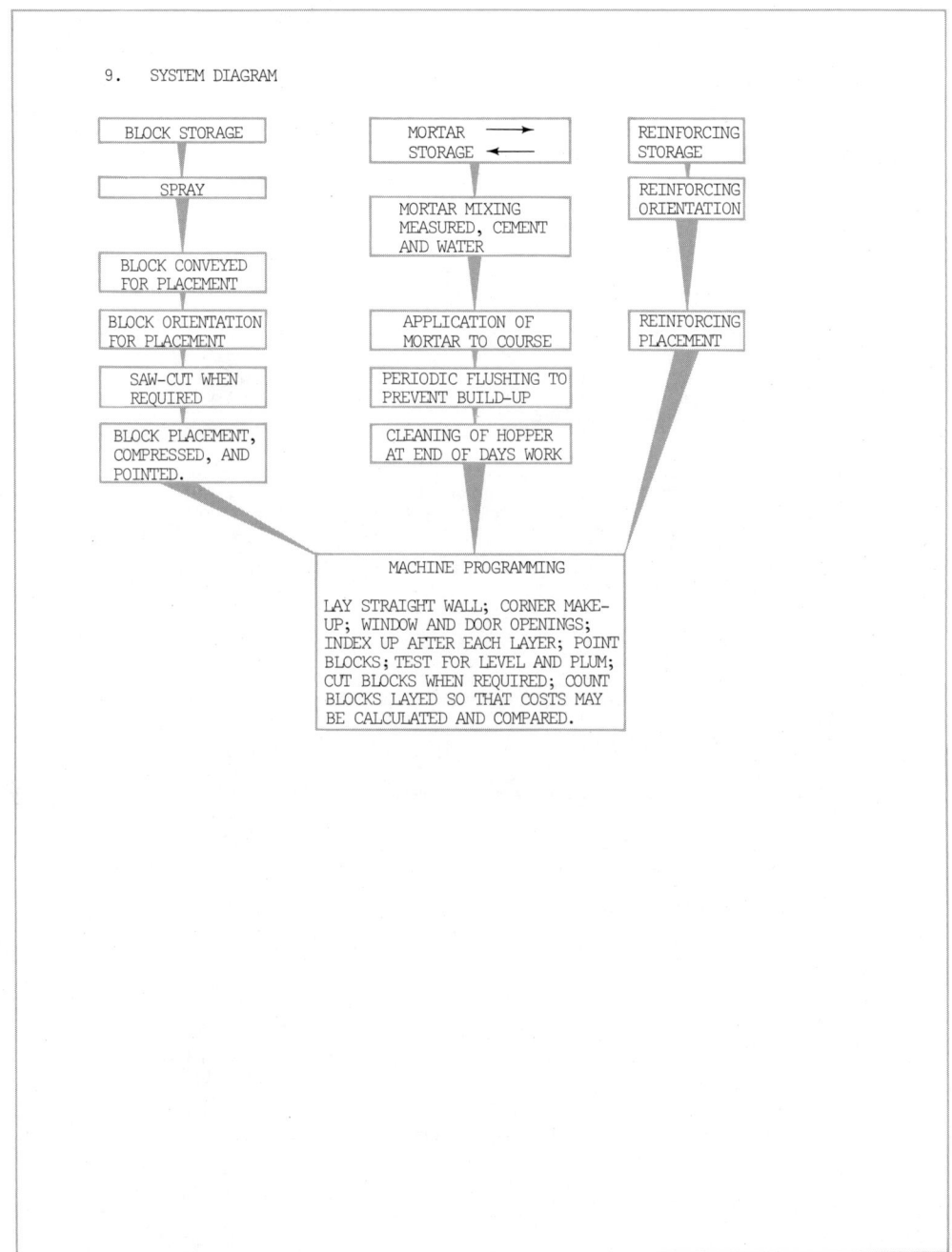

CASE 040   NATIONAL EQUIPMENT COMPANY

NATIONAL EQUIPMENT COMPANY

300 Third Avenue
Arlington, Massachusetts
HI 3 - 5800* Area Code 617

November 19, 19--

Professor James Perry
University of Winchester
Department of Engineering Design
Room 220, Bancroft Hall
Winchester, Massachusetts 01890

Dear Professor Perry:

    This will serve as your authorization to design a micro hematocrit centrifuge, with all rights including any patents resulting from your investigation to become the property of National Equipment Company. It is the mutual understanding between University of Winchester and National Equipment Company that this cooperative effort will serve two purposes: (1) University of Winchester engineering students will have an opportunity to work with a local manufacturer under an actual development and engineering project, (2) National Equipment Company may receive as a result of this effort a design that can be manufactured as a saleable product.

    There is a marketing need estimated to be between 500-2000 units per year for an inexpensive semi-portable micro hematocrit centrifuge to be used in physicians' offices and clinical laboratories. In addition, by having portable characteristics the centrifuge can be taken to a patient's home or a nursing home for running a micro capillary method for blood cell volume determination. To accomplish this test the following procedure is followed: A micro sample of blood is taken directly from the finger tip, which has been punctured and through capillary action the tube is filled to approximately 85-90% of the tube length, (2) The capillary tube is sealed, using a heat seal or molding clay or a plastic sealing cap. (3) The sample is then placed in the centrifuge and spun for a given period in order to pack the red cells at the outer end of the tube, (4) The percentage of the red cell volume is then determined by using a reading device.

    The following design criteria have been set up for the hematocrit centrifuge:

1. The unit shall have a capacity for spinning a quantity of six standard 1.5 mm diameter x 75 mm long capillary tubes. Centrifuging to be in the horizontal plane with the tubes having their longitudinal axis radially arranged around the hub of the tube holder, i.e. the spokes of a wheel.

2. Required speed of the tube holder to be in the +7000 rpm range. This results in a G level sufficient to pack the red cells in 3-4 minutes. Convenience of removing and inserting the tubes to be considered important.

3. The motor to operate on 115 volts 60 cycles and to have sealed bearings requiring no attention over the life of the machine.

Professor James Perry
University of Winchester

4. The centrifuge to be designed with convenient protective enclosure for the tube holder and readily accessible. Air flow through the protective housing to be arranged in such a way as to cool the capillary tubes and motor so that at no time, even during prolonged use, will the ambient temperature of the blood sample exceed 36°C. The construction to consider stability and portability. Unit must not creep during operation.

5. Simplicity of control is necessary. A combined starter switch and running timer (10 minutes maximum) is desirable so that the machine shuts off automatically at the end of a run cycle. A push button brake, either mechanical or electric for deceleration in approximately 30 seconds is to be considered desirable.

6. The reading device incorporated in the tube holding device is a desirable feature. Simplicity here is essential.

7. Eye appeal consistent with apparatus in a doctor's office is very important.

8. The noise level shall not exceed normal conversational level. Normal conversational standard has been determined by the Telephone Company.

9. The manufacturing cost including packaging in lots of 1,000 and excluding tool amortization should not exceed $40.00. Tooling costs cannot exceed $5,000.

10. The selling price is expected to be approximately $120.00.

The coordinator for National Equipment Company will be Ernest R. Pallen Chief Engineer. Delivery is to be arranged with no liability expressed or implied by either party.

Sincerely yours,

NATIONAL EQUIPMENT COMPANY

Ralph W. Schmidt

Ralph W. Schmidt
Executive Vice-President

RWS:o

CASE 041   AMERICAN SHIPYARDS, INC.*

AMERICAN SHIPYARDS INC.

GALVESTON, TEXAS

April 29, 19--

Mr. John F. Ericsson, Jr., President
International Maritime Designers, Inc.
Medford, Massachusetts 02155

Dear Mr. Ericsson:

As you probably know we are one of the largest shipbuilders in the United States dedicated almost exclusively to the construction of merchant vessels.

Upon request from the United States Department of Defense we are at present considering a novel type of vessel hull design proposed by Blohm & Voss AG, Hamburg, West Germany, based on hull shapes containing only flat plates. This polyhedral hull design has many advantages as far as manufacturing is concerned. First of all no bending or forming of the hull plates is necessary, also no bending of frame structural members is required, since all transverse sections become polygonal. Lofting time is therefore greatly reduced since surface fitting becomes a simple geometric problem. The only drawback of polyhedral vessels seems to be their greater drag which makes the operation costs somewhat higher.

We are interested in contracting the services of your company for the design of a polyhedral hull to approach in size and performance characteristics those of our well known "Excelsior" type vessel whose plans are enclosed here.

The Government has requested that preliminary designs be submitted by May 22nd. These designs should include the following:

1. Plans for a polyhedral hull which should not depart from the "Excelsior" displacement by more than 10%. Overall dimensions, such as length and beam should remain unchanged.

---

* Case printed by permission of Prof. Ernesto E. Blanco, Tufts University.

Mr. John F. Ericsson, Jr.
April 29, 19--
Page (2)

    Plans should contain the same number of transverse sections and waterlines as those of our vessel and should also include the development of one complete side of the hull.

    It would be desirable to have not more than 22 flat surfaces forming the entire hull surface.

2. Proposed hull computations should include the Displacement; Block Coefficient; and Fineness Coefficient for the eight values of draft. The mathematical function relating the hull's Drag Coefficient to the Draft should be determined by curve fitting techniques and a nomogram relating Drag Coefficient to Effective Horsepower at a range of speeds from 0 to 12 knots should also be submitted.

3. The proposed polyedraul hull should be compared in performance to the "Excelsior" hull by plotting horsepower versus speed at four levels of draft for each vessel.

    Performance curves should be submitted on standard chart paper.

    Please inform us whether you are interested in a contract for the work previously specified and I will instruct our Legal Department to negotiate mutually agreeable terms.

Sincerely,

M. D. Farragut

MDF/hl

Michael D. Farragut
President

INTERNATIONAL MARITIME DESIGNERS, INC.
MEDFORD, MASSACHUSETTS.

May 6, 19--

Mr. Michael D. Farragut, President
American Shipyards, Inc.
Galveston, Texas

Dear Mr. Farragut:

We were pleased to receive your request for a design proposal relating to a polyhedral hull. For the last three years we have been following the work of Blohm and Voss with great interest and during that period we have carried extensive experimentations which confirm the advantages of polyhedral hulls for cargo vessels.

Since our mutual interests lie within this area I have instructed our Design Department to start analyzing your "Excelsior" hull for polyhedral lofting pending an agreement between our Legal Departments. The deadline is close but I feel sure that our interest in this area and experienced supervision will enable us to satisfy your requirements in the short time available.

Sincerely,

J.D. Ericsson, Jr.
President

JDE/mu

—MEMORANDUM—

May 6, 19--

To: Mr. H. Dupuy De Lôme, Chief Engineer
From: J.D. Ericsson, President
Subject: American Shipyards, Inc. Polyhedral hull.

Dear Henri:

    Enclosed please find our correspondence with American Shipyards. As you will see their request fits very nicely with our "Narwhal" research project. I do not think it will be difficult for your personnel to comply with the requirements, although some additional experimentation will be necessary as usual.

    I want you to draw all "Narwhal" personnel into this project and work overtime if necessary. Let me know if you need any assistance. I want to impress American Shipyards with our competence in polyhedral lofting.

Jack

- MEMORANDUM -

May 6, 19 - -

To: Supervisory Personnel of Project "Narwhal"
From: H. Dupuy De Lôme
Subject: American Shipyards Polyhedral hull.

Gentlemen:

Upon request from Mr. Ericsson, all "Narwhal" personnel are to be transferred to the "American" Project starting on May 8th. The deadline of May 22 is very close and the work on polyhedral lofting to fit American's "Excelsior" hull should start as soon as possible.

All work on the new hull should proceed as specified in the project schedule to be issued. Please bear in mind that our reputation depends on this project.

A meeting will be held with the supervisory personnel listed below at 2:00 P.M. on May 7th, to set up project schedules.

Henri

cc: Mr. Chapman ——————— Lofting Room
    Mr. Comstock ——————— Structural Department
    Mr. Hughes ——————— Research Department
    Mr. Taylor ——————— Propulsion Department
    Mr. Warnock ——————— Towing Tank Laboratory

- MEMORANDUM -

May 7, 19--

To: Supervisory Personnel, Project "Narwhal"
From: H. Dupuy De Lôme
Subject: Schedule for Project "American"

Gentlemen:

As agreed during our meeting of May 6th the <u>delivery schedule</u> for the various phases of Project "American" will be as follows:

<u>PHASE 1</u> (May 13 & 14) Deliver the polyhedral lofting layout of hull approximating the shape of the "Excelsior" type vessel. Lofting should include a) Tentative surface fittings on "canary" sketch paper. b) A finished drawing on vellum showing transversal sections, sheer plan, and profile of water lines for one half of the proposed hull. c) The development of half of the proposed hull showing the relative positions between the planes that form the hull (on separate vellum).

<u>PHASE 2</u> (May 15 & 16) a) Deliver computations for Displacement at values of Draft of 17'; 22'; 27'; 32'; & 37'. Plot Displacement Vs. Draft on 8 1/2" x 11" standard arithmetical graph paper. b) Deliver computations for Block Coefficient and Fineness Coefficient in tabular form for each value of draft.

The material presented at this date will be examined and an empirical set of values of Drag Coefficients Vs. Draft will be determined by the Project Engineers to be used for further hull analysis work.

<u>PHASE 3</u> (May 20 & 21) a) Submit a plot of Drag Coefficient Vs. Draft on a standard 8 1/2" x 11" arithmetical graph paper. b) Through a curve fitting technique determine an empirical mathematical expression relating Drag-Coefficient to Draft for your hull. A suitable Log-Log or semi-Log standard chart should be used for curve fitting and delivered at this time.

PHASE 3 (continued)

All computation work should be presented on a separate sheet.

PHASE 4 (May 22 & 23) a) Deliver a nomograph to solve for Horsepower (HP) at Speeds (V) between 0 and 12 knots for values of Draft (D) between 17 and 37 feet. Draft scale should be made as an adjacent scale, relating on the same stem Draft (D) and Drag Coefficient (D.C.) for the proposed hull. b) On two separate standard arithmetical charts plot performance curves relating Horsepower Vs. Speed for the following values of Draft: 22'; 27'; 32'; & 37'. One chart should Exhibit the performance on the proposed hull, and the other that of the "Excelsior" type vessel so that their performance can be easily compared.

The finished project to be delivered at this date should be presented ready to be sent to the client, therefore, its elaboration should be as complete as possible.

- BACKGROUND INFORMATION -

The United States, leading manufacturing nation in the world, has been **steadily losing ground in the field of shipbuilding. The reasons given for this** situation are many, among them; high labor and operation costs; lack of modernization in the industry; lack of managerial agressiveness, etc., but whatever the reasons, the fact is that the United States today is a lower rate shipbuilder. A far cry from this Nation's position during WW II when we were producing ships faster than Hitler could sink them.

At present some experimentation on special type ships is being carried out mostly under the auspices of Government agencies. Among these special designs is a simplification of hull manufacture consisting of an approximation of the "ideal" hull shape by using combinations of flat surfaces. This type of hull is appropriately called polyhedral and its manufacturing advantages are derived from the fact that no bending or forming of hull plates or frame structural members is necessary.

Polyhedral ship hulls are nothing new, many small craft have been built out of plane surfaces long before the present serious attempts in this direction; what is new is the consideration being given to this type of design for the manufacture of commercial cargo vessels.

The slowly curving, smoothly faired and graceful lines, characteristic of ship shapes are dictated by the desire to reduce drag or frictional resistance of the hull in the water. Centuries of empirical studies and more recent attempts **by naval architects to formulate a rational approach to hull design based on fluid** dynamics theory have culminated in the present design of ships. However, in spite of the intensive research in this field the design of ship hulls is still largely based on empirical work. The ultimate shape of a hull being determined through exhaustive tests in towing tanks and largely empirical formulations.

The optimization of propulsion power requirements for a given speed requires minimum drag on the ship's hull; but today for certain low speed vessels, like cargo ships, it is found that the manufacturing complexities of a smoothly shaped hull are hardly justified by the gain in fuel economy. Savings in a ship's initial investment can easily outweigh the extra fuel costs over the life of a properly designed polyhedral vessel.

The firm of Blohm & Voss A.G. of Hamburg, West Germany has for several years been experimenting with polyhedral hulls for container ships. Their work has been closely followed in other nations and their designs are being seriously considered for future cargo ships.

The ideal design for a polyhedral hull is one in which the hull shape is the simplest that can approximate the "ideal" shape, but in practice bows are rather short to maximize cargo space having large penetration angles, and sterns have blunt, approximately bulbous, transom lines formed by several planes. The number of hull planes is kept below 20 to minimize lofting costs and fabrication complexities. If such a design can be made to perform in a way that its operation costs above those of a regular ship are offset by its initial investment savings the design becomes acceptable.

Modular ships could then be fabricated easily since only an addition to the midship section would be needed to increase tonnage and such sections could be stockpiled.

- PARAMETERS OF HULL DESIGN -

Efforts to define the shape of a ship's hull in rational terms have led to the adoption of a series of definitions and mathematical relationships some of which are expressed as follows:

<u>OVER-ALL LENGTH</u> = Maximum Center Line Length of a Vessel.

CASE 041 (CONT)

LENGTH BETWEEN PERPENDICULARS = Length of the waterline of the vessel from bow tip to stern.

EXTREME BREADTH = Width of hull over outside plating at widest section.

(D) DRAFT = Vertical distance from the bottom surface of the keel to the waterline. When the ship's trim is altered the draft may vary from bow to stern.

(d) DISPLACEMENT = Is the weight of water displaced by a vessel. The displacement in tons of 2,240 lb. is equal to the volume of the immersed section of the hull in cubic feet divided by 35 when floating in sea water. When floating in fresh water the displacement is equal to the volume in cubic feet divided by 36.
(35 cubic ft. of sea water, or 36 cu. ft. of fresh water = 1 ton of 2240 lb.)

(D.W.) DEAD WEIGHT = Is the payload capacity of a vessel in tons of 2240 lb. It is also the difference between the displacement of a vessel when fully loaded and when she is in light condition with no cargo, fuel, ballast, stores, passengers, or crew aboard, but with water in the boilers to the steaming level.

(F.C.) FINENESS COEFFICIENT OF WATER PLANE = Is the ratio of the area of the water plane to the circumscribing rectangle. For ships with fine ends the value may go up to 0.9.

(B.C.) BLOCK COEFFICIENT = Is the ratio of the volume of the displacement to the volume of a block having the same length, breadth, and draft. It varies from 0.45 for torpedo boats to 0.90 for barges.

(W.A.) WETTED AREA = Is the area of the immersed surface of a ship's hull.

(D.C.) DRAG COEFFICIENT = Is an empirical factor obtained from the analysis of towing tank data. It expresses the quality of the shape of a hull. Drag coefficients increase with draft for any hull. The exact relationship between drag coefficient and draft depends on hull shape.

$$D.C. = f(D) \qquad (eq.\ 1)$$

(F_D) DRAG FORCE = Is the force that opposes the ship's hull motion through the water. The drag force on any immersed body is the product of the Drag Coefficient times the square of the velocity.

$$\therefore F_D = D.C. \times V^2 \qquad (eq.\ 2)$$

(H.P.) HORSE POWER = As recalled from physics power is the product of force times velocity. $\therefore H.P. = F \times V$

$$\text{Horsepower} = \frac{1\ lb.\text{-}ft./sec}{550} = \frac{lb.\text{-}ft./min}{33,000} \qquad (eq.\ 3)$$

The quality of a hull can be measured by the power required to propel it at a given speed, the lower the power demands for a given speed the better the hull design.

According to Newton's First Law, if the hull is not accelerating, the force required to move it through the water is equal to the Drag Force on the hull.

The equation for the Horsepower required to propel a ship at a given speed can be obtained by the substitution of Eq. 2 into Eq. 3 giving

$$H.P. = D.C. \times V^3 \quad \text{(eq. 4)}$$

(Eq. 1) D.C. = f (D) has to be determined empirically to be able to solve for H.P. for a given speed (V) and draft (D).

-CONVERSION FACTORS-

ONE KNOT	=	One Nautical Mile/Hour.
ONE NAUTICAL MILE	=	6,080 feet.
ONE NAUTICAL MILE	=	One minute of arc on the Earth's meridian.
ONE NAUTICAL MILE	=	1,853 meters.
ONE SHIPPING TON	=	2,240 pounds.
ONE SHIPPING TON	=	20 Hundredweight.
ONE SHIPPING TON	=	35 cubic feet of sea water.
ONE SHIPPING TON	=	36 cubic feet of fresh water.
ONE SHIPPING TON	=	1 cubic meter of sea water.

AMERICAN SHIPYARDS INC.

GALVESTON, TEXAS

"EXCELSIOR" TYPE VESSEL
Hull Characteristics

Length between perpendiculars	317 Ft.
Breadth	54 Ft.
Maximum draft (including flat keel)	37 Ft.

DRAFT (Ft.)	DISPLACEMENT (Ft.^{3})	(2,240 # Tons)	BLOCK COEFF.	FINENESS COEFF.	DRAG COEFF. (lbs. $\frac{\text{min}^2}{\text{Ft.}^2}$)
2	9,018	257	---	---	---
7	57,393	1,636	---	---	---
12	110,043	3,136	---	---	---
17	164,988	4,702	0.719	0.752	0.179
22	222,228	6,333	0.730	0.773	0.309
27	282,978	8,065	0.740	0.786	0.473
32	350,073	9,977	0.754	0.850	0.673
37	421,938	12,025	0.769	0.848	0.907

CASE 041 (CONT) 351

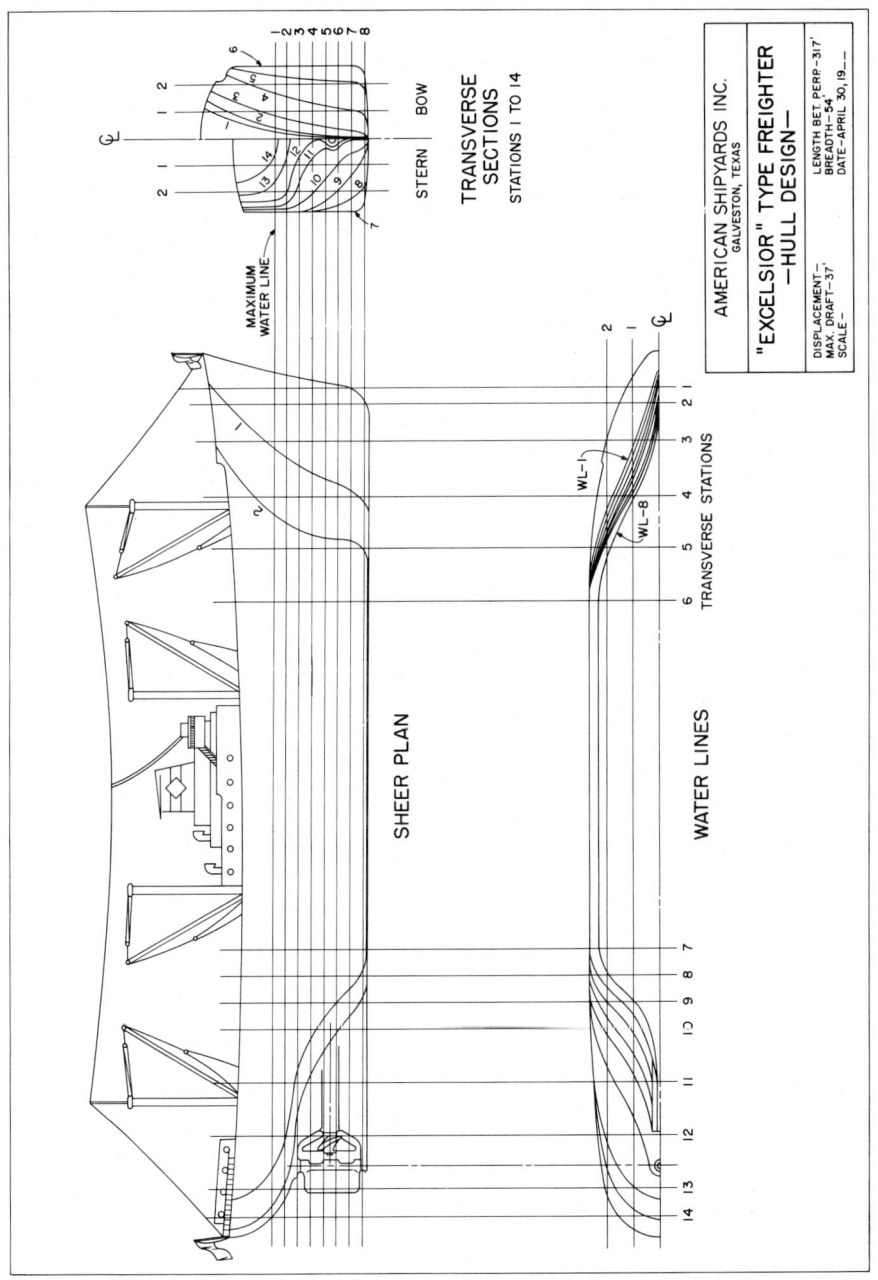

CASE 042   CHADWICK-WORKS, INC.

ENGINEERING DESIGN ASSOCIATES

1250 COLLEGE AVENUE
SOMERVILLE, MASSACHUSETTS
02144

(617)628-5000
CABLE:

M E M O R A N D U M

DATED: February 14, 19 - -

TO: INVENTIVE DESIGN GROUP

FROM: H.A. SWEET, Chief Project Engineer

SUBJECT: DESIGN AND DEVELOPMENT OF NEW TOOL LINE FOR CHADWICK - WORKS, INC.

    This memo may be considered as authorization to begin work immediately on the Chadwick-Works account. Our company has recently signed a contract with C-W to design a new commercial tool line that may be marketed through Hardware Stores. The attached materials have been collected to give the group a complete picture of C-W as we see them. All designs, including prototypes, must be completed by March 7, 19 - - to be presented to C-W management at 2:00 PM.

    Needless to say, the successful innovation of a novel tool line for C-W is important to our company in that our reputation depends upon accepted creative design.

CASE 042 (CONT)   353

(1)

THE COMPANY

Chadwick-Works, Inc. is a quality manufacturer of light industrial and retail consumer tools specializing in jigs and fixtures, gages, saws (wood and metal), wrenches (fixed and adjustable), drills (both power and hand operated), pliers, and custom items. The company commenced operation in 1948 as a four man shop specializing in jigs and fixtures to order and has had a continuous growth since that time to its present position as one of the largest and best known tool manufacturers in northern New England.

Customers who have entrusted C-W with tool purchases since their early days include General Electric, Pratt & Whitney, Sears, Raytheon, Sylvania, as well as a large number of community hardware stores.

In 1957 C-W moved to its present facilities in Northbridge and has enlarged its plant to 350,000 square feet of manufacturing, inventory, and office space. Its employees number 175 which include 125 plant supervisors, machinists, and production specialists; ten salesmen, promotion and purchasing personnel, secretarial staff, engineering, and administrative personnel.

The financial structure of the company is sound and they are in a position of being able to contract for large orders without in any way impairing their credit standing. This enables C-W to handle complete projects for customers as a single source, relieving them of the necessity of issuing and processing multiple purchase orders on one project.

FACILITIES LIST

PRESS EQUIPMENT - A wide range of presses from 5 to 400 ton capacity including hydraulic, double action, double crank, horn, straight side and inclinable types. Roll and automatic feeds and straighteners. Maximum deep drawing capacity 13-1/2".

    1 400 - ton Baldwin Southwark Hydraulic Press.
    (5' right to left, 4' front to back, 42" daylight)
    (Maximum stroke of cushion 15"; Maximum stroke of Press 36")

(2)

1	Bliss #30 Inclinable - 200 - ton capacity.
1	Hamilton Double Crank - 125 - ton capacity.
2	Double Crank Presses, 120 Ton and 135 Ton.
1	Toledo Gap Press - 125 ton capacity.
3	Double Action, 100 ton Range Presses with Long Strokes
1	Press, 75 ton 8" Stroke with Bed Adjustment up to 46" Die Height
1	Press, Bliss, 74 Ton, 4" Stroke OBI.

Plus many smaller **Presses** from 5 to 75 tons capacity.

TOOL AND DIE-MAKING EQUIPMENT- Complete tool-making facilities, including lathes, millers, shapers, radial drill surface grinders, Blanchard grinders and hardening equipment.

CUTTING EQUIPMENT- Shearing, cutting-off machines, lathes and abrasive cut-offs. Shearing capacity 10' x 1/4".

1	10' x 1/4" Cincinnati Gap Shear.
1	6' x 14 ga. Niagara Shear.
1	4' x 14 ga. Niagara Shear.
1	#216D 48" x 1/8" Nigara Circle Shear.
	Internal Circle Shears 1/8" x 4' D capacity.
1	High Speed Abrasive Cut-off Saw.
1	Abrasive Cut-off Machine.

METAL SHOP EQUIPMENT- A complete line of box and pan, apron press brakes; maximum capacity 12' x 1/4". Complete modern equipment for arc, gas, heliarc and butt welding, brazing and silver soldering. All equipment for flanging and **burring**, for the manufacture of special shapes.

1	Press Brake, 150 T, 4" Stroke OBI.
1	Verson Press Brake 10' x 1/4" capacity.
1	Press Brake, 8' x 3/16" capacity.
2	Press Brakes, 4" x 1/8" capacity.
1	8' Apron Brake
1	**Wiedemann** R41P 15 Ton 20 Station Turrett Punch Press with 10" **adjustable** rack gauge (table).
2	Complete Whistler setups
1	SCIAKY Spot Welder, 150 KVA - will qualify for certified aluminum welding up to two 1/8" thickness.
5	Spotwelders, 40 to 70 KVA range.
1	400 Amp. Heliarc Welder.
1	300 Amp. Arc Welder.
2	Aircomatic Welders.
1	300 Amp. G.E. Arc Welder.
4	Sets Gas Welding, Brazing, Silver Soldering.
2	High Speed Riveting Machines.
1	Multi-Drill & Tapping Head.
	Time Saver Deburring Machine.

(3)

FINISHING EQUIPMENT - Complete painting and finishing facilities including 15-foot water wash spray booths and automatically controlled gas-fired ovens, capable of producing a wide variety of finishes including wrinkle finish, hammertones, and area high-bake finishes. Equipment for solvent degreasing and caustic metal preparation.

- 2 Double-end Polishing and Buffing Jacks.
- 1 Variable Speed Polishing Jack
- 1 Walker Turner Sanding Belt.
- 1 Binks 15' Water Wash Spray Booth -3- man capacity.
- 1 Conveyor Oven - Heat Range 70 to 500 degrees F. Reeves Drive, 18' to 30" per minute. Carriers 30" long, 68 in number, with 4 hooks each.

BONDERIZING EQUIPMENT

CHROME - FLASH PLATING EQUIPMENT

AUTOMATIC SPRAY EQUIPMENT

TUMBLING - DEBURRING EQUIPMENT

ASSEMBLY EQUIPMENT:

Complete facilities for final assembly and packaging of large and small items, including tapping riveting and packaging machinery.

(5)

THE PROBLEM

Although the company has enjoyed a reputation of quality tools at a competitive price, it now finds that the hardware market for its line of hand and light (portable) power tools has begun to fall off during 1968. This has, in a large measure, contributed to the 18% sales decline discussed earlier.

Despite a qualified engineering staff of designers and draftsmen, the company has been unable to produce a new-product tool line in the past ten years. Its last success in this area was a grommet punch and fastener presented in 1957 when the camping (tent) industry was beginning to flourish. This line has fallen off now due to the popularity of all metal camping vehicles. It is obvious that C-W needs several new product lines to maintain its position in the tool field. John Smith, a hardware salesman with C-W for fourteen years, reports that "Sears recently dropped our tool line and took on Bonney because of their obvious human-factors styling and recent new pop-rivet line. This was a great loss since Sears was one of our biggest volume accounts, even though it sold under the Craftsman name. To sell the consumer hardware line, I must have new-products to show the store buyer."

C-W management engaged the services of Engineering Design Associates, specialists in new-product design, on February 10, 19-- as consultants to undertake the design of a new-product line in the consumer tool field. Top level management will make a decision to either drop this area of the market and retrench into industrial tools or gamble on a new-product when it reviews the designs presented by EDA on March 7, 19--.

PRESENT POSITION

For the first nine months of 1968, sales declined 18% from those of the similar period of 1967. The drop importantly reflected a downward trend in new orders during 1967 and an increased order cancellation rate in the first six months of 1968, Despite further cost-reduction efforts, results were severely penalized by the lower

(6)

volume, greater materials costs, higher wages rates, and charges for subcontract work committed during 1967. Net before taxes fell 58%. After taxes at 39.5% (including a provision for the tax surcharge), against 46.6%, net income dropped 53%. Earnings were equal to $1.00 a share, compared with $2.03 a share for the year-earlier interim. Results for 1968 exclude a special credit equal to $0.43 a share from the sale of former factory land and buildings.

INCOME STATISTICS (MILLIONS $)

Year Ended Dec. 31	Net Sales	%Oper. Inc. of Sales	Oper Inc.	Depr.& Amort.	Net Inc.	*Inventories
1968	----	-------	---	------	-----	--------
1967	$93.33	9.9	$9.22	$1.64	$5.27	$30.83
1966	93.10	10.5	9.80	1.50	5.33	29.98
1965	82.76	11.5	9.51	1.40	5.20	28.30
1964	67.22	10.7	7.18	1.15	3.87	21.66
1963	61.10	12.5	7.61	1.21	3.56	18.21
1962	59.62	14.4	8.58	1.14	4.07	18.45
1961	51.05	11.8	6.03	1.04	3.01	17.25
1960	58.23	16.0	9.30	0.87	4.87	18.84
1959	37.77	15.9	6.01	0.64	3.53	14.08
1958	30.88	12.6	3.90	0.64	2.05	10.03

* Approximately 65% of inventory is in the consumer tool line with the remaining 35% in industrial tools.

CASE 043   WRIGHT BICYCLE SHOP, INC.

THE WRIGHT BICYCLE SHOP, INC.
1908 KILL DEVIL HILL
DAYTON, OHIO 01896
AREA CODE 617   628-5000, ext. 242

April 8, 19 - -

ANNOUNCEMENT OF DEISGN COMPETITION

The Wright Bicycle Shop of Dayton, builders of quality bicycles since 1901, seeks unique designs for improvements in man-powered land transport (hereafter known as the MPLT program).

Based on the fact that the bicycle has not been improved appreciably since 1880 and the following statements by leaders in physical fitness and technological change, the Wright Shop is interested in creative solutions to the problem of manual land transportation:

"MAN HAS A PHYSICAL AND PSYCHOLOGICAL NEED TO USE HIS MUSCLES."

"A BICYCLE IS A VERY EXPENSIVE INVESTMENT, OF THE ORDER OF A YEAR'S INCOME TO MANY OF THE PEOPLE WHO MOST NEED IT."

"IT ABSORBS TOO MUCH EFFORT TO PROPEL IT, CHIEFLY BECAUSE OF HIGH AERODYNAMIC DRAG."

"POUND PER POUND ITS COST IS ABOUT THREE TIMES THAT OF A TYPICAL LIGHT CAR."

"IT SEEMS LIKELY THAT THE LESSONS OF LOS ANGELES WILL SOON BE TAKEN TO HEART SO THAT THE PROVISION OF MORE AND MORE FACILITIES FOR MOTOR VEHICLES WILL NO LONGER BE CONSIDERED TO OFFER A SOLUTION TO THE PROBLEM OF TRAFFIC—CLOGGED AND AIR POLLUTED CITIES."

The Wright Shop will accept proposals and contractual agreements on the design described on the following pages from Consultant Design Companies until April 22, 19 - -. Contracts for design studies will be awarded on April 24, and 26, 19 - - with final designs and prototypes to be presented in competition on May _____, 19 - -.

-2-

INTRODUCTION

These are the chief objects of the competition:

1. To stimulate fresh thinking on the broad subject of man-powered land transport (MPLT).

2. To provide state-of-the-art information to the Wright Shop on optimization of land transportation and related research activities.

3. To produce for the Wright Shop optimum designs for man-powered transport.

BACKGROUND

Man-powered land transport, which at the present time is confined almost entirely to that provided by the bicycle, has taken on new social significance in this automobile age. In affluent countries, the bicycle has for many been a means of obtaining needed exercise in a pleasant and useful manner, while for others it has enabled traffic-choked streets to be navigated more rapidly than by any other method. In most developing countries, the bicycle is virtually the sole means of medium and long-distance transport available to a large proportion of the population.

Despite this widespread use, bicycle development has been almost at a standstill during the whole of this century. This situation is in contrast to the ferment portrayed by the bicycling handbooks of the 1880's and 1890's, when almost all the details of modern designs evolved. It is the view of Mr. Alex Moulton, himself responsible for the only successful fresh approach to bicycle design in recent years, that the motorcar siphoned off the best brains from bicycle development.

The machine which has been evolved for the everyday rider is, despite this long period of stagnated design, far from an optimum. It absorbs too much effort to propel it, chiefly because of high aerodynamic drag. It liberally deposits oil and dirt on its rider, especially in wet weather. The most popular form of brake, the rim brake, suffers a drastic drop in efficiency in wet weather. The bicycle's reliability is poor by any standard and repairs and maintenance take an inordinate amount of time and skill. In a collision with another vehicle, or even in a fall, there is a high probability of injury to the rider. Yet despite these shortcomings and the lack of any significant research, development of design effort by any of the large, traditional bicycle manufacturers, the ordinary roadster is far from cheap. Pound per pound its cost is about three times that of a typical light car.

These considerations have prompted the present competition.

-3-

COMPETITION

The competition is being offered for the design of a complete vehicle or system, not for ideas, gimmicks or research only. Wright leaves the limits of the problem to be set by each consultant group entering. We are sure that some groups will be interested in improved versions of the racing or touring bicycle, and we shall welcome their efforts. However, we feel that we have already mentioned two problems more important to our society-those of transport in the developing countries, and of transport in the big Western Cities. We should like to expand on the problems involved in these two areas in the hope that more people may be attracted to offer new designs.

In the developing countries the bicycle is the first and the most desired of Western consumer goods. It contributes to raising the standard of living by providing mobility of the people and produce. It has these major disadvantages:

1. A bicycle is a very expensive investment, of the order of a year's income to many for people who most need it.

2. Bicycles, being made of materials such as ball bearings, steel tubes and pneumatic tires, have to be imported into countries which above all things need to stimulate their own small industry.

3. Bicycle-maintenance requirements are high in terms of skill and frequency.

4. Although large and unwieldy loads are often carried in uncomfortable and unsafe conditions, the bicycles sold in most developing countries are poorly equipped for load carrying.

5. Existing bicycles are rather unsuited to rutted tracks of hard-baked **earth**, mud, or sand, which are the conditions of a **large** proportion of the roads in these countries.

An acceptable design solution for developing countries would therefore be a vehicle which uses the maximum of indigenous materials, can be constructed and **maintained** with local skills, and **can carry goods on the vehicle rather than on the rider's head or the handlebars** as is now usually the case. Obviously some qualities have to be sacrificed to reach an optimum, and it **is the way the designer juggles the various competing demands, as well as his ingenuity and skill**, which govern the value of his solution.

A very different problem is found in the cities of the West, rapidly becoming **clogged with traffic.** It seems likely that the lessons of Los Angeles will soon be taken to heart so that **the provision of more and more facilities for motor vehicles will no longer be considered to offer a solution to the problem.** Advanced planners are, however, considering various automated-transportation schemes, from mass transit, very much as we know it today, to **arrangements** where people drive their own vehicles, probably battery powered, on to various types of powered "**guideways**." That people should move about under their own efforts is not apparently being considered; even walking quite short distances will, in many cases, be almost impossible. Before long it will be realized that man has a physical and a psychological need to use his muscles. **But even without this** consideration, it is likely that some form of **man-powered** transport would be found to have an important place in any urban transportation system from the engineering-economic viewpoint. We hope that some entries to this competition will point to interesting possibilities of solution.

-4-

Here are some of the desirable characteristics of an urban vehicle propelled by man-power.

1. It should enable "city clothes" to be worn by men and women in virtually all weathers.

2. It should be easy and comfortable to propel. The aerodynamic drag should be reduced. Rolling friction might be reduced by further lessening the unsprung mass. The possibility of an arrangement to store energy from braking or from manual work developed during waiting periods might be looked at again.

3. There may have to be a weight penalty, compared with existing bicycles, but it should still be possible for a woman to move the vehicle unaided.

4. There must be capability for carrying a small quantity of luggage--say a brief case or handbag and a couple of average shopping bags, or perhaps a baby--within the vehicle.

5. The rider should be protected from injury in the event of minor collisions or spills.

6. The vehicle should conform to general requirements for lighting, driver visibility, warning signals, braking capability, and so forth.

Within these fairly rigid specifications (which any entrant may nevertheless change to suit his view of the problem) there are endless possibilities for "trading-off" among first cost, weight, propulsion efficiency, manoeuvrability and so forth to reach an optimum solution - reached either intuitively or analytically.

These, then, are merely suggestions of approaches to problems and solutions. We would prefer that entrants adopt the type of approach used by designers in the aerospace industry, since their methods have been outstandingly successful in solving extremely difficult problems. That is, the particular problem addressed should be stated, specifications, including a "mission", drawn-up, and a cost-benefit analysis made of all the significant factors in each design variation. However, such an approach will not be considered essential to a successful entry.

CASE 044   CARRY-ALL CO.

CARRY—ALL COMPANY
1250 COLLEGE AVENUE
MEDFORD, MASSACHUSETTS 02155
AREA CODE 617   776-2100, Ext. 242

April 3, 19__

ANNOUNCEMENT OF DESIGN COMPETITION

The CARRY—ALL COMPANY of Medford, Massachusetts, manufacturers of materials handling equipment for both light and heavy industry hereby announces design competition for a specific system to transport luggage and parcels from commercial airliners to terminal facilities. The company's present position (see exhibit "A") is such that its design man - power is taxed to the limit in keeping pace with present materials handling needs in industry and has decided to subcontract design work on a competition basis for materials handling in the commercial aircraft area.

The company feels (based on initial research) that an efficient, reliable, and economic system of fast transport of luggage and parcels from commercial aircraft to terminal facilities represents a real need and the area in which it wished to diversify. Not only are present systems a deterrent to high speed passenger transportation but will become relatively incompatible considering the newly designed DC—10 with a passenger capacity of 550 and speed in excess of 600 mph scheduled for service in the early 1970's and the near future supersonic transport to travel at a speed of 2000 mph with a passenger load of near 900.

The CARRY—ALL company will accept proposals on the design described above from Consultant Design Companies until April 19, 19__. Contracts for design studies will be awarded on April 21, 19__ with final designs and prototypes presented in competition on May 17, 19__.

CARRY—ALL PRODUCTS: Belt conveyers—Trolley and monorail systems
Wire and linked rod conveyors—Magnetic rails
Skate wheel units

EXHIBIT "A"

-2-

CARRY-ALL COMPANY

POSITION: The company manufactures **material handling equipment for use in both light** and heavy industry. This equipment has a wide variety of industrial applications in the automation area as well as government-defense contract work. Emphasis in new product development is in high speed materials handling in the electronics industry for purposes of automated components assembly.

FUNDAMENTAL POSITION: The CARRY-ALL COMPANY manufactures **material handling equipment which includes belt** conveyors, trolley and monorail systems, gravity chutes, wire mesh and linked rod conveyors, magnetic rails, and skate wheel conveyors for both light and heavy industry. The standard line, marketed under the Diamond Seal trademark has been considered an efficient and reliable **product** by industry for twenty years.

Defense work accounted for about 6% of 1964-65 sales with outlets including submarines, space craft, rocket and satellite launching systems, wind tunnel conveyor equipment and some research facilities.

Industries served include petroleum, petrochemical, chemical, food and fertilizers, paper and pulp pharmaceutical, railroad loading platforms, ships and trucks, processing **industries and mechanical goods manufacturing.**

Sales in the United States and Canada are handled by more than 170 independent distributors. Sales were made to 19 countries in South America, Africa and Asia in 1963-64.

CARRY-ALL and SCHICK LIMITED, 50% owned, makes skate wheel conveyors in England for sale in Europe.

The owned plant at Medford, Massachusetts, was expanded to approximately 140,000 square feet through a 54,000 addition completed in 1964-65. Certain machinery and equipment, plus a branch facility established at Toronto, Ontario, during 1963-64 are leased. Employees: about 520.

RECENT DEVELOPMENTS:

Sales in the fiscal year ending June 30, 19 - -, rose 14% from the **year-earlier** level, to another new peak, with a reduction in military business more than offset by growing commerical demand for established and newer lines of material handling equipment. Margins widened on the larger volume, and operating income increased 18%. The provision for depreciation was up more sharply, but interest expense was smaller and, despite the absence of the gain in 1963-64 of $35,000 on redemption of debentures, pretax net expanded 15%. After a lower tax charge, at 40% against 50% the advance in final net was 17%.

-3-

For the six months ending December 31, 19 --, there was a 26% **year-to-year** expansion in sales, but profit-margins narrowed and net income fell 17%. Earnings receded to $0.34 a share on the company's larger number outstanding stock, from $0.43 in 1964 period.

Recent new product developments include materials handling components for the electronics industry and the company **expects** to announce in 19 -- a novel system to handle luggage and parcels at high speeds from **commercial** airliner to terminal facilities.

EARNINGS - DIVIDEND RECORD

Yr. End June 30	(a) Net Sales	(b) Comb. Earnings	Share Divs.	(c) Stock Range
1965	$10.88	$1.05	Nil	13 3/4-8
1964	9.52	0.96	Nil	10 5/8-7 3/4
1963	6.89	0.37	Nil	11 3/8-4 1/8
1962	6.32	0.16	Nil	12-3 5/8
1961	5.90	0.01	Nil	31 1/4-10 3/8
1960	5.02	0.48	Nil	24 1/4-10
1959	2.14	0.11	Nil	11 1/4-5 1/2
1958	1.35	0.10	Nil	------------
1957	0.82	0.07	Nil	------------

(a) In millions of dollars. (b) Adjusted for 10-for-1 split in 1959 and stk. divs. of 100% in 1960 and 5% each in 1964-65 and 1965-66. (c) Calendar year; bid prices.

CASE 045   AMERICAN-ENTERPRISES, INC.

AMERICAN ENTERPRISES, INC.
1830 BOSTON AVENUE
MEDFORD, MASSACHUSETTS, 02155

AREA CODE 617   776-2100, ext. 242

April 3, 19--

ANNOUNCEMENT OF DESIGN COMPETITION

AMERICAN-ENTERPRISES of Medford, Massachusetts ranks among the 200 largest industrial companies in the United States, and is a leading producer in the fields of construction products, cement, paints, inks, dyes, adhesives, resins, chemicals, household products, atmospheric control equipment and metallurgical products. The company supplies more than 500 essential products through its 11 products divisions. In November of 19--, A-E acquired Low Voltage Electric Company a leading manufacturer of electrical construction products such as switch and outlet boxes, conduit and cable fittings, lighting fixture supports, etc. Low Voltage's line of more than 2,000 products is considered to be the most complete in the industry.

In January of 19-- A-E considered acquiring Relbmar Motors, a manufacturer of motor cars who was having difficulty competing with the big three. After several months of research into the competitive automotive market, management decided not to attempt this acquisition due to Relbmar's commitment to the piston engine and a conventional vehicle that is not at present competing favorably in sales. Such an acquisition would be an investment in tooling, manufacturing facilities, and assembly lines geared to produce a conventional motor car. Management feels, however, that a motor vehicle geared to present transportation needs--suburban, urban, and intercity travel--using a unique form of power--designed for people--produced at low competitive cost, can be of a lucrative area for A-E to consider manufacturing.

At a meeting of the Board of Directors of A-E on March 20, 19-- it was agreed that initial designs of a novel motor vehicle hereafter referred to as the MINICAR will be placed in design competition among design consultant companies. It was felt that present automotive companies cannot produce, at the present time, a novel car due to their sizable investment in manufacturing facilities geared to producing the piston engine, metal body, four-wheel, over-styled, high-horsepower vehicles. The board feels that competing creative design companies can produce for A-E a vehicle that meets today's as well as the near future's demands on tape controlled highways, limited parking facilities, air pollution, traffic problems, and the high price of cars.

A-E will accept proposals and contractual agreements on the design described above from Consultant Design Companies until April 17, 19--. Contracts for design studies will be awarded on April 21, 19-- with final designs and prototypes to be presented in competition on May _____, 19--.

EXHIBIT "A"   -2-

                A M E R I C A N - E N T E R P R I S E S ,   I N C .

I. VITAL STATISTICS

    1. Company - American - Enterprises, Inc. 1830 Boston Ave., Medford, Mass. 02155
    2. Industry - Chemicals, metals, manufacturing.
    3. Years in Business - Founded in September of 1930.
    4. Years dividends paid - Since 1940.
    5. Exchange - Over the counter.
    6. Common shares outstanding - 11,380,686.
    7. Products - Construction products, cement, paints, inks, dyes, **adhesives**,
                   resins, chemicals, household products, atmospheric control
                   equipment, and metallurgical products. The company supplies
                   more than 500 **essential** products through its eleven products
                   divisions.
    8. Price of stock as of February 5, 19-- $35.50, bid, $37.75 offered.

II. PAST HISTORY TO DATE

    1. Company President - Robert E. Gordon (age-early forties).
        Chairman of the Board - Grover M. Cleveland (age-middle forties).
    2. Ranks among the 200 **largest industrial companies in the U.S. and is
        the leading producer in its field.**
    3. The rate of American-Enterprises' growth is indicated by the fact
        that 100 shares in 1951 valued at $1,400 would now (15 years later)
        be equal to 938.5 shares (because of splits) and worth about $56,800-
        dividend income has gone from $1.00 to the equivalent of $9.39 per
        share in this time. Shareholders have increased in cash dividend
        income, along with stock splits at frequent intervals (i.e. 2 for 1
        in 1952; 1955; 5 for 4 in 1956; 3 for 2 in 1957; and 5 for 4 in 1959).
    4. American-Enterprises has acquired the following companies since 1941;
            United Brick & Tile Co.   Dragon Cement Co.
            Master Builders Co.   Presstite Engineering Co.
            Metals Disintegrating Co.   Guardite, Inc.
            O-Cedar Corp.   Sinclair & Valentine Co.
            Lamar Pipe & Tile Co.   Booty Resineers, Inc.
            Concrete Products of America   Niagara Concrete Pipe, Ltd.
            Universal Concrete Pipe Co.   Sierra Metals Corp.
            Standard Lime & Cement Co.   Southern Dyestuff Corp.
            Stoner-Mudge, Inc.   Marietta Concrete Corp.
            Concrete Conduit Co.   Concrete Metals & Const. Co.
            Southern **Cement** Co.   Steel City Electric Co.

    5. Approximately 50% of the company's dollar sales are represented by products
        sold to the construction industry and 50% to the chemical and metal field.
    6. The company shows an average sales growth of 55% per year over the past
        ten years.

-3-

III. FUTURE

1. Company's continued growth and expansion would depend on the country's interest in research, state's road building and construction program, government's defense spending and expansion in space vehicles.

2. Company is reasonably immune to seasonal decline in commodities because of its diversification, government controlled price fixing, outmoding of products due to its continued research, and cyclic union grievances.

3. One of the secrets of American-Enterprises' success is the very strict qualifications it requires of a company before acquiring it. A primary requisite for an acquisition is that its potential for growth must be at least equal to that of American-Enterprises as a whole.

4. American-Enterprises had developed and holds patents on a new high-temperature super-alloy which is acceptable for jet engine production. This new alloy will be used by leading companies in the manufacture of gas turbine aircraft engines.

5. A $30 million expansion aimed directly at increasing productivity and plant capacity was approved on November 17, 19-- by the board of directors of the company. A major portion of the expansion will be carried out during the 1970's and is expected to be financed from retained earnings.

INDEX

Activity, 178
   dummy, 178
   lead-time, 181
   real, 178
   specification, 179
Altimeter, 150
Anthropometric data, 139
   seated male, 148
   standing male, 147
Anthropometry, 140, 142
Artificial horizon, 150
Auditory display, 140
   signals, 145
Authority, dependence on, 29

Backward node, 179
Bar chart, 195
Bioengineering, 135
Biomechanics, 135
Booz, Allen & Hamilton, Inc., 1, 177
Brainstorming, definition of, 21
Break-even chart, 207
Break-even point, 206
Burgeson, John W., 194

Chronocyclegraph, 137
Coermann, R. R., 158, 159
Color coding, instruments, 157
   resistor, 156
Commercialization, 5
Complete (PERT), 179
Controls, human factors, 139, 140
   characteristics of, 142
     data, 143

Convex polygon, area of technical feasibility, 212
Copyright, definition of, 220
   Office, 220
Creative process, 25
   decision, 25
   illumination, 26
   incubation, 26
   irritation, 25
   preparation, 25
   verification, 26
Creativity, barriers to, 28
   definition of, 15
   organizational barriers, 28
   personal barriers, 28
Critical path, definition of, 179
   managing, 187
   method, 176
   history, 176
CPM/PERT, 241
Criticism, 237

Dantzig, George B., 211
Da Vinci, Leonardo, 136
Decision matrix, 42, 133
Decisions, subjective, 52
Dependency, 180
Design attributes, attitudes, 35
   creativity, 35
   knowledge, 34
   skills, 34
Design method, analysis, 36
   conceptualization, 36
   product loop, 36

Design method (*continued*)
    production, 36
    synthesis, 36
Design process, analysis, 37, 45
    conceptualization, 37, 42
    consumption, 38, 48
    definition, 37
    distribution, 38, 47
    experiment, 38, 45
    goal, 37, 39, 40
    ideation, 37, 42
    manufacture, 38, 47
    need, 37, 39
    research, 37, 39, 40
    solution, 38, 47
    task specifications, 37, 39
Design report, 244
Design review, 239
    preliminary, 240
    specification, 239
Dial array, 151
Disclosure (patent), 224, 225
Distribution channels, 210
Dreyfuss, Henry, 135, 143, 144, 146–148
Ductility, 118

Earliest start, 184
Edison, Thomas A., 243
Einstein, Albert, 15
Elastic limit (proportional limit), 117
Elongation, 117
Engineering design, definition of, 33
Engineering psychology, 135
Entrepreneur, 242
Environment, 140, 144
Ergonomics, 135
Estimate of demand, 209
Event (PERT), 180
    end, 180
    hanging, 180
    node, 182
    predecessor, 182, 185
    start, 180
    successor, 183, 185
Exoskeletons, 161
Exponential reliability law, 55

Facts, old *or* light, 26
    recent *or* hard, 26
Failure rate, 56
Fatigue, 118

Finishes and coatings, factors, 130
    methods, 130
Float (*see* Slack)
Forward node, 180
Fuel gage, 149
Functional fixedness, 28
Functional utility, 28

Gantt, Henry L., 174
Gantt chart, 174
Gilbreth, Frank B., 136
Gilbrethian variables, 137
Gillette, King, 230
    safety razor, 230
Goldman, D. E., 158, 159

Habit transfer, 29
Hardness, Brinell, 118
    Rockwell, 118
Human body analog, 158
Human factors, 135, 139
Human factors analog, electrical, 159
    mechanical, 159

IBM 1130, 194
Idea, diagram, 21
    matrix, 21
    screening, 4
Idea sources, external, 3
    internal, 3
Ideas, association of, 16
    criticism of, 23
    how to sell, 243
Ideation, definition of, 16
Imagineering, definition of, 16
Impact strength, 119
Indenture, level of, 181
Interface, 181
Invention, article of manufacture, 224
    classes, 224
    composition of matter, 224
    machine, 224
    priority of, 224
    process, 224
    promotion, 234
    record keeping, 225
    three-legged stool, 231
Iso-profit line, 216

Job (*see* Activity)

Krendel, Ezra S., 160

Latest start, 181
Linear programming, definition of, 211
Localization of estimate, 181

Man-amplifier (manipulator), 162
Man-machine-environment system, 138
Man-machine system, 137
Manufacturing, 8
Market targets, 209
Market testing, 5, 11
Material flow, 8
Materials, composition and density, 122–125
    joinability of, 131
    mechanical properties, 127
    metal forming, 128
    plastic forming, 129
    rubber forming, 129
    selection of, 132
McCrory, R. J., 33
Memory barrier, 26
Microwave Associates, 202
Microwave switch body and cover, 202–205
Milestone, 181
Mind, conscious, 28
    subconscious, 28
Mock-up, 46
Model, adequate, 46
    dissimilar, 46
    distorted, 46
    true, 46
Modulus of elasticity, 117
Morphology, 48
    chart, 50
Morse, Richard S., 242

Network, (PERT), 181
    overall, 182
    skeleton, 183
    summary, 184
New products, committee, 13
    description, 1
    idea search, 4
    life cycle, 2
    planning department, 12
    terms, 13

Optimization, definition, 51
Osborn, Alex F., 15
Overspecialization, 29

Parallel (PERT), 182
Parameters, 48
Patent, claims, 229
    definition of, 221
    diligence, 227
    good records, 227
    self-interrogation, 226
    steps to obtain, 227
    witnesses, 227
Patent application, 222
    claims, 222
    drawing, 222
    specification, 222
    steps, 229
Patent infringment, 234
Patent Office, 220
    Official Gazette, 234
    prosecution, 229
Patent search, 228
Patent system, 220
Patentability, conditions for, 223
Percentile, 140
Phase (VE), evaluation, 201
    implementation, 201
    information, 201
    speculative, 201
Pilot production, 5
Plant operations, 8
Plastic versus metal properties, 121
Plastics, favorable considerations, 120
    thermoplastic resins, 119
    thermosetting resins, 119
    unfavorable considerations, 120
Polaris program, 178
Practical mindedness, 29
Presentation, check list, 251
    planning and preparing, 247
Price, penetrating, 210
    skimming, 210
Pricing, 206
    competitive maturity, 209
    market maturity, 209
    policies, 208
    technical maturity, 209
Probabilities, event, 195
    job (activity), 195
Probability analysis, 194

Process development, 7, 8
Product development, 5, 7
Product manager, 9
PERT, history, 177
PERT/CPM, project management, 197
PERT Network, 176
   preparation, 189
Proprietary information, 232, 233
Protection (patent), 224
Prototype, 46
   development, 7

Rating factor, 44, 133
Recall mechanism, 27
Reduction to practice, 224
Redundancy, 57
Reliability, definition, 53
Reproduction processes, 246
Research and development, 5
Responsible (PERT), 182
Ridicule, fear of, 30
Robot, industrial, 166
   unimate, 166
Role-playing, 19

Schedule date, 182
Scientific method, existing knowledge, 35
  hypothesis, 35
  knowledge loop, 35
  proof, 35
Semi-critical path (PERT), 182
Shear strength, transverse, 119
Sheet metal, properties, 125
Slack (PERT), 183
   negative, 183
   positive, 183
   secondary, 183
SAE, metal coding system, 116
Stainless steel, fabrication properties, 126
State-of-the-art, 35
Steinbeck, John, 61
Stiffness, 118
Stress versus strain curve, 117

Subnet, 183
Surplus path, 184
Symbolic language, 153
Synectics, definition of, 24

Taylor, Frederick W., 174
Telephone dial, 152
Tensile test, 116
Test review, 240
Tichauer, E. R., 137, 138
Time, activity duration, 178
  earliest activity completion, 184
  earliest expected, 184
  latest allowable, 184
  most likely, 179
  optimistic, 179
  pessimistic, 179
Torsion, 119
Trade secrets, 233
Trademark, definition of, 221
Trade-off, 53

Ultimate strain, 117
Ultimate strength, 117
Unimate, 166
   applications, 169
   specifications, 167
Unimation, Inc., 166

Value Engineering (VE), case history, 202
   definition of, 199
   process, 200
Visual display, 139, 140
   data, 146
   media, 248
Visualization, functional, 20
Von Gierke, H. E., 159

Wear-out period, 56
Weighting factor, 44
Weighting value, 133

Yield point (yield strength), 117